Intelligent Systems Reference Library

Volume 148

The aim of this series is to publish a Reference Library, including novel advances and developments in all aspects of Intelligent Systems in an easily accessible and well structured form. The series includes reference works, handbooks, compendia, textbooks, well-structured monographs, dictionaries, and encyclopedias. It contains well integrated knowledge and current information in the field of Intelligent Systems. The series covers the theory, applications, and design methods of Intelligent Systems. Virtually all disciplines such as engineering, computer science, avionics, business, e-commerce, environment, healthcare, physics and life science are included. The list of topics spans all the areas of modern intelligent systems such as: Ambient intelligence, Computational intelligence, Social intelligence, Computational neuroscience, Artificial life, Virtual society, Cognitive systems, DNA and immunity-based systems, e-Learning and teaching, Human-centred computing and Machine ethics, Intelligent control, Intelligent data analysis, Knowledge-based paradigms, Knowledge management, Intelligent agents, Intelligent decision making, Intelligent network security, Interactive entertainment, Learning paradigms, Recommender systems, Robotics and Mechatronics including human-machine teaming, Self-organizing and adaptive systems, Soft computing including Neural systems, Fuzzy systems, Evolutionary computing and the Fusion of these paradigms, Perception and Vision, Web intelligence and Multimedia.

More information about this series at http://www.springer.com/series/8578

Miloš Savić · Mirjana Ivanović
Lakhmi C. Jain

Complex Networks in Software, Knowledge, and Social Systems

 Springer

Miloš Savić
Faculty of Sciences, Department of
    Mathematics and Informatics
University of Novi Sad
Novi Sad
Serbia

Lakhmi C. Jain
Centre for Artificial Intelligence, Faculty of
    Engineering and Information Technology
University of Technology Sydney
Sydney, NSW
Australia

Mirjana Ivanović
Faculty of Sciences, Department of
    Mathematics and Informatics
University of Novi Sad
Novi Sad
Serbia

ISSN 1868-4394   ISSN 1868-4408   (electronic)
Intelligent Systems Reference Library
ISBN 978-3-030-08195-9   ISBN 978-3-319-91196-0   (eBook)
https://doi.org/10.1007/978-3-319-91196-0

Printed on acid-free paper

This Springer imprint is published by the registered company Springer International Publishing AG
part of Springer Nature
The registered company address is: Gewerbestrasse 11, 6330 Cham, Switzerland

Foreword

We are living in the information age being surrounded by diverse types of complex networks. The study of complex networks has gained a significant research interest in recent years, mostly because of their ubiquitous presence in nature and society, leading to an inter-disciplinary research field involving researchers from all major scientific disciplines.

This monograph deals with three types of complex networks describing the structure of software systems, the semantic web ontologies, and the self-organized social structure of research collaboration. In Chaps. 1 and 2, the authors give an overview of fundamental concepts, metrics, methods, and models important in studying real-world complex networks. As the main research contribution of the monograph, they propose and empirically validate several novel methods to analyze complex networks in which nodes are enriched with the domain-independent structural metrics and the metrics from a particular domain (i.e., software metrics, ontology metrics, and metrics of research performance, respectively).

Software networks are directed graphs that represent the dependencies among software entities present in a complex software system. One software system can be represented by several software networks reflecting its structure at different granularity levels. For example, the design structure of an object-oriented software system is typically described by three different kinds of software networks that depict dependencies among methods, classes, and packages (i.e., modules or namespaces). The applications of software networks are numerous, including the analysis of software systems using graph-based methods, computation of software design metrics, program comprehension and visualization, reverse engineering of software systems, identification of key software components, identification of design flaws in source code, analysis of change impact, and prediction of defects in software systems.

The authors give a comprehensive overview of previous empirical studies of software networks in Chap. 3. In the same chapter, they introduce a novel methodology to examine coupling and cohesion in software systems, found on enriched software networks. The authors also propose domain-independent graph clustering evaluation metrics for measuring the cohesiveness of software entities indicating their benefits over commonly used software cohesion metrics. The case studies presented here show that the proposed methodology has both theoretical and practical relevance. It enables a deeper understanding of phenomena that are commonly considered as indicators of poorly designed software systems (i.e., high coupling, low cohesion, and large cyclic dependencies). Additionally, it can be utilized for software engineering practitioners to identify keys, distinctive features of highly coupled software entities, software entities involved in cyclic dependencies, and software entities causing low cohesion providing valuable information for software development, testing, and maintenance activities.

The ontology formally describes the concepts and relationships in a domain of discourse. Ontologies have a prominent role in the development of the semantic web where they serve as shared and agreed-upon knowledge models enabling information reuse and interoperability. Ontologies and networks are very closely related—an ontology is a set of axioms inducing a semantic network of ontological entities present in the ontology. In Chap. 4, the authors show that modular semantic web ontologies represented by enriched ontology networks can be studied and evaluated in the same way as software systems represented by enriched software networks.

The last four chapters of the monograph are devoted to co-authorship networks. Co-authorship networks are social networks in which nodes represent researchers and links do research collaborations among them. In Chap. 5, the authors first discuss several graph-based representations of research collaboration and several ways to quantify its strength. In Chap. 6, they focus on the author name disambiguation problem appearing when extracting a co-authorship network from a bibliographic database in which authors are not uniquely identified. They provide a comprehensive overview of existing heuristic and machine learning approaches to solving the author name disambiguation. Then, the authors propose a novel supervised network-based method for disambiguating author names in bibliographic data.

Research collaboration is one of the fundamental determinants of contemporary science. The study of co-authorship networks is thus crucial for understanding the social structure and evolution of research communities. In Chap. 7, the authors give a thorough overview of existing empirical studies of co-authorship networks and identify their common structural and evolutionary properties. In Chap. 8, they propose a novel methodology based on enriched co-authorship networks to analyze the structure and evolution of research collaboration. The accompanying case study shows that the proposed methodology enables an in-depth analysis of research collaboration and its relationships with other indicators of research performance.

In my opinion, researchers and students interested in complex networks may benefit a lot from this monograph in two ways. First, the monograph provides a comprehensive and up-to-date overview of studies of complex networks from three important domains. Second, it introduces new methods to study complex networks enriched with domain-dependent metrics that are empirically validated with relevant and interesting case studies. The monograph may be also useful for researchers and practitioners in software engineering, ontology engineering, and scientometrics since it gives a network-based perspective on important issues from those three disciplines. I have recognized the significance of the original research contributions presented in the monograph and thus expect that they will motivate further research directions and novel applications.

Seoul, Korea   Prof. Sang-Wook Kim
                                                                                                                              Hanyang University

Preface

A wide variety of complex natural, engineered, conceptual, and social systems of high technological and scientific importance can be represented by networks—structures that describe relations, dependencies, and interactions between constituent parts of a complex system. Well-known examples of complex networked systems include technological systems such as Internet, power grids, telecommunication, and transportation networks; social systems such as academia, corporations, markets, and online communities; biological systems such as brain, metabolic pathways, and gene regulatory networks; and ecological systems such as food chains. In order to understand, control, or improve a complex system composed out of a large number of inter-related parts, it is necessary to quantify, characterize, and comprehend the structure and evolution of underlying complex networks.

The focus of this monograph is on complex networks from three domains: (1) networks extracted from source code of computer programs that represent the design of software systems, (2) networks extracted from source code of semantic web ontologies that describe the structure of shared and reusable knowledge, and (3) networks extracted from bibliographic databases that reflect scientific collaboration. In the monograph, we present novel methods for analyzing *enriched* software, ontology, and co-authorship networks, i.e., complex networks in which nodes are enriched with both domain-dependent metrics (software, ontology, and metrics of research performance, respectively) and domain-independent metrics used in complex network analysis.

The monograph is intended primarily for researchers, teachers, and students interested in complex networks and data analysis and mining. Additionally, it may also be interesting for researchers dealing with software engineering, ontology engineering, and scientometrics since it addresses topics from those disciplines within the framework of complex networks.

The monograph consists of three major parts entitled "Introduction", "Software and Ontology Networks: Complex Networks in Source Code", and "Co-authorship Networks: Social Networks of Research Collaboration".

Part I. In Chap. 1, we make an introduction to complex networks and outline our main research contributions presented in this monograph. The next chapter, Chap. 2, presents fundamental complex network measures, algorithms, and models. Those two chapters contain the necessary theoretical background and preliminaries used in the rest of the monograph.

Part II. The second part of the monograph is devoted to software and ontology networks. Those two types of complex networks, although representing two different kinds of complex man-made systems, have one important thing in common —they show dependencies between entities present in a system described in a formal language. In Chap. 3, after presenting an overview of the literature investigating software networks, we propose and empirically evaluate a novel methodology to study the structure of enriched software networks. In Chap. 4, we apply the same methodology to study the design of a large-scale modularized ontology.

Part III. The last part of the monograph is focused on co-authorship networks. This part contains four chapters. In Chap. 5, we discuss different models of co-authorship networks, different schemes to quantify the strength of research collaboration, different types of co-authorship networks, and their main applications. Chapter 6 is devoted to the extraction of co-authorship networks from bibliographic databases. We start with an overview of existing approaches to the author name disambiguation problem and their actual utilization in empirical studies analyzing co-authorship networks. In the same chapter, we study the performance of various string similarity metrics for identifying name synonyms in bibliographic records. We present a novel network-based method to disambiguate author names and investigate the impact of author name disambiguation to the structure of co-authorship networks. A comprehensive overview of studies dealing with the analysis of co-authorship networks is given in Chap. 7. Finally, in Chap. 8, we propose a novel methodology to study the structure and evolution of enriched co-authorship networks and demonstrate it on a case study in the domain of intra-institutional research collaboration.

Novi Sad, Serbia   Miloš Savić
Novi Sad, Serbia   Mirjana Ivanović
Sydney, Australia   Lakhmi C. Jain

Contents

Part I Introduction

1   Introduction to Complex Networks   3
   1.1 Complex Networks   3
   1.2 Software Networks   7
   1.3 Ontology Networks   8
   1.4 Co-authorship Networks   9
   1.5 Research Contributions of the Monograph   10
   References   12

2   Fundamentals of Complex Network Analysis   17
   2.1 Basic Concepts   17
   2.2 Complex Network Measures and Methods   21
        2.2.1 Connectivity of Nodes   21
        2.2.2 Distance Metrics   27
        2.2.3 Centrality Metrics and Algorithms   28
        2.2.4 Node Similarity Metrics   35
        2.2.5 Link Reciprocity   38
        2.2.6 Clustering, Cohesive Groups and Community Detection
               Algorithms   39
   2.3 Basic Complex Network Models   45
   References   53

**Part II Software and Ontology Networks: Complex Networks
           in Source Code**

3   Analysis of Software Networks   59
   3.1 Preliminaries and Definitions   61
   3.2 Structure of Software Networks   63
   3.3 Evolution of Software Networks   69

3.4   Analysis of Enriched Software Networks   72
       3.4.1   Metric-Based Comparison Test   73
       3.4.2   Analysis of Strongly Connected Components
                and Cyclic Dependencies   76
       3.4.3   Analysis of Coupling Among Software Entities   78
       3.4.4   Graph Clustering Evaluation Metrics as Software
                Cohesion Metrics   82
       3.4.5   Analysis of Cohesion of Software Entities   86
3.5   Experimental Dataset   88
3.6   Results and Discussion   91
       3.6.1   Strongly Connected Components and Cyclic
                Dependencies   93
       3.6.2   Degree Distribution Analysis   104
       3.6.3   Analysis of Highly Coupled Software Entities   117
       3.6.4   Correlations Between Cohesion Metrics   125
       3.6.5   Analysis of Package and Class Cohesion   128
3.7   Conclusions   133
References   135

4 Analysis of Ontology Networks   143
4.1   Preliminaries and Definitions   145
4.2   Related Work   148
4.3   Analysis of Enriched Ontology Networks: A Case Study   151
4.4   Results and Discussion   155
       4.4.1   Strongly Connected Components and Cyclic
                Dependencies   156
       4.4.2   Correlation Based Analysis of Ontology Modules   160
       4.4.3   Degree Distribution Analysis   162
       4.4.4   Highly Coupled Ontological Entities   165
       4.4.5   Cohesiveness of Ontology Modules   167
4.5   Conclusions   171
References   173

Part III   Co-authorship Networks: Social Networks of Research
           Collaboration

5 Co-authorship Networks: An Introduction   179
5.1   Co-authorship Networks as Undirected Graphs   181
5.2   Co-authorship Networks as Directed Graphs   182
5.3   Co-authorship Networks as Hypergraphs   184
5.4   Types of Co-authorship Networks   185
5.5   Applications of Co-authorship Networks   186
References   189

6   Extraction of Co-authorship Networks   193
    6.1   Bibliographic Databases   194
    6.2   Extraction of Co-authorship Networks from
          People-Article-Centered Bibliography Databases.   196
    6.3   Initial-Based Name Disambiguation Approaches.   197
    6.4   Heuristic Name Disambiguation Approaches   200
    6.5   Comparison of String Similarity Metrics for Name
          Disambiguation Tasks.   203
          6.5.1   Analyzed String Similarity Metrics   203
          6.5.2   Dataset   206
          6.5.3   Evaluation Methodology   207
          6.5.4   Results and Discussion   208
    6.6   Machine Learning Name Disambiguation Approaches   211
          6.6.1   Author Grouping Methods   212
          6.6.2   Author Assignment Methods   215
    6.7   Name Disambiguation Approach Based on Reference
          Similarity Network Clustering   218
          6.7.1   Experimental Evaluation   221
    6.8   Author Identification in Massive Bibliography Databases   225
    6.9   Impact of Name Disambiguation on Co-authorship Network
          Structure: A Case Study   227
    References   230

7   Analysis of Co-authorship Networks   235
    7.1   Empirical Studies of Field Co-authorship Networks   236
    7.2   Co-authorship Networks of Computer Science Authors   245
          7.2.1   Co-authorship Networks of Topical Computer
                  Science Communities.   249
          7.2.2   Co-authorship Networks of Computer Science
                  Conferences   252
    7.3   Co-authorship Networks of Mathematicians   255
    7.4   Journal Co-authorship Networks   258
    7.5   National Co-authorship Networks   260
    7.6   Summary   264
    References   268

**8   Analysis of Enriched Co-authorship Networks: Methodology
    and a Case Study**   277
    8.1   Methodology   278
    8.2   Case Study.   287

8.3   Network Analysis: Results and Discussion   291
       8.3.1   Network Structure   291
       8.3.2   Identification of Research Groups   296
       8.3.3   Collaborations Among Research Groups   298
       8.3.4   Comparison of Research Groups   305
       8.3.5   Gender Analysis of Research Groups   308
       8.3.6   Network Evolution   310
8.4   Conclusions   314
References   316

About the Authors

Dr. Miloš Savić is an Assistant Professor at the Department of Mathematics and Informatics, Faculty of Sciences, University of Novi Sad, Serbia, where he received his B.Sc., M.Sc., and Ph.D. degrees in computer science. His research interests are related to complex network analysis with focus on social, information, ontology, and software networks. He is co-author of 30 research papers published in international journals and proceedings of international conferences. During his studies, he received the faculty award "Aleksandar Saša Popović" for outstanding research work in the field of Computer Science. He is also a Senior Teaching Associate at the Petnica Science Center. From 2014, he serves as an Editorial Assistant for the *Computer Science and Information Systems* (ComSIS) journal.

Dr. Mirjana Ivanović holds the position of Full Professor at Faculty of Sciences, University of Novi Sad. She is a member of the University Council for Informatics. She is author or co-author of 14 textbooks, several monographs, and more than 350 research papers on multi-agent systems, e-learning, and intelligent techniques, most of which are published in international journals and conferences. She is/was a member of Program Committees of more than 230 international conferences, participant of numerous international research projects and principal investigator of more than 15 projects. She delivered several keynote speeches at international conferences, and visited numerous academic institutions all over the world as visiting researcher and teacher. Currently, she is Editor-in-Chief of the Computer Science and Information Systems journal.

Dr. Lakhmi C. Jain, Ph.D., ME, BE(Hons), Fellow (Engineers Australia) is with the Faculty of Education, Science, Technology and Mathematics at the University of Canberra, Australia and University of Technology Sydney, Australia. He founded the KES International for providing a professional community the opportunities for publications, knowledge exchange, cooperation, and teaming. Involving around 5,000 researchers drawn from universities and companies worldwide, KES facilitates international cooperation and generates synergy in teaching and research. KES regularly provides networking opportunities for professional community through one of the largest conferences of its kind in the area of KES. www.kesinternational.org.

His interests focus on the artificial intelligence paradigms and their applications in complex systems, security, e-education, e-healthcare, unmanned air vehicles, and intelligent agents.

Acronyms

ABST	Package Abstractness
AEC	Afferent–Efferent Coupling
AEXPR	Average Expression Complexity
AP	Average Population of Classes
AVGODF	Average Out-Degree Fraction
AXM	Number of Axioms
BA	Barabási-Albert
BET	Betweenness Centrality
BFS	Breadth First Search
CBO	Coupling Between Objects
CC	Clustering Coefficient
CC	Cyclomatic Complexity (Chap. 3)
CCD	Complementary Cumulative Distribution
CCN	Class Collaboration Network
CDF	Cumulative Distribution Function
CIG	Computer Intelligence in Games
CK	Chidamber–Kemerer
CLO	Closeness Centrality
COLL	Number of Collaborators
COMP	Internal Connectedness
COND	Conductance
CR	Class Richness
CRIS	Current Research Information System
CSCW	Computer Supported Cooperative Work
CUTR	Cut ratio
DBE	Department of Biology and Ecology
DC	Department of Chemistry, Biochemistry, and Environmental Protection
DD	Dominant Department
DEG	Degree

DEN	Internal Density
DFS	Depth First Search
DG	Department of Geography, Tourism, and Hotel Management
DIT	Depth of Inheritance Tree
DMI	Department of Mathematics and Informatics
DP	Department of Physics
EAND	Eager Associative Name Disambiguation
EB	Edge Betweenness
EC	Evolutionary Computation
ECIS	European Conference on Information Systems
ECOLL	Number of External Collaborators
ECST	Enriched Concrete Syntax Tree
ECTEL	European Conference on Technology Enhanced Learning
EM	Expectation–Maximization
ER	Erdős–Renyi
EVC	Eigenvector Centrality
EXP	Expansion
FODF	Flake Out-Degree Fraction
FS-UNS	Faculty of Sciences—University of Novi Sad
GCE	Graph Clustering Evaluation
GDN	General Dependency Network
GMO	Greedy Modularity Optimization
GWCC	Giant Weakly Connected Component
HCI	Human–Computer Interaction
HDIFF	Halstead Difficulty
HITS	Hyperlink-Induced Topic Search
HITSA	HITS Authority Score
HITSH	HITS Hub Score
HK	Henry–Kafura Complexity
HM	Hitz–Montazeri
HVOL	Halstead Volume
ICIS	International Conference of Information Systems
ICN	International Collaboration Network
IM	InfoMap
IMDb	Internet Movie Database
IN	In-degree
IR	Information Retrieval
IRI	Internationalized Resource Identifier
IS	Information Systems
KS	Kolmogorov–Smirnov
LAND	Lazy Associative Name Disambiguation
LCC	Loose Class Cohesion
LCOLL	Number of Local Collaborators
LCOM	Lack of Cohesion in Methods
LIS	Library and Information Science

LOC	Lines of Code
LV	Louvain (community detection algorithm)
MAXODF	Maximum Out-Degree Fraction
MCL	Markov Cluster Algorithm
MCN	Method Collaboration Network
MLE	Maximum Likelihood Estimation
MR	Mathematical Reviews
MWU	Mann—Whitney U
NCLASS	Number of Classes
NEC	Number of External Classes
NINST	Number of Instances
NMI	Normalized Mutual Information
NOC	Number of Children
NUMA	Number of Attributes
NUME	Number of Entities
NUMM	Number of Methods
OCN	Ontology Class Network
ODF	Out-Degree Fraction
OMN	Ontology Module Network
ONGRAM	Ontology Graphs and Metrics
OO	Object-Oriented
OON	Ontology Object Network
OSN	Ontology Subsumption Network
OUT	Out-degree
OWL	Web Ontology Language
PCN	Package Collaboration Network
PMF	Probability Mass Function
PNAS	Proceedings of the National Academy of Sciences
PR	Page Rank
PROF	Productivity (Fractional Counting)
PRON	Productivity (Normal Counting)
PROS	Productivity (Straight Counting)
PS	Probability of Superiority
RDF	Resource Description Framework
REC	References to External Classes
RR	Relationship Richness
RS	Radicchi Strong
RSNC	Reference Similarity Network Clustering
SBM	Stochastic Block Model
SCC	Strongly Connected Component
SCG	Static Call Graph
SICRIS	Slovenian Current Research Information System
SIGIR	Special Interest Group on Information Retrieval
SIGMOD	Special Interest Group on Management of Data
SLAND	Self-training Lazy Associative Name Disambiguation

SNA	Social Network Analysis
SNEIPL	Software Networks Extractor Independent of Programming Language
SOM	Spectral Optimization of Modularity
SQALE	Software Quality Assessment based on Lifecycle Expectations
SRCI	Serbian Research Competency Index
SVM	Support Vector Machine
SW	Small-World
SWEET	Semantic Web for Earth and Environmental Terminology
TCC	Tight Class Cohesion
TDNS	Top Degree Node Set
TEXPR	Total Expression Complexity
TF-IDF	Term Frequency—Inverse Document Frequency
TOT	Total Degree
UNS	University of Novi Sad
W3C	World Wide Web Consortium
WBET	Weighted Betweenness Centrality
WCC	Weakly Connected Component
WCLO	Weighted Closeness Centrality
WCOLL	Strength of Collaboration
WCRE	Working Conference on Reverse Engineering
WDEG	Weighted Degree Centrality
WECOLL	Strength of External Collaboration
WLCOLL	Strength of Local Collaboration
WMC	Weighted Methods Per Class
WS	Watts–Strogatz
WT	Walktrap
WWW	World Wide Web

Part I
Introduction

Chapter 1
Introduction to Complex Networks

Abstract Complex networks are graphs describing complex natural, conceptual and engineered systems. In this chapter we present an introduction to complex networks by giving several examples of technological, social, information and biological networks. Then, we discuss complex networks that are in the focus of this monograph (software, ontology and co-authorship networks). Finally, we briefly outline our main research contributions presented in the monograph.

1.1 Complex Networks

Generally speaking, complex systems are systems that are difficult to understand, model and predict. The term system usually denotes a set of interrelated and interdependent components (real or abstract) which together form an integrated whole. The main characteristic of complex systems is that their behavior cannot be inferred from the behavior of individual components. Complex systems are characterized by a large number of constituent components that may interact in many different ways. A variety of complex systems can be described by networks that show relations, dependencies and interactions among their constituent components. Networks are composed out of nodes (also called vertices, sites or actors) and links (also called edges, bonds or ties) connecting nodes. From a mathematical and computer science point of view, complex networks are graph data structures describing large and complex real-world systems. Examples of complex networks are numerous and can be found in almost any scientific discipline. Mark Newman classified complex networks into four broad categories: technological networks, social networks, information networks and biological networks [42, 46]. However, this classification should not be taken too rigorously since many networks may belong to more than one of the Newman's categories.

Technological networks represent engineered, man-made complex systems. Well known examples of technological networks forming the backbone of technological societies are networks of computers and related devices (e.g. Internet), power grids, telephone networks (networks of landlines and wireless links that transmit phone calls) and transportation networks such as airport, railway and road networks. Other

© Springer International Publishing AG, part of Springer Nature 2019
M. Savić et al., *Complex Networks in Software, Knowledge,
and Social Systems*, Intelligent Systems Reference Library 148,
https://doi.org/10.1007/978-3-319-91196-0_1

examples include: networks representing software systems such as call graphs [51] and class collaboration networks [52], networks depicting the structure of oil refineries [4] and networks of electronic circuits [15].

Social networks describe interactions and relations between social entities. Social entities can be individuals, social groups, institutions or even whole nations. There are many types of social links: links determined by individual opinions on other individuals, links describing transfers of material resources, links denoting collaboration, cooperation and coalition, links resulting from behavioral interactions, links imposed by formal relations within formally organized social groups, and so on. The study of social networks dates back to 1930s. The first social networks documented in the literature are sociograms made by psychotherapist Jacob L. Moreno published in his book "Who shall survive" (1934). Moreno's first sociograms visualizing sitting preferences within a group of schoolmates clearly show the presence of gender homophily and community structures in small social groups.

Another well-known example of an early documented social network constructed by direct observations of social interactions is the Zachary's karate club network [66]. This social network encompasses 34 members of an university karate club. Wayne Zachary (an anthropologist from the Temple University, Philadelphia, USA) connected two club's members by a weighted link if they frequently interacted outside regular club activities. Link weights in the network denote the strength of social interactions. This network is particularly interesting because the club split into two groups due to a conflict between two club's members (the chief administrator and the instructor). Zachary applied the Ford–Fulkerson algorithm to the network before the split and showed that the algorithm gives accurate assignments of the members into the groups formed after the split (only one assignment was incorrect). Nowadays, the Zachary's karate club network is one of benchmark networks used to test the ability of community detection algorithms to infer ground truth communities.

Social network analysis has a very long tradition within social sciences [12]. The topics studied by social scientists within the framework of social networks are quite extensive including occupational mobility, political and economic systems, decision making within groups, coalition formation, diffusion of innovations, sociology of science and corporate interlocking [62]. However, the main characteristic of traditional social network analysis studies is that they are based on small-scale social networks assembled through interviews, questionnaires or direct observations of social ties. With the rise of information technologies, Internet and massive amounts of data about human activities, it becomes possible to construct and analyze large-scale social networks. Examples include the network of collaboration among movie actors extracted from the IMDb database, networks of research collaboration extracted from massive bibliographic databases, the world trade network extracted from the Comtrade database and networks of personal interactions extracted from e-mail server logs, phone call logs or social networking sites such as Facebook and Twitter.

Information networks show connections between data items. The best known example of an information network is the World Wide Web (WWW) network. The nodes in the WWW network represent web pages. Two nodes A and B are connected by a directed link $A \rightarrow B$ if the web page represented by A contains at least one hyperlink to the web page represented by B. It should be emphasized that it is practically impossible to form the complete WWW network due to the decentralized and highly dynamic nature of the WWW, i.e. there is no central authority containing a complete list of web pages. However, a large subsets of the WWW network can be sampled using computer programs known as crawlers. Crawlers employ graph traversing techniques to collect web pages starting from an initially given set of web pages. Obviously, web pages that are not reachable from any of the pages contained in the initial set will not be present in the sampled network. WWW pages are documents and the WWW network belongs to a subset of information networks known as citation networks. The set of nodes in a citation network corresponds to a set of documents, while links represent references between those documents. Other examples of frequently investigated citation networks include citation networks of scientific papers, citation networks of patents and citation networks of legal documents.

Another important class of information networks are linguistic networks. Two dominant sub-categories within linguistic networks are semantic and structural linguistic networks. Links in semantic linguistic networks represent semantic relationships (e.g., synonyms or antonyms) between words or concepts. Structural linguistic networks are formed from large textual corpora taking into account the inner structure of words, position of words within a text or similarity between words or sentences. Examples of structural linguistic networks are word co-occurrence networks [14] and sentence similarity networks [70].

Tabular datasets can also be transformed into information networks [57]. A tabular dataset D is a set of data instances (or data points), where each data instance is described with a feature vector containing a fixed number of features or attributes. Features can be either categorical, ordinal or numerical. To transform D into an information network we have to define a function quantifying the distance (or similarity) between data instances. Then, the nodes of the network are data instances themselves, while two nodes A and B are connected if their distance satisfies some predefined criteria. For example, we can construct a k-nearest neighbor network in which a directed link from A to B exists if B is among the top k nearest data points to A. Another example is an ε-radius network in which A and B are connected by an undirected link if the distance between A and B is smaller than or equal to ε. It is also possible to derive an information network in which nodes represent features of data instances. As an example we can mention feature correlation networks that have practical applications in unsupervised feature selection [53].

Recommender networks are information networks that have a huge practical importance in e-commerce systems [42]. Recommender networks are bipartite graphs displaying preferences of individuals towards some items (e.g. products like books and films). They can be exploited in collaborative filtering systems to infer unexpressed preferences and produce recommendations to end users.

Various biological systems also can be represented by networks [19, 42, 46]. Examples include protein-protein interaction networks, gene regulatory networks, gene co-expression network, metabolic networks and neural networks. The nodes in a protein-protein interaction network are proteins and two proteins are connected together if it has been experimentally demonstrated that they bind together to build or activate a protein complex. Gene regulatory networks describe how cells control the expression of genes. On the other hand, two genes are connected in a gene co-expression network if their expression profiles are highly correlated. The nodes in a metabolic network represent metabolites. Two metabolites are connected by an undirected link if they participate in the same metabolic reaction. Neural networks contain neurons and synapses that attempt to mimic biological problem solving mechanisms [29]. They are typically designed to learn from experience using a dense interconnection of neurons using synapses. A number of neural networks were proposed in the past using various schemes such as feedforward connection, learning schemes and so on. Biological networks also include networks describing interactions between species such as food webs.

In the past years, researchers empirically analyzed a large number of large-scale complex networks from different domains [19]. The burst started in 1998 with Watts and Strogatz [63] who discovered that three large and sparse real-world networks (the collaboration graph of film actors, the power-grid network of Western United States and the neural network of the worm *C. Elegans*) exhibit the small-world property (a small distance between two randomly chosen nodes) and a high level of local clustering (highly dense ego-networks, i.e. highly dense sub-graphs induced by a randomly chosen node and its adjacent nodes). This discovery was significant because the classic theory of random graphs, used until then in modeling complex networked structures, cannot explain the presence of these two qualities together in one large and sparse graph.

Analysis of statistical properties of networks that represent large portions of the World Wide Web [3, 33] and the Internet at the physical level [22] led to the discovery that their degree distributions (probability $P(k)$ that a randomly chosen node has exactly k incident links) follow power-laws of the form $P(k) \sim k^{-\gamma}$, a property that the Erdős-Renyi model of random graphs [10, 20, 21] does not predict. Networks obeying the previously mentioned connectivity pattern are also known as *scale-free networks* [6]. A vast majority of nodes in a scale-free network are loosely coupled, but scale-free networks also contain a small fraction of nodes (called hubs or preferential nodes) whose degree of connectedness is unexpectedly high and tend to increase as networks evolve [6]. Two important consequences of the scale-free networked organization are (1) the "robust, yet fragile" property [2, 11] (scale-free networks are robust to random failures, but they experience an extreme disintegration when failures occur in hubs) and (2) the absence of an propagation threshold in spreading processes [49].

Newman's studies of complex networks from different domains revealed another two important characteristics of real-world networks: assortativity mixing patterns (a tendency that hubs either establish or avoid connections among themselves) [44, 45] and community organization (the presence of highly dense sub-graphs in a sparse graph) [23, 47]. Leskovec et al. [35] investigated the evolution of large-scale networks from different domains observing two evolutionary phenomena: denisification power-laws (the number of links grows super-linearly as a power of the number of nodes during evolution, which means that the average node degree increases in time) and shrinking diameters (the diameter of a network decreases as it grows).

After previously mentioned grounding works in the field of complex network analysis, researchers have analyzed a variety of complex biological, social, technological and conceptual systems represented as networks (good overviews can be found in [1, 9, 19, 46]). These studies initiated a new theory of complex networks (also known as *network science*) whose focus is on the analysis techniques and mathematical models which can reveal, reproduce and explain frequently observed structural and evolutionary characteristics of real-world networks.

The focus of this book is on three types of complex networks:

1. software networks – technological networks extracted from the source code of computer programs that represent the design of software systems.
2. ontology networks – information-technological networks extracted from the source code of semantic web ontologies that describe the structure of shared and reusable knowledge.
3. co-authorship networks – social networks extracted from bibliographic databases that show collaboration among scientists.

1.2 Software Networks

Modern software systems consist of many hundreds or even thousands of inter-related software entities appearing at different levels of abstraction. For example, complex software systems written in Java consist of packages, packages group related classes and interfaces, while classes and interfaces declare or define related methods and class attributes. Interactions, dependencies, relationships, or collaborations between software entities form various types of *software networks* which provide different granularity views of corresponding software systems. In the literature, software networks are also known as software collaboration graphs [41], software architecture maps [60] and software architecture graphs [30]. Depending on the level of abstraction, specific software networks, such as package, class and method collaboration networks [27], can be distinguished. Additionally, different coupling types between software entities of the same type determine different subtypes of software networks [64]. Due to the terminological and type diversity, we use the generic term "software network" to refer to any architectural (entity-level) graph representation of real-world software systems.

The applications of software networks span multiple fields such as empirical analysis of design complexity of software systems, their reverse engineering and computation of software metrics [54]. Links in software networks may denote various relationships between software entities such as coupling, inheritance and invocation. This means that software networks can be used to compute software metrics related to software design quality. The primary goal of a reverse engineering activity is to identify the main components of the system and dependencies among them in order to create a representation of the system at a higher level of abstraction [16]. A typical reverse engineering activity starts with the extraction of fact bases [32]. Source code is the most popular, valuable and trusted source of information for fact extraction because other artifacts (e.g. documentation, release notes, information collected from version management or bug tracking systems) may be missing, outdated, or unsynchronized with the actual implementation. Fact extraction is an automatic process during which the source code is analyzed to identify software entities and their mutual relationships. Therefore, software networks can be viewed as fact bases used in reverse engineering, architecture recovery and software comprehension activities. Architecture recovery techniques usually perform software network partitioning [17, 38] or cluster software entities according to feature vectors that can be constructed from software networks [5, 39]. Graphical representations of software entities and dependencies between them have long been accepted as comprehension aids to support reverse engineering processes [34]. Moreover, nodes in software networks can be enriched with software metrics information in order to provide visual, polymetric views of analyzed software systems [34]. Additionally, software networks can be exploited to identify and remove "bad smells" from source code [48], to support static concept location in source code [55], program comprehension during incremental change [13] and software component retrieval [28], to identify design patterns in source code [37], and to predict defects in software systems [8, 59, 69].

1.3   Ontology Networks

The term ontology has a very broad meaning. In information and computer sciences ontology means a specification of a conceptualization [26]. An ontology formally describes concepts and relations present in a domain of discourse, and as such models a certain part of reality. In the context of the Semantic Web vision [7, 56], ontologies are formal specifications of shared and reusable knowledge that support automated data-driven reasoning, data integration and interoperability of computer programs processing web accessible resources. The traditional World Wide Web can be viewed as a collection of inter-linked documents created by humans for humans. The Semantic Web is an extension of the World Wide Web based on the concept of structured inter-linked data that can be "consumed" by both humans and autonomous software agents [7].

Information always involves multiple inter-related entities in some context. Therefore, if we want to build computer programs able to perform tasks involving publicly available data then we have to specify the structure of data and their surrounding context. The sentence "John Doe is the doctoral adviser of Richard Roe" is an example of information that involves two entities in the context of university education. If a person does not have basic knowledge about the structure of academic organizations then he/she is unable to understand the sentence and possibly act upon it. Even worse situation is with computer programs that do not have any inherent cognitive capabilities. However, if we formally specify the structure of academic organizations by stating relevant concepts (such as Professor and PhDStudent), relations (such as AdvisorOf), and their mutual associations in the form of "subject-predicate-object"

(Professor-AdvisorOf-PhDStudent) then a software agent would be able to interpret the symbols of the triplet "John Doe-AdvisorOf-Richard Roe", i.e. it can infer that John Doe is an university professor and that Richard Roe is a PhD student. A semantic network of concepts, relations and data fragments naturally emerges from a sequence of triplets like the two previously given reflecting knowledge present in a particular domain.

Ontology networks show dependencies among ontological entities present in an ontology. Ontological descriptions contain axioms that define associations among ontological entities and axioms that specify non-relational properties of ontological entities. For example, the transitivity of a relation is one of non-relational properties that can be used to infer new non-explicitly stated associations among ontological entities. Since ontological networks reflect associations stated in ontological axioms they can be used to evaluate quality and complexity of ontological descriptions [67]. Similar to software networks, ontology networks can be exploited in a variety of reverse engineering and comprehension activities such as automated ontology modularization [18, 58], ontology summarization [68] and ontology visualization [31].

1.4   Co-authorship Networks

Collaboration between researchers is one of the key features of modern science. One of the most productive mathematicians of all time Paul Erdős has written over 1500 papers with over 500 other researchers [25]. This enormously high productivity inspired the concept of the Erdős number [24], which is defined to be one for his co-authors, two for co-authors of his co-authors, and so on. The Erdős number for a scientist is actually the length of the shortest path connecting him/her to Erdős in the appropriate *co-authorship network*. The nodes in a co-authorship network represent researchers – people who published at least one research paper. Two researchers are connected by an undirected link if they authored at least one paper together, with or without other co-authors. Additionally, link weights can be introduced in order to express the strength of research collaboration.

Co-authorship networks can be viewed as ordinary social networks restricted to people doing science: links in co-authorship networks denote collaboration and imply a strong academic bond. It has long been realized that the analysis of co-authorship networks can help us to understand the structure and evolution of corresponding academic societies [43]. Additionally, those networks can be used to automatically determine the most appropriate reviewers for a manuscript [50] and to predict future research collaborations [36, 65].

Co-authorship networks are not the only type of complex networks relevant to empirical analysis of scientific practice. Other two important types of scientometrics networks are citation and affiliation networks. Citation networks show citations among scientific papers, thus representing the structure of scientific knowledge. Affiliation networks are bipartite graphs that capture the affiliation of researchers to institutions. Affiliation networks can be combined with co-authorship networks in order to study scientific collaboration at institutional and country levels. On the other hand, a study that combines analysis of co-authorship and citation networks can reveal correlations and intersections between authorship and citation [40], as well as the influence of research collaboration on citation practices [61].

1.5 Research Contributions of the Monograph

In this monograph we propose several network-based methods to analyze complex software, knowledge and social systems. In Chap. 3 we present a novel network-based methodology to examine coupling and cohesion in software systems. The main idea of the methodology is to enrich nodes of software networks with both software metrics and domain-independent metrics used in complex network analysis in order to determine keys and distinctive characteristics of software entities that are commonly considered as indicators of poorly designed software solutions, i.e.

- highly coupled software entities,
- software entities involved in cyclic dependencies, and
- software entities causing low cohesion.

The proposed methodology is empirically evaluated on package, class and method collaboration networks of five large-scale open-source software systems written in Java (Tomcat, Lucene, Ant, Xerces and JFreeChart). Contrary to previous studies, our empirical findings indicate that high coupling in real software systems cannot be accurately modeled by power-law distributions, further implying that the power-law scaling exponent is not a reliable metric of software design quality. Our experimental results also indicate that highly coupled classes/methods tend to be more voluminous, more internally complex and more functionally important than loosely coupled classes/methods. Highly coupled classes tend to be significantly less cohesive than loosely coupled classes. Second, they do not tend to be localized in class inheritance hierarchies. Additionally, our results show that extremely highly coupled classes/methods are caused dominantly by their extensive internal reuse, not by an extensive aggregation of other classes/methods. This result implies that extreme high coupling in software systems can indicate only negative consequences of extensive internal reuse, not negative consequences of extensive internal aggregation. It is observed that package collaboration networks exhibit a core-periphery structure with a strongly connected core encompassing the most coupled and functionally the most important packages. The analysis of strongly connected components in class collaboration networks shows that they densify with size. This densification can be modeled by a power-law whose scaling exponent can be used as an indicator of software design quality with respect to the principle of avoiding cyclic dependencies. Finally, it is noticed that poorly cohesive packages (resp. classes) are dominantly caused by classes (resp. methods) having a high internal complexity and efferent coupling. This result suggests that improving cohesion in software systems, besides leading to a more coherent, understandable and extensible modular software structure, may also positively influence software quality attributes associated to code complexity (understandability and testability) and efferent coupling (error-proneness and external reusability).

In Chap. 4 we demonstrate that our methodology to analyze enriched software networks can also be applied to enriched ontology networks in order to

1. evaluate the coupling and cohesion of ontological entities, and
2. assess the design quality of modular semantic web ontologies.

The empirical evaluation is conducted on the enriched ontology module, class and subsumption network of the SWEET ontology (a large modular ontology describing the terminology related to earth and environmental sciences). Similarly as for software systems, it is observed that

1. the coupling of ontological entities cannot be accurately modeled by power-laws as frequently reported in the literature,
2. highly coupled ontology modules tend to be more voluminous and more functionally important compared to loosely coupled modules, and
3. high ontology module coupling tend to be dominantly caused by an extensive internal reuse of ontology modules.

Obtained results also indicate that the modular structure of the SWEET ontology exhibits both good and bad qualities from an ontology engineering perspective. The SWEET ontology modules tend to have a good degree of internal cohesion implying that the SWEET ontology mainly consists of self-contained ontology modules. However, the SWEET ontology module network has a giant strongly connected component encompassing the most important and the most reused SWEET modules, which implies that the modular structure of SWEET does not enable an easy comprehension and efficient external reuse of core SWEET modules.

In Chap. 6 we discuss the problem of extracting co-authorship networks from bibliographical records stored in different types of bibliographic databases. The extraction of co-authorship networks from bibliographic databases in which authors are uniquely identified is a straightforward task. In all other cases, the name disambiguation problem appears manifesting in two forms: name homonymy (many different individuals may have the same name) and name synonymy (a single individual may appear under different names in bibliographic records). After a discussion of existing approaches for solving the name disambiguation problem and their actual utilization in empirical studies of co-authorship networks, we investigate the performance of various string similarity metrics for identifying name synonyms in bibliographic records. The obtained results indicate that the Jaccard n-gram string similarity metric exhibits the best performance for matching full-full author name pairs (name pairs in which both names include full forename and full surname), while the TF-IDF n-gram string similarity metric gives the highest accuracy when matching full-short author name pairs (name pairs in which one name is fully given and the other name includes full surname and forename reduced to initial letters). In the case of short-short name pairs (the forename in both names is reduced to initial letters), the best accuracy is achieved by both the Jaccard *n*-gram and TF-IDF *n*-gram distance. Then, we propose a novel supervised network-based name disambiguation method. The method is based on reference similarity networks, i.e. networks in which nodes represent bibliographic references with two references being directly connected if they have a high string similarity. The method requires a training set containing pairs of co-referent bibliographic references. The optimal string similarity metric and the corresponding reference similarity threshold are determined in the training phase by a greedy algorithm. Then, unique authors are identified by clustering reference similarity networks using community detection techniques. In the same chapter, we also present a case study investigating the impact of name disambiguation on the structure of co-authorship networks.

After giving a comprehensive overview of empirical studies of co-authorship networks (Chap. 7), in Chap. 8 we propose a novel methodology to study the structure and evolution of enriched co-authorship networks. The nodes of an enriched co-authorship network are annotated with attributes indicating demographic characteristics of researchers (e.g. gender, age, academic position, and so on) and research evaluation metrics that quantify various determinants of research performance. The first step of the proposed methodology is to use different community detection techniques and graph clustering evaluation metrics to determine research groups within the analyzed co-authorship network. After that, non-parametric statistical tests are employed to derive directed graphs of superiority relations between research groups with respect to researcher evaluation metrics attached to the nodes of the examined co-authorship network. Non-parametric statistical tests are also utilized to perform gender-based analysis of research groups, as well as to investigate differences between

1. researchers from the core of the network and researchers located at the periphery of the network, and
2. researchers involved in inter-group research collaborations and researchers whose collaboration is bounded to their own research groups.

Regarding the evolution of enriched co-authorship network, we present an algorithm for mining attachment preferences that takes into account attributes associated to nodes. The proposed methodology is illustrated on an enriched co-authorship network reflecting intra-institutional research collaboration.

References

1. Albert, R., Barabási, A.L.: Statistical mechanics of complex networks. Rev. Mod. Phys. **74**(1), 47–97 (2002). https://doi.org/10.1103/RevModPhys.74.47
2. Albert, R., Jeong, H., Barabasi, A.: Error and attack tolerance of complex networks. Nature **406**(6794), 378–382 (2000). https://doi.org/10.1038/35019019
3. Albert, R., Jeong, H., Barabási, A.L.: Diameter of the world wide web. Nature **401**, 130–131 (1999). https://doi.org/10.1038/43601

4. Andrade Jr., J.S., Bezerra, D.M., Ribeiro Filho, J., Moreira, A.A.: The complex topology of chemical plants. Phys. A Stat. Mech. Appl. **360**(2), 637–643 (2006). https://doi.org/10.1016/j.physa.2005.06.092
5. Anquetil, N., Fourrier, C., Lethbridge, T.C.: Experiments with clustering as a software remodularization method. In: Proceedings of the Sixth Working Conference on Reverse Engineering, WCRE '99, pp. 235–255. IEEE Computer Society, Washington, DC, USA (1999). https://doi.org/10.1109/WCRE.1999.806964
6. Barabasi, A.L., Albert, R.: Emergence of scaling in random networks. Science **286**(5439), 509–512 (1999). https://doi.org/10.1126/science.286.5439.509
7. Berners-Lee, T., Hendler, J., Lassila, O.: The semantic web. Sci. Am. **284**(5), 34–43 (2001)
8. Bhattacharya, P., Iliofotou, M., Neamtiu, I., Faloutsos, M.: Graph-based analysis and prediction for software evolution. In: Proceedings of the 34th International Conference on Software Engineering, ICSE '12, pp. 419–429. IEEE Press, Piscataway, NJ, USA (2012)
9. Boccaletti, S., Latora, V., Moreno, Y., Chavez, M., Hwang, D.: Complex networks: structure and dynamics. Phys. Rep. **424**(45), 175–308 (2006). https://doi.org/10.1016/j.physrep.2005.10.009
10. Bollobás, B.: Random graphs. Cambridge University Press, Cambridge (2001)
11. Bollobás, B., Riordan, O.: Robustness and vulnerability of scale-free random graphs. Internet Math. **1**(1), 1–35 (2003). https://doi.org/10.1080/15427951.2004.10129080
12. Borgatti, S.P., Mehra, A., Brass, D.J., Labianca, G.: Network analysis in the social sciences. Science **323**(5916), 892–895 (2009). https://doi.org/10.1126/science.1165821
13. Buckner, J., Buchta, J., Petrenko, M., Rajlich, V.: Jripples: a tool for program comprehension during incremental change. In: Proceedings of the 13th International Workshop on Program Comprehension, IWPC '05, pp. 149–152. IEEE Computer Society, Washington, DC, USA (2005). https://doi.org/10.1109/WPC.2005.22
14. Cancho, R.F., Solé, R.V.: The small world of human language. Proc. R. Soc. Lond. Ser B Biol Sci. **268**(1482), 2261–2265 (2001). https://doi.org/10.1098/rspb.2001.1800
15. Cancho, RFi, Janssen, C., Solé, R.V.: Topology of technology graphs: small world patterns in electronic circuits. Phys. Rev. E **64**, 046119 (2001). https://doi.org/10.1103/PhysRevE.64.046119
16. Chikofsky, E.J., Cross II, J.H.: Reverse engineering and design recovery: a taxonomy. IEEE Softw. **7**(1), 13–17 (1990). https://doi.org/10.1109/52.43044
17. Chiricota, Y., Jourdan, F., Melançon, G.: Software components capture using graph clustering. In: Proceedings of the 11th IEEE International Workshop on Program Comprehension, IWPC '03, pp. 217–226. IEEE Computer Society, Washington, DC, USA (2003). https://doi.org/10.1109/WPC.2003.1199205
18. Coskun, G., Rothe, M., Teymourian, K., Paschke, A.: Applying community detection algorithms on ontologies for identifying concept groups. In: Kutz, O., Schneider, T. (eds.) Workshop on Modular Ontologies, vol. 230, pp. 12–24. IOS Press (2011). https://doi.org/10.3233/978-1-60750-799-4-12
19. Costa, L.d.F., Oliveira, O., Travieso, G., Rodrigues, F.A., Villas Boas, P., Antiqueira, L., Viana, M.P., Correa Rocha, L.: Analyzing and modeling real-world phenomena with complex networks: a survey of applications. Adv. Phys. **60**(3), 329–412 (2011). https://doi.org/10.1080/00018732.2011.572452
20. Erdős, P., Rényi, A.: On random graphs I. Publ. Math. Debr. **6**, 290–297 (1959)
21. Erdős, P., Rényi, A.: On the evolution of random graphs. Publ. Math. Inst. Hung. Acad. Sci. **5**, 17–61 (1960)
22. Faloutsos, M., Faloutsos, P., Faloutsos, C.: On power-law relationships of the internet topology. ACM SIGCOMM Comput. Commun. Rev. **29**, 251–262 (1999). https://doi.org/10.1145/316194.316229
23. Girvan, M., Newman, M.E.J.: Community structure in social and biological networks. Proc. Natl. Acad. Sci. **99**(12), 7821–7826 (2002). https://doi.org/10.1073/pnas.122653799
24. Goffman, C.: And what is your Erdős number? Am. Math. Mont. **76**(7), 149 (1969)

25. Grossman, J.W.: Paul Erds: the master of collaboration. In: Graham, R.L., Nettil, J., Butler, S. (eds.) The Mathematics of Paul Erds II, pp. 489–496. Springer, New York (2013). https://doi. org/10.1007/978-1-4614-7254-4_27

26. Gruber, T.R.: A translation approach to portable ontology specifications. Knowl. Acquisition **5**(2), 199–220 (1993). https://doi.org/10.1006/knac.1993.1008

27. Hylland-Wood, D., Carrington, D., Kaplan, S.: Scale-free nature of Java software package, class and method collaboration graphs. Technical report, TR-MS1286, MIND Laboratory, University of Maryland, College Park, USA (2006)

28. Inoue, K., Yokomori, R., Yamamoto, T., Matsushita, M., Kusumoto, S.: Ranking significance of software components based on use relations. IEEE Trans. Softw. Eng. **31**(3), 213–225 (2005). https://doi.org/10.1109/TSE.2005.38

29. Jain, L.C.: Soft Computing Techniques in Knowledge-Based Intelligent Engineering Systems. Springer, Berlin (1997)

30. Jenkins, S., Kirk, S.R.: Software architecture graphs as complex networks: a novel partitioning scheme to measure stability and evolution. Inf. Sci. **177**, 2587–2601 (2007). https://doi.org/ 10.1016/j.ins.2007.01.021

31. Katifori, A., Halatsis, C., Lepouras, G., Vassilakis, C., Giannopoulou, E.: Ontology visualization methods – a survey. ACM Comput. Surv. **39**(4) (2007). https://doi.org/10.1145/1287620. 1287621

32. Kienle, H.M., Müller, H.A.: Rigi - an environment for software reverse engineering, exploration, visualization, and redocumentation. Sci. Comput. Progr. **75**(4), 247–263 (2010). https:// doi.org/10.1016/j.scico.2009.10.007

33. Kumar, R., Raghavan, P., Rajagopalan, S., Tomkins, A.: Extracting large-scale knowledge bases from the web. In: Proceedings of the 25th International Conference on Very Large Data Bases (VLDB '99), pp. 639–650 (1999)

34. Lanza, M., Ducasse, S.: Polymetric views - a lightweight visual approach to reverse engineering. IEEE Trans. Softw. Eng. **29**(9), 782–795 (2003). https://doi.org/10.1109/TSE.2003.1232284

35. Leskovec, J., Kleinberg, J., Faloutsos, C.: Graph evolution: densification and shrinking diameters. ACM Trans. Knowl. Discov. Data **1**(1) (2007). https://doi.org/10.1145/1217299.1217301

36. Liben-Nowell, D., Kleinberg, J.: The link prediction problem for social networks. In: Proceedings of the Twelfth International Conference on Information and Knowledge Management, CIKM '03, pp. 556–559. ACM, New York, NY, USA (2003). https://doi.org/10.1145/956863. 956972

37. Lucia, A.D., Deufemia, V., Gravino, C., Risi, M.: Design pattern recovery through visual language parsing and source code analysis. J. Syst. Softw. **82**(7), 1177–1193 (2009). https:// doi.org/10.1016/j.jss.2009.02.012

38. Mancoridis, S., Mitchell, B.S., Rorres, C., Chen, Y., Gansner, E.R.: Using automatic clustering to produce high-level system organizations of source code. In: Proceedings of the 6th International Workshop on Program Comprehension, IWPC '98, pp. 45–52. IEEE Computer Society, Washington, DC, USA (1998). https://doi.org/10.1109/WPC.1998.693283

39. Maqbool, O., Babri, H.: Hierarchical clustering for software architecture recovery. IEEE Trans. Softw. Eng. **33**(11), 759–780 (2007). https://doi.org/10.1109/TSE.2007.70732

40. Martin, T., Ball, B., Karrer, B., Newman, M.E.J.: Coauthorship and citation patterns in the physical review. Phys. Rev. E **88**, 012814 (2013). https://doi.org/10.1103/PhysRevE.88.012814

41. Myers, C.R.: Software systems as complex networks: structure, function, and evolvability of software collaboration graphs. Phys. Rev. E **68**(4), 046116 (2003). https://doi.org/10.1103/ PhysRevE.68.046116

42. Newman, M.: Networks: An Introduction. Oxford University Press Inc., New York (2010)

43. Newman, M.E.J.: The structure of scientific collaboration networks. Proc. Natl. Acad. Sci. **98**(2), 404–409 (2001). https://doi.org/10.1073/pnas.98.2.404

44. Newman, M.E.J.: Assortative mixing in networks. Phys. Rev. Lett. **89**, 208701 (2002). https:// doi.org/10.1103/PhysRevLett.89.208701

45. Newman, M.E.J.: Mixing patterns in networks. Phys. Rev. E **67**, 026126 (2003). https://doi. org/10.1103/PhysRevE.67.026126

46. Newman, M.E.J.: The structure and function of complex networks. SIAM Rev. **45**, 167–256 (2003). https://doi.org/10.1137/S003614450342480
47. Newman, M.E.J., Girvan, M.: Finding and evaluating community structure in networks. Phys. Rev. E **69**(2), 026113 (2004). https://doi.org/10.1103/PhysRevE.69.026113
48. Oliveto, R., Gethers, M., Bavota, G., Poshyvanyk, D., De Lucia, A.: Identifying method friendships to remove the feature envy bad smell. In: Proceedings of the 33rd International Conference on Software Engineering, ICSE '11, pp. 820–823. ACM, New York, NY, USA (2011). https://doi.org/10.1145/1985793.1985913
49. Pastor-Satorras, R., Vespignani, A.: Epidemic spreading in scale-free networks. Phys. Rev. Lett. **86**, 3200–3203 (2001). https://doi.org/10.1103/PhysRevLett.86.3200
50. Rodriguez, M.A., Bollen, J.: An algorithm to determine peer-reviewers. In: Proceedings of the 17th ACM Conference on Information and Knowledge Management, CIKM '08, pp. 319–328. ACM, New York, NY, USA (2008)
51. Ryder, B.G.: Constructing the call graph of a program. IEEE Trans. Softw. Eng. **5**(3), 216–226 (1979). https://doi.org/10.1109/TSE.1979.234183
52. Savić, M., Ivanović, M., Radovanović, M.: Characteristics of class collaboration networks in large Java software projects. Inf. Technol. Control **40**(1), 48–58 (2011). https://doi.org/10.5755/j01.itc.40.1.192
53. Savić, M., Kurbalija, V., Ivanović, M., Bosnić, Z.: A Feature Selection Method Based on Feature Correlation Networks, pp. 248–261. Springer International Publishing, Cham (2017). https://doi.org/10.1007/978-3-319-66854-3_19
54. Savić, M., Rakić, G., Budimac, Z., Ivanović, M.: A language-independent approach to the extraction of dependencies between source code entities. Inf. Softw. Technol. **56**(10), 1268–1288 (2014). https://doi.org/10.1016/j.infsof.2014.04.011
55. Scanniello, G., Marcus, A.: Clustering support for static concept location in source code. In: Proceedings of the 19th International Conference on Program Comprehension (ICPC 2011), pp. 1 10 (2011). https://doi.org/10.1109/ICPC.2011.13
56. Shadbolt, N., Berners-Lee, T., Hall, W.: The semantic web revisited. IEEE Intell. Syst. **21**(3), 96–101 (2006). https://doi.org/10.1109/MIS.2006.62
57. Silva, T.C., Zhao, L.: Network Construction Techniques, pp. 93–132. Springer International Publishing, Cham (2016). https://doi.org/10.1007/978-3-319-17290-3_4
58. Stuckenschmidt, H., Schlicht, A.: Structure-based partitioning of large ontologies. In: Stuckenschmidt, H., Parent, C., Spaccapietra, S. (eds.) Modular Ontologies, pp. 187–210. Springer, Berlin (2009). https://doi.org/10.1007/978-3-642-01907-4_9
59. Tosun, A., Turhan, B., Bener, A.: Validation of network measures as indicators of defective modules in software systems. In: Proceedings of the 5th International Conference on Predictor Models in Software Engineering, PROMISE '09, pp. 5:1–5:9. ACM, New York, NY, USA (2009). https://doi.org/10.1145/1540438.1540446
60. Valverde, S., Cancho, R.F., Solé, R.V.: Scale-free networks from optimal design. EPL (Europhys. Lett.) **60**(4), 512–517 (2002). https://doi.org/10.1209/epl/i2002-00248-2
61. Wallace, M.L., Larivire, V., Gingras, Y.: A small world of citations? the influence of collaboration networks on citation practices. PLoS ONE **7**(3), e33339 (2012). https://doi.org/10.1371/journal.pone.0033339
62. Wasserman, S., Faust, K., Iacobucci, D.: Social Network Analysis: Methods and Applications. Cambridge University Press, Cambridge (1994)
63. Watts, D.J., Strogatz, S.H.: Collective dynamics of 'small-world' networks. Nature **393**(6684), 409–10 (1998). https://doi.org/10.1038/30918
64. Wheeldon, R., Counsell, S.: Power law distributions in class relationships. In: Proceedings of the Third IEEE International Workshop on Source Code Analysis and Manipulation, pp. 45–54 (2003). https://doi.org/10.1109/SCAM.2003.1238030
65. Yan, E., Guns, R.: Predicting and recommending collaborations: an author-, institution-, and country-level analysis. J. Inf. **8**(2), 295–309 (2014). https://doi.org/10.1016/j.joi.2014.01.008
66. Zachary, W.: An information flow model for conflict and fission in small groups. J. Anthropol. Res. **33**, 452–473 (1977)

67. Zhang, H., Li, Y.F., Tan, H.B.K.: Measuring design complexity of semantic web ontologies. J. Syst. Softw. **83**(5), 803–814 (2010). https://doi.org/10.1016/j.jss.2009.11.735
68. Zhang, X., Cheng, G., Qu, Y.: Ontology summarization based on RDF sentence graph. In: Proceedings of the 16th International Conference on World Wide Web, WWW '07, pp. 707–716. ACM, New York, NY, USA (2007). https://doi.org/10.1145/1242572.1242668
69. Zimmermann, T., Nagappan, N.: Predicting defects using network analysis on dependency graphs. In: Proceedings of the 30th International Conference on Software Engineering, ICSE '08, pp. 531–540. ACM, New York, NY, USA (2008). https://doi.org/10.1145/1368088.1368161
70. Zupanc, K., Savić, M., Bosnić, Z., Ivanović, M.: Evaluating coherence of essays using sentence-similarity networks. In: Proceedings of the 18th International Conference on Computer Systems and Technologies, CompSysTech'17, pp. 65–72. ACM, New York, NY, USA (2017). https://doi.org/10.1145/3134302.3134322

Chapter 2
Fundamentals of Complex Network Analysis

Abstract Complex network analysis is a collection of quantitative methods for studying the structure and dynamics of complex networked systems. This chapter presents the fundamentals of complex network analysis. We start out by presenting the basic concepts of complex networks and graph theory. Then, we focus on fundamental network analysis measures and algorithms related to node connectivity, distance, centrality, similarity and clustering. Finally, we discuss fundamental complex network models and their characteristics.

2.1 Basic Concepts

The term network denotes a graph representation of some real-world system. Graphs are one of the fundamental structures studied in mathematics and computer science and they have broad applications in many fields of contemporary science, including biology, social sciences, economics and engineering. A graph describes a set of objects and relations that exist among these objects.

Definition 2.1 (*Graph*) A graph $G = (V, E)$ is a set V of nodes (vertices) and a set E of links (edges) connecting pairs of nodes.

There are two basic types of graphs: undirected and directed graphs. The main difference between undirected and directed graphs is that in undirected graphs links are bidirectional, while in directed graphs they have a direction.

Definition 2.2 (*Undirected graph*) An undirected graph G is a pair (V, E) where E is a set of two-elements subsets of V, i.e. $E = \{\{a, b\} \mid a, b \in V\}$.

Let $e = \{a, b\}$ be a link in an undirected graph G connecting two nodes a and b. Then we say that (1) a and b are adjacent or neighbors, (2) a and b are directly connected, (3) both a and b are incident with e, and (4) a and b are end-points of e. The link between a and b is also written as ab (or ba). We also use $a \leftrightarrow b$ to denote

© Springer International Publishing AG, part of Springer Nature 2019
M. Savić et al., *Complex Networks in Software, Knowledge, and Social Systems*, Intelligent Systems Reference Library 148, https://doi.org/10.1007/978-3-319-91196-0_2

the link connecting a and b. Definition 2.2 corresponds to so-called *simple graphs*, i.e. graphs that do not contain multiple links between two nodes (parallel links) nor links connecting a node to itself (loops). Throughout this chapter and the rest of the monograph, unless otherwise stated, we assume that graphs and networks are simple.

Definition 2.3 (*Subgraph, induced subgraph*) A graph $G' = (V', E')$ is a subgraph of a graph $G = (V, E)$, denoted by $G' \subseteq G$, if $V' \subseteq V$ and $E' \subseteq E$. If V' is a proper subset of V ($V' \subset V$) and E' is a proper subset of E ($E' \subset E$) then G' is a proper subgraph of G. If E' contains all links of G that have endpoints in V' then we say that G' is the subgraph of G induced by V'.

Nodes a and b are indirectly connected in a graph if there is a path from a to b.

Definition 2.4 (*Path, distance*) A path from a node a to a node b in an undirected graph G is a subgraph $G' = (V', E')$ of G such that:

- $V' = \{a, v_1, v_2, v_3, \ldots, v_{n-1}, v_n, b\}$, and
- $E' = \big\{\{a, v_1\}, \{v_1, v_2\}, \{v_2, v_3\} \ldots, \{v_i, v_{i+1}\}, \ldots, \{v_{n-1}, v_n\}, \{v_n, b\}\big\}$.

The number of edges in G' is the *length* of the path between a and b. The *distance* between a and b is the length of the shortest path between a and b.

If there is a path from a to b in G then we say that b is reachable from a and vice versa. If every node in G can be reached from every other node then G is a *connected* graph. In other words, every pair of nodes in a connected graph is either directly or indirectly connected. If a graph is not connected then it consists of more than one connected component. Connected components of an undirected graph are its maximally connected subgraphs.

Definition 2.5 (*Union of graphs*) A graph $C = (V_c, E_c)$ is the union of graphs $A = (V_a, E_a)$ and $B = (V_b, E_b)$, $C = A \cup B$, if $V_c = V_a \cup V_b$ and $E_c = E_a \cup E_b$.

Definition 2.6 (*Intersection of graphs*) A graph $C = (V_c, E_c)$ is the intersection of graphs $A = (V_a, E_a)$ and $B = (V_b, E_b)$, $C = A \cap B$, if $V_c = V_a \cap V_b$ and $E_c = E_a \cap E_b$.

Definition 2.7 (*Connected components*) The connected components of a graph G are the subgraphs $(C_i)_{i=1}^{k}$ of G such that

- C_i is connected for each i.
- $\bigcup C_i = G$.
- $C_x \cap C_y = \emptyset$ for each distinct x and y, where $\emptyset$ denotes an empty graph having zero nodes and zero links.

The connected components of an undirected graph G can be determined using basic graph traversal algorithms such as breadth-first search (BFS) or depth-first search (DFS). Additionally, BFS can be employed to find the shortest path between two arbitrary nodes in G. If two nodes belong to different connected components then there is no path connecting them. Consequently, the distance between unreachable nodes is either undefined or treated as infinite.

Definition 2.8 (*Giant connected component*) If G has a component that encompasses a vast majority of nodes then this component is called a giant connected component.

A formal definition of a giant connected component can be given assuming that there is an infinite process governing the evolution of G [76]. In practice, the largest component is considered giant if its size (the number of nodes contained in the component) is considerably higher than the size of the second largest component [74].

Definition 2.9 (*Directed graph*) A directed graph (or digraph) G is a pair (V, E) where E is a set of ordered pairs of nodes, i.e. $E = \{(a, b) \mid a, b \in V\}$.

We also use $a \rightarrow b$ to denote a link $e = (a, b)$ from a node a to a node b in a directed graph G. The link $e = a \rightarrow b$ means that a is directly connected to b in the sense that e emanates from a pointing to (or referencing) b. The node a is called the source node, while the node b is called the destination node. We say that a and b are end-points of e when it is not important which of them is the source or destination node.

Let x and y be two nodes in a directed graph G such that x is not directly connected to y, i.e. G does not contain a link $x \rightarrow y$. The node x is indirectly connected to y if there is a *directed* path from x to y. A path from x to y exists if y can be reached from x following an alternating sequence of nodes and links $(x, e_0, v_1, e_1, v_2, \ldots, v_k, e_k, y)$ satisfying the following conditions:

- $e_0 = x \rightarrow v_1$,
- $e_i = v_i \rightarrow v_{i+1}$ for each i in $[1, k-1]$,
- $e_k = v_k \rightarrow y$.

The distance from x to y is equal to the length of the shortest directed path from x to y. It should be noticed that the distance from x to y is not necessarily equal to the distance from y to x due to the directed nature of links. Even more, there may not be a path from y to x when there is a path from x to y. The shortest path between two nodes in a directed graph can be determined by the BFS algorithm.

There are two kinds of connected components in directed graphs: weakly connected and strongly connected components.

Definition 2.10 (*Undirected projection of directed graph*) Let G be a directed graph. The undirected projection of G, denoted by G^U, is an undirected graph having the same set of nodes as G. Two nodes a and b are connected in G^U if at least one directed link between a and b is present in G.

Definition 2.11 (*Weakly connected component*) A subgraph $W = (V_w, E_w)$ of a directed graph $G = (V, E)$ is its weakly connected component if V_w forms a connected component in the undirected projection of G. E_w contains all links from E whose end points belong to V_w, i.e. W is induced by V_w.

Definition 2.12 (*Strongly connected component*) A subgraph $S = (V_s, E_s)$ of a directed graph $G = (V, E)$ is its strongly connected component if for every two nodes a and b from V_s there is a path from a to b and a path from b to a. V_s is the maximal subset of V with respect to inclusion which means that any proper super-graph of S is not strongly connected. E_s contains all links from E whose end points belong to V_s, i.e. S is induced by V_s.

The weakly connected components of a directed graph G can be obtained by applying BFS or DFS to the undirected projection of G. On the other side, strongly connected components can be identified in linear time with respect to the number of nodes and links using the Tarjan algorithm [89], which is an extension of the DFS algorithm.

If we assign real values to links then we obtain so-called *weighted* graphs. Both directed and undirected graphs may be weighted graphs. Weighted graphs are par-ticularly useful for representing relations or interactions that may exhibit different levels of strength.

Definition 2.13 (*Weighted graph*) A graph $G = (V, E)$ is a weighted graph if there is an associated function $w : E \to \mathbb{R}$ that assigns numerical weights to the links in E. The weight of a link e ($e \in E$) is denoted by $w(e)$.

The meaning of link weights strongly depends on the domain of the corresponding network. For example, link weights in co-authorship networks can be used to describe the strength of research collaboration. Unless otherwise stated, we assume that link weights are real positive values since this is a common situation in most application cases. The length of a path between two nodes in a weighted graph is measured as the sum of link weights along the path. The notion of node distance generalizes accordingly. Shortest paths in weighted graphs with positive link weights can be efficiently determined using the Dijkstra algorithm [30]. It should be emphasized that for networks in which link weights reflect the proximity of nodes (e.g., the strength of research collaboration in co-authorship networks) an algorithm for determining shortest paths should be applied after inverting link weights (i.e., $w(e) \to 1/w(e)$).

It is often useful, when defining and discussing graph measures and algorithms, to represent a graph by its adjacency matrix.

Definition 2.14 (*Adjacency matrix*) The adjacency matrix A of an unweighted undi-rected graph $G = (V, E)$ is a $n \times n$ square matrix where n is the number of nodes ($n = |V|$). $A_{ij} = 1$ if nodes i and j ($i, j \in V$) are directly connected ($a \leftrightarrow b \in E$), or zero otherwise. In the case that G is a directed graph then $A_{ij} = 1$ if there is a link from a to b ($a \to b \in E$), or zero otherwise.

For undirected graphs, A is a symmetric matrix with respect to the main diagonal, i.e. $A_{ij} = A_{ji}$. Additionally, $A_{ii} = 0$ if G does not contain loops. In the case that G contains parallel links then A_{ij} is equal to the number of links connecting i and j in the case of undirected graphs, or the number of links from i to j in the case of directed graphs. For weighted graphs, A_{ij} is equal to the weight of the link connecting i and j, or zero otherwise.

2.2 Complex Network Measures and Methods

Throughout this section, unless otherwise stated, we assume that $G = (V, E)$ denotes an arbitrary non-weighted undirected graph, without loops and parallel links, that contains n nodes labeled by numbers from 1 to n. Two basic structural quantities describing G are

1. the number of nodes (n, the cardinality of V), and
2. the number of links (the cardinality of E).

The number of links, denoted by l, is a positive integer in the range $[0, \max(l)]$ where $\max(l) = n(n - 1)/2$ is the maximal number of links that the nodes in G can form. For directed graphs, without parallel links and loops, $\max(l) = n(n - 1)$. The ratio of l and $\max(l)$ can be used to determine whether G is a sparse graph or a dense graph.

Definition 2.15 (*Graph density*) The density of G, denoted by $D(G)$, is equal to $l/\max(l)$.

Definition 2.16 (*Sparse graph*) G is a sparse graph if $l \ll \max(l)$.

Definition 2.17 (*Dense graph*) G is a dense graph if $l \approx \max(l)$.

Definition 2.18 (*Complete graph*) G is a complete graph if $D(G) = 1$. In other words, G is complete when each node is directly connected to every other node.

Most real-world large-scale networks are actually sparse graphs [61, 74]. Consequently, their adjacency matrices are dominantly filled with zeros. Therefore, computer programs performing analysis of large-scale networks usually rely on the adjacency list graph data structure in which

- G is represented as an array A of nodes, and
- to each node a is attached a list L containing those nodes that are directly connected to a in the case of undirected graphs or the nodes to which a points in the case of directed graphs. Each element in L actually represents one link in G, and it may contain some additional information (e.g. a weight in the case of weighted graphs).

2.2.1 Connectivity of Nodes

The most basic structural characteristic of a node in G is its degree.

Definition 2.19 (*Node degree*) The degree of a node i in G, denoted by k_i, is the number of links incident with i, i.e. $k_i = \sum_{j=1}^{n} A_{ij}$.

Definition 2.20 (*Isolated node*) A node i is an isolated node if $k_i = 0$.

If G does not contain parallel links then k_i is equal to the number of nodes to which i is directly connected. By the first theorem of graph theory, the average degree of G, denoted by $\langle k \rangle$, is equal to $2l/n$. The density of G can also be expressed in terms of $\langle k \rangle$:

$$D(G) = \frac{2l}{n(n-1)} = \frac{\langle k \rangle}{n-1} \tag{2.1}$$

Therefore, G is sparse when $\langle k \rangle \ll n - 1$, while it is dense when $\langle k \rangle \approx n - 1$.

Definition 2.21 (*Regular graph*) G is regular if all of its nodes have the same degree.

The connectivity of nodes in a regular graph can be described by one number – the average degree. For a non-regular graph the connectivity of nodes can be expressed by its degree distribution.

Definition 2.22 (*Degree distribution*) The degree distribution of G is given by the probability mass function $P(k) = P\{D = k\}$, where D is a random variable representing the degree of a randomly chosen node. In other words, $P(k)$ is the fraction of nodes in G having degree equal to k.

Definition 2.23 (*Complementary cumulative degree distribution*) The complementary cumulative degree distribution function $CCD(k)$ is the probability of observing a node with degree greater than or equal to k, that is, $CCD(k) = \sum_{i=k}^{\infty} P(i)$, where $P(i)$ is the degree distribution of G. Equivalently, $CCD(k)$ is the fraction of nodes in G whose degree is greater than or equal to k.

The connectivity of a node in a directed graph is expressed by two degrees: in-degree (the number of in-coming links) and out-degree (the number of out-going links). The degree of the node is then the sum of its in-degree and out-degree, and it is called total degree in order to emphasize the directed nature of links. Consequently, there are three degree distributions summarizing the connectivity of nodes in directed graphs: in-degree, out-degree and total-degree distributions.

Definition 2.24 (*Node in-degree*) The in-degree of a node i in a directed graph G, denoted by $k_{in}(i)$, is the number of links pointing to i, i.e. $k_{in}(i) = \sum_{j=1}^{n} A_{ji}$.

Definition 2.25 (*Node out-degree*) The out-degree of a node i in a directed graph G, denoted by $k_{out}(i)$, is the number of links emanating from i, i.e. $k_{out}(i) = \sum_{j=1}^{n} A_{ij}$.

The connectivity of nodes in weighted graphs can also be characterized by node strength.

Definition 2.26 (*Node strength*) The strength of a node i in a weighted graph G, denoted by s_i, is the sum of weights of links incident with i.

One of the frequently observed characteristics of large-scale real-world networks is that their degree distributions follow power-laws [4, 13, 29, 74]. Power-law degree distributions imply the existence of *hubs* – a small fraction of nodes with an extremely high degree, much higher than the average node degree.

Definition 2.27 (*Power-law*) Discrete probability distributions of the form $P(k) = Ck^{-\gamma}$, where C and γ are constants, are said to follow a power-law. The constant γ is called the scaling exponent of the power-law. Smaller γ implies slower decay of $P(k)$ causing a more skewed distribution. The constant C is determined by the normalization requirement that $\sum_{k=k_{min}}^{\infty} P(k) = 1$.

Definition 2.28 (*Scale-free network*) Networks whose degree distributions follow a power-law in the tail, $P(k) \sim Ck^{-\gamma}$, are known as scale-free networks.

If $P(k)$ follows the power-law $P(k) \sim Ck^{-\gamma}$ then the plot of $P(k)$ on logarithmic scales appears as a straight line of slope $-\gamma$ ($\log P(k) \sim -\gamma \log k$). The complementary cumulative degree distribution of a scale-free network also appears as a straight line on log-log plots but with slope $-(\gamma - 1)$ rather than $-\gamma$ [74, 75]. For $P(k) \sim Ck^{-\gamma}$ we have that $CCD(k) = \sum_{i=k}^{\infty} P(i) \sim \frac{C}{\gamma-1} k^{-(\gamma-1)}$. This can be easily seen if we treat (and thus approximate) discrete random variable D, that represents the degree of a randomly chosen node, as a continuous random variable. Then the sum of $P(i)$ can be expressed as the integral of a power function which is also a power function (with the exponent increased by one).

Researchers often rely on linear regression methods applied to log-log transformed empirically observed degree distributions when checking whether studied networks belong to the class of scale-free networks. However, this method for testing power-laws in empirical data was shown to be biased: (1) besides power-laws many other theoretical distributions appear as straight lines on log-log plots, and (2) linear regression leads to inaccurate estimates of power-law scaling exponents [28, 45]. Clauset et al. proposed a statistically robust method for fitting a power-law distribution to a given degree sequence (an array containing degrees of all nodes in a network) and checking its plausibility with respect to alternative theoretical distributions [28]. Their method consists of the three following steps:

1. The scaling parameter of a power-law (α) is determined using the maximum likelihood estimation (MLE) with respect to a lower bound of the power-law behavior (denoted by k_m) in the input degree sequence. The MLE for α in the case of a discrete power-law probability distribution is given by

$$\hat{\alpha} = 1 + n \left[\sum_{i=1}^{n} \ln \frac{k_i}{k_m - 0.5} \right]^{-1} \tag{2.2}$$

where n is the number of nodes whose degree is higher than or equal to k_m, and k_i is the degree of node i. k_m is determined by minimizing the weighted Kolmogorov-Smirnov (KS) statistic. The KS statistic is the maximum distance between the cumulative distribution function (CDF) of the input degree sequence and the fitted power-law model. The weighted variant of the KS statistic, denoted by KS^*, adds weights to distances in order to obtain uniform sensitivity across the whole range of degree values:

$$KS^* = \max_{k \geq k_m} \frac{|S(k) - P(k)|}{\sqrt{P(k)(1 - P(k))}} \qquad (2.3)$$

where $S(k)$ is the empirically observed CDF and $P(k)$ is the CDF of the power-law model.

2. A large number of synthetic datasets containing values following a power-law distribution are generated using the estimated values of k_m and α in order to compute the goodness of the fitted power-law model. For each synthetic dataset, the parameters of the power-law model are determined in the same way as for empirical data and the value of KS^* statistic is recorded. The quality of the power-law fit (p-value) for empirical data is the probability that a randomly selected synthetic dataset has a higher value of KS^* compared to the value of KS^* obtained for empirical data. If the obtained p-value is lower than 0.1 then the power-law hypothesis is rejected, i.e. power-law is not considered as a plausible statistical model for the empirically observed degree distribution. The number of synthetic dataset has to be higher than 2500 in order to ensure that the p-value is accurate up to 2 decimal digits.

3. The parameters of alternative distributions are also determined by appropriate MLEs. The power-law fit is compared to the fits of alternative distributions using the likelihood ratio test. This test checks the null hypothesis that two theoretical distributions are equally far from an empirically observed distribution.

One important feature of social systems is the presence of homophily. Homophilic behavior in social systems means that an individual tends to establish social interactions with individuals being similar to him/her by one or more social characteristics such as age, ethnicity, professional vocation, income, educational level, and so on. The presence of homophily can be spotted not only in social networks but also in a variety of real-world networks from other domains. For example, research papers in a citation network tend to be more connected to papers in the same research field than to papers in other research fields. Considering degree as the most basic structural characteristic of nodes, three types of networks can be distinguished:

- *Assortative networks* – networks in which hubs (nodes with a high degree) tend to be connected among themselves.
- *Disassortative networks* – networks in which hubs tend to avoid connections to other hubs.
- *Non-assortative networks* – networks that are neither assortative nor disassortative. Such networks are also called uncorrelated networks.

Newman [71, 73] proposed a measure known as the assortativity index to quantify the level of degree-based node mixing in a network.

Definition 2.29 (*Assortativity index*) The assortativity index of G, denoted by $A(G)$, is the Pearson correlation coefficient between random variables X and Y, where X and Y are degrees (total-degrees when G is directed or strengths when G is weighted) of end-points of a randomly selected link.

$A(G)$ takes values in the range $[-1, 1]$. A positive value of $A(G)$ means that G is an assortative network. A higher value implies a stronger assortative mixing pattern. A negative value of $A(G)$ means that G is disassortative. A lower value implies a stronger disassortative mixing pattern. Finally, $A(G)$ close to zero means that G is an uncorrelated network. The assortativity index can be generalized to any numerical characteristic of nodes, as well as to any pair of numerical characteristics of nodes.

Definition 2.30 (*Generalized assortativity index*) Given two node metrics M_1 and M_2, the generalized assortativity index of G with respect to M_1 and M_2, denoted by $A(G, M_1, M_2)$, is the Pearson correlation coefficient between random variables X and Y, where X is the value of M_1 and Y is the value of M_2 for end-points of a randomly selected link.

Thus, for directed networks we can measure in-degree assortativity $A(G, k_{in}, k_{in})$, out-degree assortativity $A(G, k_{out}, k_{out})$, in-out-degree assortativity $A(G, k_{in}, k_{out})$, and out-in-degree assortativity $A(G, k_{out}, k_{in})$ which are different instances of the generalized assortativity index for indicated node metrics.

Another way to determine whether a network exhibits an assortative mixing is to compute the slope of the function $k_{nn}(k)$ that denotes the average degree of the nearest neighbors of nodes with degree k [78]. If $k_{nn}(k)$ increases with k then the network is assortative, while for disassortative networks $k_{nn}(k)$ is a decreasing function of k.

Newman also proposed a measure to quantify network assortativty with respect to discrete or enumerative node attributes [73]. Let G be a directed network whose nodes are divided into k categories. We can define a mixing matrix e such that e_{xy} is equal to the fraction of links connecting nodes of type x to nodes of type y ($1 \leq x, y \leq k$). For undirected networks we have that $e_{xy} = e_{yx}$. Then, the assortativity index of G with respect to the predefined categorization of nodes can be quantified as

$$r = \left(\sum_{i=1}^{k} e_{ii} - \sum_{i=1}^{k} a_i b_i \right) \Big/ \left(1 - \sum_{i=1}^{k} a_i b_i \right) \tag{2.4}$$

where a_i is the fraction of links in which the source node is of type i and b_i represents the fraction of links in which the destination node belongs to the category i. $\sum_{i=1}^{k} e_{ii}$ is the fraction of links connecting nodes that are in the same category, while $\sum_{i=1}^{k} a_i b_i$ would be the fraction of such links if all links in the network were formed randomly. The above stated formula gives $r = 1$ for perfectly assortative networks ($\sum_{i=1}^{k} e_{ii} = 1$ and $\sum_{i=1}^{k} a_i b_i = 0$), $r > 0$ for assortative networks, $r = 0$ for non-assortative networks ($e_{xy} = a_x b_y$), and a negative value for dissasortative networks.

The presence of a strong degree-based homophily in a network indicates that it exhibits a core-periphery structure. A network with a core-periphery structure is composed of core high-degree nodes densely connected among themselves and sparsely connected peripheral low-degree nodes [21, 84]. The k-core decomposition technique [5, 86] can be used to identify nested cores in a complex network and quantify the connectivity of nodes with respect to identified cores.

Definition 2.31 (*K-core*) An induced subgraph K of a graph G is the k-core of G if K is the largest subgraph in which each node has degree equal to or higher than k.

The k-core of G can be obtained by successively removing nodes having degree lower than k until all nodes in the remaining graph have degree higher than or equal to k (when a node is removed from G then the degree of adjacent nodes decreases by one). K-cores are nested: all nodes from a p-core also belong to a q-core if $q < p$.

Definition 2.32 (*Shell index, coreness*) The shell index or coreness of a node x is equal to k if x belongs to the k-core of G, but not to the $(k + 1)$-core of G.

It is important to observe that nodes having a high degree may have a low shell index. For example, if a node x is connected to k nodes and all of them have degree equal to 1 then the shell index of x equals 1, not k. Therefore, the shell index measure can be used to identify two considerably different types of hubs: hubs mostly connected to other hubs have a high shell index, while hubs dominantly connected to non-hubs have a low shell index. The k-core decomposition can be efficiently performed (without deleting nodes) by the algorithm proposed by Batagelj and Zaveršnik [8, 53].

Algorithm 2.1: The Batagelj-Zaveršnik algorithm for k-core decomposition

---

input : a graph $G = (V, E)$
output: S – an array of integers such that $S[v]$ ($v \in V$) is equal to the shell index of v

$m =$ the maximal degree of nodes in V
$d =$ an array of integers of length n where $n = |V|$
$D =$ an array of m empty node sets

foreach $v \in V$ **do**
    $k =$ the degree of node v
    $d[v] = k$
    add v to $D[k]$
end

for $k = 0$ **to** m **do**
    while $D[k] \neq \emptyset$ **do**
        $x =$ remove a random node from $D[k]$
        $S[x] = k$
        foreach $v \in V : \{x, v\} \in E$ **do**
            if $d[v] > k$ **then**
                remove v from $D[d[v]]$
                add v to $D[d[v] - 1]$
                $d[v] = d[v] - 1$
            end
        end
    end
end

2.2.2 Distance Metrics

Two nodes in a graph G that belong to the same connected component can be indirectly connected via more than one path. However, shortest paths are the most important considering the efficiency of communication and information flow. Please recall that the distance between two nodes i and j, denoted by d_{ij}, is defined as the length of the shortest path connecting i and j.

Definition 2.33 (*Small-world coefficient, characteristic path length*) The small-world coefficient or characteristic path length of a graph G, denoted by $L(G)$, is the average distance between nodes in G:

$$L(G) = \frac{1}{n(n-1)} \sum_{i,j \in V, i \neq j} d_{ij} \tag{2.5}$$

The problem with the definition given above is that some distances may be undefined (or infinite) if G contains more than one connected component. One possibility to avoid this problem is to focus computation only to the largest connected component (the largest strongly connected component if G is a directed graph). This approach is usually taken when G has a giant (strongly) connected component. The second possibility is to consider only those pairs of nodes that are directly or indirectly connected. The third possibility is to take a related, well-defined measure known as *efficiency* [13]. The efficiency of a graph is defined as

$$E(G) = \frac{1}{n(n-1)} \sum_{i,j \in V, i \neq j} \frac{1}{d_{ij}} \tag{2.6}$$

where $1/d_{ij} = 0$ if there is no path connecting i and j.

Definition 2.34 (*Small-world property*) Having the small-world property means that the small-world coefficient is a small value (typically $L(G) \approx log(n)$), much smaller than the number of nodes in G.

Definition 2.35 (*Diameter*) The diameter of G is the length of the longest shortest path in G:

$$\text{Diam}(G) = \max_{i,j \in V, i \neq j} d_{ij} \tag{2.7}$$

Definition 2.36 (*Eccentricity*) The eccentricity of a node x is the maximal distance between x and any other node in G:

$$\text{Ecc}(x) = \max_{y \in V \setminus \{x\}} d_{xy} \tag{2.8}$$

Thus, the diameter of G is equal to the maximum eccentricity of any node in G.

Definition 2.37 (*Radius*) The radius of G is the minimum eccentricity of any node in G:

$$\text{Radius}(G) = \min_{x \in V} \text{Ecc}(x) \qquad (2.9)$$

2.2.3   Centrality Metrics and Algorithms

The main aim of centrality metrics is to quantify the importance of a node according to its position and connectedness within a network. Consequently, centrality metrics can be exploited to rank nodes according to some notion of structural importance and identify the most important ones. Node degree is the most simplest measure of node importance. For example, an actor having a high degree in a social network can be considered important since he is in a position to directly disseminate his opinions and articulate his interests to a large number of other actors. A large number of contacts in a social network indicates highly influential, visible and prestigious actors having a high social capital. Other fundamental notions and metrics of structural importance also originate from social network analysis. The three most fundamental node centrality metrics are betweenness centrality [23, 40, 41], closeness centrality [9, 10] and eigenvector centrality [19, 20, 79].

Definition 2.38 (*Betweenness centrality*) The betweenness centrality of a node z in a graph G, denoted by $C_b(z)$, is the extent to which z is located on the shortest paths between two arbitrary nodes different than z:

$$C_b(z) = \sum_{x,y \in V, x \neq y \neq z} \frac{\sigma(x, y, z)}{\sigma(x, y)} \qquad (2.10)$$

where $\sigma(x, y)$ is the total number of shortest paths between x and y, and $\sigma(x, y, z)$ is the total number of shortest paths between x and y passing through z.

The term $\sigma(x, y, z)/\sigma(x, y)$ represents the probability that a shortest path between x and y contains z. $C_b(z)$ takes a maximum value when z is located on all shortest paths between all other nodes. Thus, betweenness centrality can be normalized to the unit interval through division by the number of node pairs not including z, which is $T = (N - 1)(N - 2)/2$, where N is the number of nodes in G ($T = (N - 1)$ $(N - 2)$ for directed graphs). If a large fraction of shortest paths contain z then z can be considered as an important junction point of the network having a vital role to the overall connectedness of its nodes. Indeed, removing nodes in the betweenness centrality order causes an extremely quick fragmentation of real-world networks into a large number of disjoint connected components [15, 50]. In networks with a clustered or community structure, nodes having a high betweenness centrality tend to be located at the intersections of communities which means that they connect together different node groups. On the other hand, nodes with a low betweenness centrality are typically located at the periphery of the network. In social networks, betweenness centrality can be viewed as a measure of the influence an actor has over the spread of information across the network, i.e. actors with a high value of betweenness centrality are in the position to maintain and control the spread of information across the network.

The crucial observation for computing betweenness centrality is that a node z lies on a shortest path between nodes x and y if $d_{xy} = d_{xz} + d_{zy}$, where d_{xy} denotes the distance between x and y. Then, the number of shortest paths between x and y containing z can be computed as

$$\sigma(x, y, z) = \begin{cases} 0 & \text{if } d_{xy} < d_{xz} + d_{zy} \\ \sigma(x, z) \cdot \sigma(z, y) & \text{otherwise} \end{cases} \tag{2.11}$$

Therefore, to compute the betweenness centrality of z we have to determine the distance and the number of shortest paths between z and all other nodes. Both problems can be solved in a straightforward manner by simple modifications of the BFS algorithm in the case of non-weighted networks and the Dijsktra algorithm in the case of weighted networks, leading to an algorithm for computing betweenness centrality of all nodes in a network in $O(N^3)$ time and $O(N^2)$ space. Brandes proposed a much more efficient algorithm for computing betweenness centrality that works in $O(N^2)$ time and requires $O(N)$ space for sparse networks [22]. Let $\delta_{xy}(z)$ denote the probability that a shortest path between x and y contains z, i.e. $\delta_{xy}(z) = \sigma(x, y, z)/\sigma(x, y)$. The Brandes algorithm is based on the notion of the dependency of x to z defined as

$$\delta_x(z) = \sum_{y \in V} \delta_{xy}(z) \tag{2.12}$$

Then, the betweenness centrality of z is equal to

$$C_b(z) = \sum_{x \in V} \delta_x(z) \tag{2.13}$$

Brandes showed that the dependency of x on any z obeys the following recurrence relation

$$\delta_x(z) = \sum_{w \in S(z)} \frac{\sigma(x, z)}{\sigma(x, w)} (1 + \delta_x(w)) \tag{2.14}$$

where $S(z)$ denotes the set of successors of z in the BFS/Dijkstra tree. The previous recurrence relation enables betweenness centrality computation by an algorithm that has two phases for each node:

1. A modified BFS/Dijkstra algorithm to compute the distance and the number of shortest paths between a node x and all other nodes.
2. The second phase visits all nodes reachable from x in the reverse order of their BFS/Dijsktra discovery to accumulate dependencies according to the recurrence relation and update the betweenness centrality of visited nodes.

Betweenness centrality can also be defined for links and groups of nodes.

Definition 2.39 (*Link betweenness centrality*) The betweenness centrality of a link l is the extent to which l is located on the shortest paths between two arbitrary nodes

$$C_b(l) = \sum_{x,y \in V} \frac{\sigma(x, y, l)}{\sigma(x, y)} \tag{2.15}$$

where $\sigma(x, y)$ is the total number of shortest paths between x and y, and $\sigma(x, y, z)$ is the total number of shortest paths between x and y containing l.

Definition 2.40 (*Group betweenness centrality*) The betweenness centrality of a group of nodes S, $S \subset V$, is the extent to which the nodes from S are located on the shortest paths between two arbitrary nodes that are not in S:

$$C_b(S) = \sum_{x,y \in V \setminus S} \frac{\sigma(x, y, S)}{\sigma(x, y)} \tag{2.16}$$

where $\sigma(x, y)$ is the total number of shortest paths between x and y, and $\sigma(x, y, S)$ is the total number of shortest paths between x and y containing at least one node from S.

Definition 2.41 (*Closeness centrality*) The closeness centrality of a node z in a graph G, denoted by $C_c(z)$, is inversely proportional to the total distance between z and all other nodes in G:

$$C_c(z) = \frac{1}{\sum_{i \in V \setminus \{z\}} d_{zi}} \tag{2.17}$$

The intuition behind the closeness centrality measure is that a node can be considered important if it is in proximity to many other nodes. Let us consider a simple spreading process in which information reaching a node is propagated to all of its neighbors. Then, information originating at nodes having a high closeness centrality will propagate more efficiently through the network. The problems with the definition of closeness centrality arise when the network consists of more than one (strongly) connected component. If we restrict the total distance only to reachable nodes then the closeness centrality measure becomes strongly biased towards nodes that have a small set of reachable nodes [14]. An alternative solution is to consider the normalized harmonic centrality that is defined as

$$C_h(z) = \frac{1}{n-1} \sum_{i \in R(z)} \frac{1}{d_{zi}} \tag{2.18}$$

where $R(z)$ denotes the set of nodes reachable from z.

The closeness vitality index is a centrality index closely related to closeness centrality. This measure quantifies the drop in the Wiener index after a node is removed from the network. A higher drop indicates a more important node considering the cost of message passing in all-to-all communication between nodes.

Definition 2.42 (*The Wiener index*) The Wiener index of a graph G, denoted by $W(G)$, is the sum of node distances for all node pairs in G:

$$W(G) = \sum_{x,y \in V, x \neq y} d_{xy} \qquad (2.19)$$

Definition 2.43 (*The closeness vitality index*) The closeness vitality index of a node z in a graph G, denoted by $CV(z)$, is the difference of the Wiener index of G and the Wiener index of G without z:

$$CV(z) = W(G) - W(G \setminus \{z\}) \qquad (2.20)$$

The same idea can also be applied for links, i.e. we can quantify the importance of a link l by measuring the difference of $W(G)$ and $W(G \setminus \{l\})$. The closeness vitality index possesses the same weakness as the closeness centrality measure: a graph may become disconnected after a node or link is removed, leading to an undefined or infinite value of the measure.

Definition 2.44 (*Eigenvector centrality*) The eigenvector centrality of a node z in a graph G, denoted by $C_e(z)$, is proportional to the sum of eigenvector centralities of its neighbors:

$$C_e(z) = \frac{1}{\lambda} \sum_{i \in N(z)} C_e(i) \qquad (2.21)$$

where λ is a constant and $N(z)$ denotes the set of nodes directly connected to z, i.e. $N(z) = \{w : \{w, z\} \in E\}$.

The main idea of the eigenvector centrality metric is that a node can be considered important if it is directly connected to important nodes. In the case of directed graphs, $N(z)$ is the set of nodes pointing to z, i.e. z is important if it is referenced by important nodes. Since

$$\sum_{i \in N(z)} C_e(i) = \sum_{i \in V} A_{iz} C_e(i) \qquad (2.22)$$

the recursive relation defining eigenvector centrality can be stated as the eigenvector equation

$$\lambda C_e = A C_e \qquad (2.23)$$

where C_e is the vector containing eigenvector centralities of all nodes. Therefore, C_e is the eigenvector of the adjacency matrix A associated with the eigenvalue λ. If λ is the largest eigenvalue of A and A is reducible, which means that G is (strongly) connected, then C_e is both unique (up to a constant factor) and positive. The common way to compute the eigenvector corresponding to the largest eigenvalue is by the power iteration algorithm (see Algorithm 2.2). The power iteration algorithm starts with an unit vector C_e. In each iteration, A is multiplied with C_e, the resulting vector C_e^n is normalized to an unit vector and then copied into C_e. The iterations are repeated until either the desired precision is reached or the number of iterations exceeds a given maximum value.

Let $G' = (V', E')$ be an arbitrary directed graph that is not strongly connected and let V^s denote the subset of V' that contains those nodes which belong to non-trivial strongly connected components of G' (strongly connected components encompassing two or more nodes). The main problem of the eigenvector centrality measure when applied to G' is that only nodes in V^s and nodes reachable from nodes in V^s can have non-zero eigenvector centrality. The situation is even more drastic in acyclic directed graphs in which the eigenvector centrality of each node is zero (acyclic directed graphs do not contain non-trivial strongly connected components). A centrality measure known as the Katz centrality overcomes the previous problem by adding a fixed amount of importance to each node.

Algorithm 2.2: The algorithm for computing eigenvector centralities of all nodes in a graph based on the power iteration method.

---

input : a graph $G = (V, E)$
            M – the maximal number of iterations
            ε – a real value indicating the desired precision
output: C_e – a vector containing eigenvector centralities of nodes in G

$N = |V|$
$C_e = [1, 1, 1, \ldots, 1]_N$
$C_e^n = [0, 0, 0, \ldots, 0]_N$
$\Delta = \varepsilon$
$i = M$
while $\Delta \geq \varepsilon \wedge i > 0$ **do**
    $i = i - 1$
    foreach $v \in V$ **do**
        // determine the new value of eigenvector centrality of v according to
        // the current eigenvector centralities of its neighbors
        $C_e^n[v] = 0$
        foreach $w \in V : \{v, w\} \in E$ **do**
            $C_e^n[v] = C_e^n[v] + C_e[w]$
        end
    end
    // normalize C_e^n to an unit vector
    $C_e^n = C_e^n / \|C_e^n\|$
    // compute the L_1 distance between C_e^n and C_e
    $\Delta = 0$
    foreach $v \in V$ **do**
        $\Delta = \Delta + |C_e^n[v] - C_e[v]|$
    end
    // copy C_e^n to C_e for the next iteration
    $C_e = C_e^n$
end

Definition 2.45 (*Katz centrality*) The Katz centrality of a node z in a directed graph G, denoted by $C_k(z)$, is defined as

$$C_k(z) = \beta + \alpha \sum_{i \in N(z)} C_k(i) \tag{2.24}$$

where α and β are constants and $N(z)$ denotes the set of nodes pointing to z, i.e. $N(z) = \{w : (w, z) \in E\}$.

Similarly to eigenvector centrality, Katz centrality can be computed by successive approximations. To ensure the convergence, the parameter α has to be strictly less than $1/\lambda$, where λ is the dominant eigenvalue of the adjacency matrix.

The PageRank measure is another variant of eigenvector centrality for directed graphs. This measure was initially designed for ranking nodes in a web graph crawled by the Google search engine [25]. Three main principles behind the measure are: (1) the importance of a node x depends on the importance of nodes which point to x, (2) x transfers its importance to referenced nodes in equal shares, and (3) a fixed amount of importance is given to each node "for free" (as in the Katz centrality).

Definition 2.46 (*PageRank*) The PageRank of a node z in a directed graph G, denoted by $PR(z)$, is given by the following recurrence relation:

$$PR(z) = \frac{1-\alpha}{N} + \alpha \sum_{w \in V \,:\, w \to z} \frac{PR(w)}{k_{out}(w)} \tag{2.25}$$

$$= \frac{1-\alpha}{N} + \alpha \sum_{w \in V} A_{wz} \frac{PR(w)}{k_{out}(w)} \tag{2.26}$$

where $k_{out}(w)$ is the out-degree of node w, N is the number of nodes in G and α is a constant called the dumping factor (usually $\alpha = 0.85$).

Similarly to other recursively defined centrality metrics, PageRank can be computed by successive approximations staring from a vector in which all nodes have the same PageRank value (see Algorithm 2.3).

PageRank has a nice probabilistic interpretation related to random walks in graphs. Let W denote a random walker (an agent walking randomly through a graph) initially positioned at a randomly selected node of a directed graph G. Let a_t denote the node at which the walker is positioned after t random walk steps, and let S_t be the out-neighborhood of a_t ($S_t = \{b : (a_t, b) \in E\}$). The walker moves from a_t to a_{t+1} according to the following rules:

1. With probability α it moves from a_t to a randomly selected node from S_t. If $S_t = \emptyset$ then it stays at a_t ($a_{t+1} = a_t$).
2. With probability $1 - \alpha$ the walker jumps to a randomly selected node from G (the teleportation rule).

The teleportation rule ensures that the walker is never "trapped" at some node in the case that G is not strongly connected. Then, the PageRank of a node i is equal to the probability that the random walker is located at i after k random walk steps (according to the above given rules) when k tends to infinity.

The HITS (Hyperlink-Induced Topic Search) hub and authority scores introduced by Kleinberg [55] are also widely used centrality measures for directed networks. The HITS algorithm was also initially conceived for ranking web pages. In information networks, such as WWW, two types of important nodes can be distinguished: authorities and hubs. Authorities are nodes that contain information relevant to a topic of interest, while hubs are nodes that point to authorities. A node can be considered as a good hub if it points to many good authorities, while it is a good authority if it is being referenced by many good hubs. Therefore, the main idea of the HITS algorithm is to use two mutually recursive scores indicating whether a node is a good hub, a good authority or both.

Algorithm 2.3: The algorithm for computing PageRank.

---

input : a graph $G = (V, E)$
         M – the maximal number of iterations
         ε – a real value indicating the desired precision
         α – the damping factor
output: C_p – a vector containing PageRank values of nodes in G

$N = |V|$
$C_p = [1/N, 1/N, 1/N, \ldots, 1/N]_N$
$C_p^n = [0, 0, 0, \ldots, 0]_N$
$\Delta = \varepsilon$
$i = M$

while $\Delta \geq \varepsilon \wedge i > 0$ **do**
 | $i = i - 1$
 | **foreach** $v \in V$ **do**
 | | $C_p^n[v] = 0$
 | | **foreach** $w \in V : (w, v) \in E$ **do**
 | | | $C_p^n[v] = C_p^n[v] + C_p[w]/\text{out-degree}(w)$
 | | **end**
 | | $C_p^n[v] = (1 - \alpha)/N + \alpha C_p^n[v]$
 | **end**
 | $C_p^n = C_p^n/\|C_p^n\|$
 | $\Delta = 0$
 | **foreach** $v \in V$ **do**
 | | $\Delta = \Delta + |C_p^n[v] - C_p[v]|$
 | **end**
 | $C_p = C_p^n$
end

Definition 2.47 (*HITS hub and authority scores*) The authority score of a node z in a directed graph G, denoted by $A(z)$ is proportional to the sum of hub scores of nodes pointing to z. On the other side, the hub score of z, denoted by $H(z)$ is proportional to the sum of authority scores of nodes to which z points, i.e. $A(z)$ and $H(z)$ satisfy the following recurrence relations:

$$H(z) = \alpha \sum_{w \in V : z \to w} A(w) \tag{2.27}$$

$$A(z) = \beta \sum_{w \in V : w \to z} H(w) \tag{2.28}$$

where α and β are constants.

It can be shown that vectors containing hub and authority scores of all nodes in a directed network are actually the principal eigenvectors of AA^T and A^TA, respectively. Consequently, simple and straightforward modifications of Algorithm 2.2 lead to an algorithm for computing HITS hub and authority scores.

2.2.4 Node Similarity Metrics

Node similarity metrics quantify structural similarity (proximity) or dissimilarity (distance) between two nodes in a network. Those metrics have important applications in hierarchical clustering of nodes, identification of missing links and link prediction.

Let S denote a function that quantifies the similarity of nodes in an arbitrary graph $G = (V, E)$, i.e. $S : V \times V \to \mathbb{R}$. We suppose that a higher value of S indicates a higher similarity between nodes. To apply a hierarchical clustering algorithm to G we have to define a function S_C quantifying the similarity between non-overlapping groups of nodes. There are three most common approaches to define S_C having S defined:

1. *Maximum similarity approach.* The similarity between two groups of nodes C_1 and C_2 is equal to the maximum similarity between nodes in C_1 and C_2, i.e.

$$S_C(C_1, C_2) = \max_{x \in C_1, y \in C_2} S(x, y) \tag{2.29}$$

2. *Minimum similarity approach.* The similarity between C_1 and C_2 is equal to the minimum similarity between nodes in C_1 and C_2, i.e.

$$S_C(C_1, C_2) = \min_{x \in C_1, y \in C_2} S(x, y) \tag{2.30}$$

3. *Average similarity approach.* The similarity between C_1 and C_2 is equal to the average similarity between nodes in C_1 and C_2, i.e.

$$S_C(C_1, C_2) = \frac{1}{|C_1||C_2|} \sum_{x \in C_1, y \in C_2} S(x, y) \tag{2.31}$$

where $|\cdot|$ denotes the number of nodes in a group.

Then, the clustering dendrogram of G (a tree defining hierarchically nested clusters of nodes) can be obtained by the following agglomerative clustering algorithm:

1. Put each node in a separate group and form a list of groups L. Each group in L becomes a leaf node in the dendrogram.
2. Find two most similar groups C_1 and C_2 in L according to S_C. Join C_1 and C_2 into a new group C_{12}. Add C_{12} to L after removing C_1 and C_2 from L. Finally, add C_{12} to the dendrogram such that C_{12} is the parent node of C_1 and C_2.
3. Repeat the previous step if L contains more than one group of nodes.

The clustering dendrogram can also be obtained by employing the divisive hierarchical clustering approach:

1. Put all nodes in one group R that is the root of the dendrogram. Add R to an empty list of groups denoted by L.
2. For each group C in L: split C into two groups C_1 and C_2 such that C_1 and C_2 have the lowest similarity, remove C from L, add C_1 and C_2 to L, and add C_1 and C_2 to the dendrogram such that C is the parent node of C_1 and C_2.
3. Repeat the previous step if L contains at least one group having more than one node.

Node similarity metrics also have important applications related to link prediction and identification of missing links in complex networks [63, 65]. Let L_d denotes the set of node pairs that are not directly connected in G, i.e. $L_d = \{(x, y) : x, y \in V, \{x, y\} \notin E\}$. The elements of L_d can be ranked according to the similarity of paired nodes. Then, the top ranked pairs constitute the most probable links that may appear in the future. Node similarity measures can also be used as features in supervised link prediction [48, 64]. A comprehensive overview of link prediction methods can be found in [66].

Node similarity metrics can be divided into two broad categories: metrics based on structural equivalence and metrics based on regular equivalence [76]. Two nodes are structurally equivalent if they are connected to the same nodes, i.e.

$$A \text{ is structurally equivalent to } B \Leftrightarrow N(A) = N(B) \tag{2.32}$$

where $N(A)$ is the set of neighbors of A. The main idea behind node similarity metrics based on the notion of structural equivalence is that two nodes can be considered similar if they have a large number of common neighbors. The most used node similarity metrics belonging to this category are:

1. The number of common neighbors:

$$S(x, y) = |N(x) \cap N(y)| = \sum_{z \in V} A_{xz} A_{zy} = (A^2)_{xy} \tag{2.33}$$

where A is the adjacency matrix of G.

2. The Jaccard index which is the number of common neighbors normalized by the total number of neighbors of x and y:

$$S(x, y) = \frac{|N(x) \cap N(y)|}{|N(x) \cup N(y)|} \tag{2.34}$$

The Jaccard index improves the number of common neighbors by giving a higher similarity to pairs of loosely coupled nodes.

3. The cosine similarity between the rows (or columns) of x and y in the adjacency matrix A:

$$S(x, y) = \frac{A_x \cdot A_y}{|A_x| \, |A_y|} = \frac{\sum_{z \in V} A_{xz} A_{yz}}{\sqrt{\sum_{z \in V} A_{xz}^2 \, \sum_{z \in V} A_{yz}^2}} \tag{2.35}$$

This measure is also known as the Salton index. For unweighted graphs, $A_{xz}^2 = A_{xz}$ and $\sum_{z \in V} A_{xz}^2 = |N(x)|$. Thus, the cosine node similarity in unweighted graphs is equivalent to

$$S(x, y) = \frac{|N(x) \cap N(y)|}{\sqrt{|N(x)| \, |N(y)|}} \tag{2.36}$$

4. The Person correlation coefficient between the rows (or columns) of x and y in the adjacency matrix A.

5. The Adamic-Adar index [2] that assigns weights to common neighbors of x and y. The main idea of this measure is that common neighbors are not equally important for the similarity between x and y. Hubs (highly coupled nodes) should obtain comparatively smaller weights since they have a high probability to be in the set of common neighbors. Then, the similarity between x and y is calculated as the sum of weights of common neighbors by the following formula:

$$S(x, y) = \sum_{z \in N(x) \cap N(y)} 1/log(|N(z)|) \tag{2.37}$$

Node similarity metrics based on the concept of regular equivalence are founded on the principle that two nodes are similar if they are connected to similar nodes. The two most used node similarity metrics from this category are:

1. The Katz index expressing the idea that x and y are similar if x is either directly or indirectly connected to a node z that is similar to y. The Katz index is computed by the following formula:

$$S(x, y) = \sum_{l=1}^{\infty} \beta^l P(x, y, l) \tag{2.38}$$

where β is a parameter of the measure and $P(x, y, l)$ is the number of paths between x and y of length l. The parameter β controls the importance of paths between x and y with respect to their length. Smaller values of β make shorter paths more important. The value of β should be smaller than the dominant eigenvalue of the adjacency matrix A in order to ensure the convergence. It can be shown that the matrix containing the values of the Katz similarity index for all pairs of nodes is equivalent to $(I - \beta A)^{-1} - I$, where I is the identity matrix.

2. The SimRank measure introduced by Jeh and Widom [52]. The basic idea behind this measure is that two nodes can be considered similar if their neighbors are similar. SimRank is defined by the following recurrence relation:

$$S(x, y) = \frac{C}{|N(x)|\,|N(y)|} \sum_{p \in N(x), q \in N(y)} S(p, q) \qquad (2.39)$$

where C is a constant. In the case of directed graphs $N(x)$ is the set of in-neighbors of x. It can be observed that two nodes without common (in-)neighbors may have a high SimRank similarity. SimRank can be computed by successive approximations starting from $\text{SimRank}(x, x) = 1$ and $\text{SimRank}(x, y) = 0$.

2.2.5 Link Reciprocity

In a directed graph, a link connecting a node A to a node B is reciprocated if the graph also contains a link connecting B to A. The tendency of node pairs to form reciprocal links can be quantified by the link reciprocity metric.

Definition 2.48 (*Link reciprocity*) The link reciprocity of a directed graph G, denoted by $r(G)$, is the fraction of reciprocal links in G, or equivalently the probability that a randomly selected link is reciprocated. Therefore, the link reciprocity of G can be computed by the following formula:

$$r(G) = \frac{\sum_{i=1}^{n} \sum_{j=1}^{n} A_{ij} A_{ji}}{l} \qquad (2.40)$$

where n and l are the number of nodes and links in G, respectively, and A denotes the adjacency matrix of G.

Obviously, $r(G) = 0$ implies that G does not contain reciprocal links, while $r(G) = 1$ implies that all links are reciprocated. However, $r(G)$ does not tell us whether reciprocal links occur more or less frequently than it can be expected by random chance. Garlaschelli and Loffredo [42] proposed a measure of link reciprocity that takes into account the Erdős-Renyi model of random graphs as the null model (see Sect. 2.3).

Definition 2.49 (*Normalized link reciprocity*) The normalized link reciprocity of a directed graph G, denoted by $r_n(G)$, is the Pearson correlation coefficient between entries A_{ij} and A_{ji} ($i \neq j$) of the adjacency matrix A of G. Therefore, $r_n(G)$ is computed by the following formula:

$$r_n(G) = \frac{\sum_{i \neq j}(A_{ij} - \overline{A})(A_{ji} - \overline{A})}{\sum_{i \neq j}(A_{ij} - \overline{A})^2} = \frac{r(G) - \overline{A}}{1 - \overline{A}} \qquad (2.41)$$

where $\overline{A}$ is the average value of the entries in A with the main diagonal being excluded, i.e.

$$\overline{A} = \frac{\sum_{i \neq j} A_{ij}}{n(n-1)} \qquad (2.42)$$

It can be seen that $\overline{A}$ actually represents the density of G. Therefore, $\overline{A}$ can be interpreted as the probability that two nodes are connected in a directed graph of the same size in which links are formed at random. Graphs obeying $r_n < 0$ are called anti-reciprocal meaning that their link reciprocity is smaller than expected by random chance, while $r_n > 0$ indicates exactly the opposite.

2.2.6   Clustering, Cohesive Groups and Community Detection Algorithms

One of prominent characteristics of social systems is transitivity of social interactions, i.e. the tendency of neighbors of a node to be neighbors themselves. Strong homophily (assortativity) and transitivity cause the existence of highly cohesive subgraphs in complex networks. The most cohesive subgraphs that can be found in a graph are called *cliques*.

Definition 2.50 (*Clique*) A clique in a graph G is a completely connected subgraph S of G. This means that every two nodes in S are directly connected. Being a clique is an invariant property under the node removal operation: if a node is removed from a clique then the resulting subgraph is still a clique.

Cliques are perfectly dense (their density equals 1), perfectly compact (their diameter equals 1), and perfectly connected (a better connectivity is not possible). A clique C is maximal if it cannot be extended to a larger clique, i.e. C is not a part of any larger clique. Maximal cliques can be identified by the Bron–Kerbosch backtracking algorithm [26] and its variants [33].

Being a completely connected subgraph is a too strong criterion for determining whether a subset of nodes form a cohesive group. Subgraphs that are close to being completely connected can also be considered strongly cohesive. Therefore, several weakened notions of a clique were proposed in the literature [3, 67, 87]. A subset of nodes W in a graph G is said to be a k-clique if for all $x, y \in W$ the distance between x and y is less than or equal to k [3]. The notion of k-cliques is more flexible than that of cliques, but it has one major weakness: shortest paths between two nodes in W do not necessarily contain only nodes from W. Therefore, k-cliques may have diameter higher than k and, even more, they may have more than one connected component, which is not consistent with an intuitive view of cohesive node groups.

Mokken [67] proposed two alternatives for k-cliques: k-clubs and k-clans. A k-clan is a maximal k-clique whose diameter does not exceed k. On the other hand, a k-club is a maximal subgraph of diameter k. Other relaxed notions of a clique are based on the connectivity of nodes within a subgraph. W is said to be a k-plex if every node in W is connected to at least $|W| - k$ other nodes in W [87]. The notion of k-cores discussed in Sect. 2.2.1 is also a relaxed, connectivity-based notion of a clique.

The transitivity of links can be measured locally, at the level of ego-networks, as well as globally, at the level of graph partitions.

Definition 2.51 (*Ego network*) The ego network of a node x in a graph G, denoted by $Ego(x)$, is a subgraph of G induced by x and its neighbors.

The tendency of the neighbors of a node to be neighbors themselves is most commonly quantified by the clustering coefficient [90].

Definition 2.52 (*Clustering coefficient*) The clustering coefficient of node x in a undirected graph G, denoted by $CC(x)$, is equal to the probability that two randomly selected neighbors of x are directly connected in G

$$CC(x) = P(y \leftrightarrow z \mid x \leftrightarrow y, x \leftrightarrow z) = \frac{2 \left| \{ \{y, z\} \,:\, y, z \in N(x), \{y, z\} \in E \} \right|}{k_x(k_x - 1)}$$

(2.43)

where k_x denotes the degree of x and $N(x)$ is the set of neighbors of x.

It can be seen that the clustering coefficient of x is actually equal to the density of a graph that is obtained by removing x from $Ego(x)$. Therefore, $CC(x) = 0$ implies that $Ego(x)$ without x consists solely of isolated nodes, i.e. the ego network is a star graph. On the opposite side, $CC(i) = 1$ means that both $Ego(x)$ and $Ego(x)$ without x are cliques. The same approach, the density of an ego network without the ego node, can also be applied to measure local transitivity in directed graphs. In the case of directed graphs, the maximal number of links that could exist in the neighborhood of x is equal to $k_x(k_x - 1)$, where k_x denotes the total-degree of x. Then, the clustering coefficient of x is equal to

$$CC(x) = \frac{|\{(y, z) \,:\, y, z \in N(x), (y, z) \in E\}|}{k_x(k_x - 1)}$$

(2.44)

Informally speaking, a *community*, *cluster* or *module* in a graph is a subset of nodes that are more densely connected among themselves than with the rest of the graph. In other words, communities determine cohesive subgraphs which are sparsely connected to each other. A graph exhibits a community structure if the set of its nodes can be partitioned either into non-overlapping or overlapping communities. Community structures are typical for social networks, but they also frequently appear in complex networks from other domains. For example, tightly connected groups of nodes in a WWW graph often correspond to pages dealing with the same topic [36]. The identification of communities in a network, also known as *community detection*, is among the most important algorithmic problems in complex network analysis. Community detection enables us to study the structure of a network at the mesoscopic level by constructing and analyzing a reduced network in which nodes represent communities and links correspond to interactions between communities. They also enable us to make readable visualizations of extremely large networks. Additionally, roles of individual nodes and links can be investigated with respect to an underlying organization of nodes into communities. An important advance in community detection techniques was made by Newman and Girvan [44, 69] who introduced a measure called *modularity* to assess the quality of a network division into communities.

Definition 2.53 (*Graph partition*) A partition of a graph $G = (V, E)$, denoted by $P(G)$, is an assignment of its nodes to k sets of nodes, $P(G) = \{C_1, C_2, \ldots, C_k\}$, such that $V = \bigcup_{i=1}^{k} C_i$. $P(G)$ is a non-overlapping partition if $C_i \cap C_j = \emptyset$ for $1 \leq i, j \leq k, i \neq j$. Otherwise, $P(G)$ is overlapping.

Definition 2.54 (*Community membership function*) A community membership function, denoted by CM, is a function that maps a node to its community (or communities), with respect to some partition $P(G)$ which determines the division of nodes into communities.

Definition 2.55 (*Intra-community link*) A link e connecting nodes a and b is an intra-community link if a and b are members of the same community, i.e. $CM(a) = CM(b)$ ($CM(a) \cap CM(b) \neq \emptyset$ in the case of overlapping communities).

Definition 2.56 (*Inter-community link*) A link e connecting nodes a and b is an inter-community link if a and b belong to different communities, i.e. $CM(a) \neq CM(b)$ ($CM(a) \cap CM(b) = \emptyset$ in the case of overlapping communities).

Definition 2.57 (*Modularity*) Let $G = (V, E)$ be an arbitrary undirected graph and let $P(G)$ denote a partition of G into non-overlapping communities. The modularity of $P(G)$, denoted by Q, is computed according to the following formula:

$$Q = \frac{1}{2l} \sum_{i,j \in V} (A_{ij} - P_{ij})\delta(i, j) \tag{2.45}$$

where A is the adjacency matrix of G, l is the number of links in E, $\delta(i, j) = 1$ if nodes i and j are in the same community ($CM(i) = CM(j)$) or 0 otherwise, and P_{ij} is the expected number of links connecting i and j considering some null random graph model.

The configuration model of random graphs (see Sect. 2.3) is usually used as the null random graph model to derive a concrete modularity measure. This model describes random graphs having a given size and degree distribution. The expected number of links between two nodes i and j according to the configuration model is equal to

$$P_{ij} = \frac{k_i k_j}{2l} \tag{2.46}$$

where k_x denotes the degree of node x. Then, the modularity measure can be expressed as

$$Q = \sum_{c=1}^{n_c} \left[\frac{l_c}{l} - \left(\frac{d_c}{2l} \right)^2 \right] \qquad (2.47)$$

where n_c is the number of communities, l_c denotes the total number of intra-cluster links in community c and d_c is the total degree of nodes in c. The first term in the previous equation, l_c/l, is actually the fraction of links inside community c. The second term, $(d_c/(2l))^2$, represents the expected fraction of links inside c in a comparable random graph. If the first term is significantly larger than the second term then c contains more links than expected by random chance. Therefore, the comparison with a random graph leads to the modularity-based definition of community: c is a highly cohesive subset of nodes if

$$\frac{l_c}{l} \gg \left(\frac{d_c}{2l} \right)^2 \qquad (2.48)$$

The modularity measure can be easily generalized to directed, weighted and directed-weighted graphs [38, 59, 70]. Additionally, several researchers proposed extensions of the modularity measure for partitions into overlapping communities [68, 77, 88].

The modularity measure was originally proposed to select the best partitioning of a dendrogram constructed by the Girvan-Newman community detection algorithm [44, 69]. Later, a variety of other community detection algorithms incorporated the modularity measure either as the guiding principle to find the best division of a network into communities or as the primary criterion to select the best division from a set of candidates. Although widely used, the modularity measure has two weaknesses. Fortunato and Barthélemy [39] demonstrated that modularity maximization may fail to identify strong and evident communities (e.g. cliques) smaller than a certain size that depends on the number of links in the network (the resolution limit problem). Good et al. [46] showed that typically an exponential number of structurally diverse network partitions have modularity very close to the maximum value (the degeneracy problem). Therefore, it is highly important to consider other notions of communities and strongly cohesive subgraphs when performing community detection based on modularity maximization. Radicchi et al. [82] proposed definitions of two types of communities based on characteristics of internal and external degrees of nodes within community.

Definition 2.58 (*Internal node degree*) The internal degree of a node i in a graph G, denoted by $k_{int}(i)$, with respect to a community membership function CM, is equal to

$$k_{int}(i) = \sum_{j \in CM(i)} A_{ij} \qquad (2.49)$$

In other words, the internal degree of i is equal to

- the number of intra-community links incident with i in the case of undirected unweighted graphs,
- the number of intra-community links emanating from i in the case of directed unweighted graphs,
- the sum of weights (total strength) of intra-community links incident with i in the case of undirected weighted graphs, and
- the sum of weights of out-going intra-community links incident with i in the case of directed weighted graphs.

Definition 2.59 (*External node degree*) The external degree of node a node i in a graph G, denoted by $k_{ext}(i)$, with respect to a community membership function CM, is equal to

$$k_{ext}(i) = \sum_{j \notin CM(i)} A_{ij} \tag{2.50}$$

In other words, the external degree of i is equal to

- the number of inter-community links incident with i in the case of undirected unweighted graphs,
- the number of out-going inter-community links incident with i in the case of directed unweighted graphs,
- the sum of weights (total strength) of inter-community links incident with i in the case of undirected weighted graphs, and
- the sum of weights of inter-community links emanating from i in the case of directed weighted graphs.

Clearly, $k_{int}(i) + k_{ext}(i)$ is equal to the (out-)degree/strength of i since each link incident with i is either an intra-community or inter-community link.

Definition 2.60 (*Radicchi strong community*) A subgraph C of a graph G is a Radicchi strong community (or community in the strong sense) if

$$(\forall i \in C) \; k_{int}(i) > k_{ext}(i) \tag{2.51}$$

Definition 2.61 (*Radicchi weak community*) A subgraph C of a graph G is a Radicchi weak community (or community in the weak sense) if

$$\sum_{i \in C} k_{int}(i) > \sum_{i \in C} k_{ext}(i) \tag{2.52}$$

A comprehensive overview of community detection algorithms can be found in the paper by Santo Fortunato [38]. Additionally, the article by Xie et al. [91] provides a focused overview of techniques for detecting overlapping communities. Fortunato classified community detection methods into the following broad categories: traditional clustering and graph partitioning algorithms (e.g. the Kernighan-Lin algorithm, K-means, the spectral bisection method), divisive algorithms, modularity-based algorithms, dynamic algorithms and methods based on statistical inference.

The basic characteristic of divisive methods is that they build clustering dendrograms by progressively removing links that are most likely to be inter-community links. For example, the Girvan-Newman algorithm [44] uses the edge betweenness centrality measure to identify inter-community links. Other measures proposed to detect inter-community links are the edge-clustering coefficient [82] and the information centrality [37].

Modularity-based community detection methods strive to maximize the modularity measure. Modularity maximization is known to be a NP-complete problem [24], so several modularity maximization strategies and heuristics were proposed in the literature. The greedy modularity optimization algorithm, originally introduced by Newman [72] and later improved by Clauset, Newman and Moore [27], relies on a greedy hierarchical agglomeration strategy. The algorithm starts with a partition in which nodes are put in separate communities (one node into exactly one community). The change in the modularity measure after merging any two adjacent communities is computed in each iteration of the algorithm, and the merge operation which maximally increase or minimally decrease modularity is performed. Finally, the partition with the highest modularity is selected as the final result of the algorithm.

Another widely used modularity-based community detection method is the Louvain algorithm proposed by Blondel et al. [12]. The Louvain algorithm is based on a greedy multi-resolution strategy to maximize modularity. The algorithms starts with a partition in which all nodes are placed in different communities. The modularity measure is firstly optimized locally. The change in modularity after moving a node to an adjacent community is computed for each node and each adjacent community. The move operation which maximally increases modularity is performed. This process is repeated until no individual move can increase modularity. Then, the algorithm creates a network of communities. Each node of this network represents one community and two nodes are connected if they represent adjacent communities. After that, the local optimization of modularity is performed on the network of communities. The whole process is repeated until the local optimization of modularity yields no community reassignments. Other modularity-based methods include spectral, extremal and simulated annealing approaches to modularity maximization [38].

The principal idea of dynamic community detection methods is that communities can be identified by dynamical processes running on networks. For example, the widely used Walktrap algorithm [81] is based on the observation that relatively short random walks tend to get trapped into dense sub-networks due to a high density of intra-community links. In other words, if a short random walk starting at a node x reaches a node y with a high probability then x and y should be placed in the same community. Let D denote the transition matrix of a network, i.e. $D_{ij} = A_{ij}/k_i$ where A is the adjacency matrix of the network and k_i is the (out-)degree of i. D_{ij} is equal to the probability that a random walker moves from i to j in one random walk step. The Walktrap algorithm uses a node similarity measure defined as the probability that the random walker moves from one node to another in exactly k steps, where k is the parameter of the algorithm. This probability can be computed by raising D to the power of k. Then, nodes are grouped into communities relying on an agglomerative hierarchical clustering technique based on the Ward's minimum variance criterion for merging adjacent communities and the modularity measure for selecting the best partitioning. The Infomap algorithm is another widely used community detection technique based on random walks. The main idea of this algorithm is that communities can be revealed by optimally compressing a description of a random walk on the network [85].

The MCL (Markov Cluster) algorithm [31] is another popular community detection algorithm based on random walks. This algorithm identifies communities by an iterative process consisting of two alternating steps called expansion and inflation. The iterative process is performed over a stochastic matrix M. Initially, M_{xy} is equal to the probability that the random walker moves from y to x in one random walk step, i.e. $M_{xy} = D_{xy}^T$. In the expansion step, M is raised to an integer power k ($k = 2$ is usually used). In the inflation step, each entry of M is raised to a real power (usually set to 2). The idea of the inflation step is to boost the probabilities of intra-cluster random walks. The resulting matrix is rescaled to be stochastic again, and the process is repeated until M reaches a steady state. Finally, communities identified by the algorithm correspond to weakly connected components of a directed graph whose adjacency matrix is M.

Another popular and widely used community detection technique belonging to the category of dynamic methods is the label propagation algorithm [83]. The basic idea of the label propagation algorithm is that each node carries a label denoting the community to which it belongs. The algorithm starts with an initial configuration in which each node has an unique label. Then, an iterative process in which each node, in a random order, takes the most frequent label appearing in its neighborhood is repeated until labels stabilize. In this way, the nodes belonging to a dense sub-network quickly reach a consensus on a label representing the whole sub-network. At the end of the algorithm, nodes having the same label are grouped into one community.

2.3 Basic Complex Network Models

Mathematical models of complex networks help us to understand properties of real-world complex networks. If some property of a real-world network is reproducible by a theoretical model then this property can be explained by the founding principles of the model. Secondly, mathematical models of complex networks enable predictions about the evolution of real-world large-scale networks. By studying mathematical models of complex networks we additionally build our intuition about networks in the sense that we associate evolutionary principles to typically observable outcomes of those principles.

The study of mathematical models of complex graphs started with the pioneering work of Erdős and Renyi. They introduced a model of non-regular graphs that is nowadays known as the Erdős-Renyi (ER) model of random graphs [18, 34, 35]. The ER model has been extensively investigated in the mathematical literature since it enables applications of the probabilistic method in graph theory. Let $S(N, L)$ denote the set of all graphs having N nodes, labeled from 1 to N, and L links. Let us suppose that we want to prove (or disprove) that there is a graph from $S(N, L)$ having some property P. Obviously, this problem can be solved by taking a brute-force approach: we make an algorithm for generating all members of $S(N, L)$ that additionally checks whether the last constructed member satisfies P. The profound idea behind the ER model is that such kind of problems can be solved in a non-constructivist way by showing that a randomly selected graph from $S(N, L)$ exhibits P with a non-zero probability.

Definition 2.62 (*Erdős-Renyi random graph*) A random graph $G_{er}(N, L)$ is a graph containing N nodes labeled from 1 to N and L links that are randomly selected (with an equal selection probability) from the set of all possible links that can be formed among these N nodes.

A closely related but a slightly broad view of random graphs is given by the following definition originally introduced by Gilbert [43].

Definition 2.63 (*Random graph*) A random graph $G(N, p)$ is a graph containing N nodes labeled from 1 to N where each pair of nodes is connected with some fixed probability p.

$G_{er}(N, L)$ is an outcome of a stochastic process defined over the sample space $S(N, L)$, where the members of $S(N, L)$ have an equal probability of realization. On the other hand, $G(N, p)$ is an outcome of a stochastic process defined over a bigger sample space $S(N)$ that encompasses all labeled graphs having N nodes. The members of $S(N)$ are not equiprobable – the most probable realizations are those graphs having pM links, where M denotes the maximal number of links among N nodes ($M = N(N - 1)/2$). In the asymptotic limit $N \to \infty$, the models $G(N, p)$ and $G_{er}(N, pM)$ are interchangeable [13, 17]. However, the Gilbert's definition is usually preferred since it enables an easier analytic treatment of random graphs. The algorithm for sampling undirected random graphs according to the Gilbert's definition is shown in Algorithm 2.4. This algorithm can be modified in a straightforward manner for directed random graphs (both loops iterate through the whole set of nodes; directed links are created instead of undirected links).

The random graph model is typically used as the null model when analyzing real-world networks. A random graph represents a typical graph of some size, typical considering the ensemble of all graphs having that size. Therefore, structural properties of real-world networks are usually firstly compared to analytical predictions obtained using the random graph model. If an empirically observed statistic of a real-world network significantly differs from the prediction based on the random graph model then the network is "atypical", i.e. it exhibits a structural property that is not characteristic for a large majority of graphs having the same number of nodes and links. The basic properties of random graphs are [4, 13]:

1. The degree distribution of a random graph $G(N, p)$ follows the binomial distribution $B(N - 1, p)$ that can be well approximated by the Poisson distribution $\text{Pois}(\langle k \rangle)$ for large N, where $\langle k \rangle$ is the average degree.

Algorithm 2.4: The algorithm for sampling undirected random graphs.

---

input : N – the number of nodes
$\qquad$ p – the connection probability
$\qquad$ random() – a function returning a pseudo-random number in $[0, 1)$
output: $G = (V, E)$ – an undirected random graph

$V = \{1, 2, \ldots, N\}$
$E = \emptyset$

for $i = 1$ **to** $N - 1$ **do**
$\quad$ **for** $j = i + 1$ **to** N **do**
$\quad\quad$ **if** random() $\leq p$ **then**
$\quad\quad\quad$ $E = E \cup \{\{i, j\}\}$
$\quad\quad$ **end**
$\quad$ **end**
end

---

2. The small-world coefficient of a random graph grows logarithmically with the number of nodes implying that random graphs are small-worlds. The diameter of $G(N, p)$ can be approximated as

$$\text{Diameter} \approx \frac{ln(N)}{ln(p(N - 1))} = \frac{ln(N)}{ln(\langle k \rangle)} \qquad (2.53)$$

3. A link in $G(N, p)$ is created independently of previously created links. Therefore, the clustering coefficient of $G(N, p)$ is equal to p. This implies that large and sparse random graphs exhibit a low degree of local clustering. Another consequence of independent formation of links is that $G(N, p)$ is a non-assortative (degree uncorrelated) graph.

4. Characteristics of connected components of $G(N, p)$ depend on p. Namely, there is a critical connection probability $p_c = 1/(N - 1)$ that corresponds to a critical average degree $\langle k_c \rangle = 1$ which is related to the emergence of giant connected components. If the average degree of $G(N, p)$ is higher than the critical average degree then $G(N, p)$ almost surely has a giant connected component. On the other hand, if the average degree is less than critical then $G(N, p)$ almost surely does not have a giant connected component.

The basic random graph model can be extended in various ways to obtain stochastic complex network models generating more realistic artificial graphs with respect to certain structural properties. Consequently, generalized random models can be used instead of the basic random graph model as the null model of real-world networks. Two generalizations of the ER model known as the configuration model [11, 13, 74, 76] and stochastic block models [1, 49] have been shown to be especially useful in complex network analysis. The configuration model is a model of a random graph $G(D)$ with a fixed degree sequence D (an array containing the degree of each node). $G(D)$ can be generated by forming nodes with "stubs" (half-links with only one endpoint fixed) such that a node i has D_i stubs. Then, we select two stubs uniformly at random and assemble them into a link connecting nodes incident to selected stubs. As the consequence, the number of stubs decreases by two. The previous step is repeated until the number of stubs becomes zero (see Algorithm 2.5).

Algorithm 2.5: The algorithm for sampling random graphs with a given degree sequence.

---

input : N – the number of nodes

D – an array of integers such that $D[i]$ represents the degree of node i

random() – a function returning a pseudo-random number in $[0, 1)$

output: $G = (V, E)$ – an undirected random graph with degree sequence D

$V = \{1, 2, \ldots, N\}$
$E = \emptyset$
$S =$ an empty list of nodes

// form S such that each node x appears $D[x]$ times in S
for $i = 1$ **to** N **do**
    for $j = 1$ **to** $D[i]$ **do**
       | add i to S
    end
end

while S is not empty **do**
    repeat
       | $i = 1 + \lfloor \text{random}() \cdot \text{size}(S) \rfloor$
       | $j = 1 + \lfloor \text{random}() \cdot \text{size}(S) \rfloor$
    until $i \neq j \wedge \{i, j\} \notin E$
    $x = i$-th element of S
    $y = j$-th element of S
    remove x from S
    remove y from S
    $E = E \cup \{\{x, y\}\}$
end

Stochastic block models (SBMs) are random graph models with an accompanying probabilistic configuration of nodes into groups. Those models enable sampling of random graphs that have a clustered (community) structure. The parameters of a SBM are:

1. The number of nodes,
2. The number of node groups (communities), denoted as q,
3. A probability vector $m = (m_1, m_2, ..., m_q)$, where m_i represents the probability that a randomly selected node belongs to a group i.
4. A square $q \times q$ stochastic block matrix B, where B_{ij} is the probability that a node from a group i is connected to a node from a group j.

Algorithm 2.6 shows the procedure for generating random graphs according to a given stochastic block model. SBMs are equivalent to the basic model of random graphs for constant stochastic block matrices. SBMs generate networks with a strong community structure if

$$(\forall i, j, i \neq j) \; B_{ii} \gg B_{ij} \tag{2.54}$$

A special case of SBM models in which $B_{ii} = p$ and $B_{ij} = q$ for each i and j ($i \neq j$) is known as the planted partition model. The planted partition model with $p \gg q$ generates assortative networks, while for $q \gg p$ the model produces disassortative networks with respect to predefined groups.

Algorithm 2.6: The algorithm for sampling random graphs according to a given stochastic block model.

---

input : N – the number of nodes
  q – the number of node groups
  m – a probability vector used to assign nodes to groups
  B – an affinity matrix
  random() – a function returning a pseudo-random number in $[0, 1)$
output: $G = (V, E)$ – an undirected random graph
  g – an array indicating the membership of nodes to groups

$V = \{1, 2, \dots, N\}$
$E = \emptyset$

for $i = 1$ **to** N **do**
  | $g[i] = $ the smallest number k such that random() $\leq \sum_{i=1}^{k} m[i]$
end

for $i = 1$ **to** $N - 1$ **do**
  | **for** $j = i + 1$ **to** N **do**
  | | createLink $=$ random() $\leq B[g[i]][g[j]]$
  | | **if** createLink **then**
  | | | $E = E \cup \{\{i, j\}\}$
  | | **end**
  | **end**
end

The rise of network science started with the study by Watts and Strogatz [90] on small-world networks. The authors examined the structure of three complex networks from different domains (the collaboration network of movie actors extracted from IMDb, the electrical power grid of the western United States and the neural network of the nematode worm *C. elegans*). They observed that analyzed networks possess the small-world property, but exhibit a considerably higher degree of local clustering than comparable random graphs. To explain that phenomenon, Watts and Strogatz proposed and investigated characteristics of a complex network model that interpolates between ring lattices and random graphs. Ring lattices are regular graphs obtained by placing nodes on a ring and connecting each node to its k-nearest neighbors, $k/2$ on each side (k should be an even number). Ring lattices exhibit a high degree of local clustering, i.e. the clustering coefficient of a ring lattice tends to $3/4$ as k grows. Secondly, sparse ring lattices are not small-world graphs. Compared to random graphs, ring lattices have exactly the opposite characteristics of local clustering and shortest paths. In the Watts-Strogatz (WS) model, we start from a ring lattice. Then, each link connecting a node n_i to a neighbor n_j, $i < j$, is rewired with a probability p to a randomly selected node sampled in a way to avoid parallel links and loops (see Algorithm 2.7). For $p = 0$ the WS model generates ring lattices, while for $p = 1$ it produces pure random graphs. Watts and Strogatz showed that the rewiring procedure for $0.01 < p < 0.1$ results in graphs having at the same time the small-word property and a high degree of local clustering.

Algorithm 2.7: The Watts-Strogatz algorithm for sampling small-world random graphs exhibiting a high degree of local clustering.

---

input : N – the number of nodes
        k – the number of nearest neighbors ($k \ll N$)
        p – the rewiring probability for links
        random() – a function returning a pseudo-random number in $[0, 1)$
output: $G = (V, E)$ – an undirected random graph exhibiting the small-world property in
        the Watts-Strogatz sense

$V = \{1, 2, \ldots, N\}$
$E = \emptyset$

for $i = 1$ **to** N **do**
   for $j = 1$ **to** $k/2$ **do**
      $d = i + j$
      if $d > N$ **then**
         $d = d - N$
      end
      $E = E \cup \{\{i, d\}\}$
   end
end

foreach $\{i, j\} \in E : i < j$ **do**
   if random() $\leq p$ **then**
      repeat
         $d = 1 + \lfloor$random()$\cdot N \rfloor$
      until $d \neq i \wedge \{i, d\} \notin E$
      $E = E \setminus \{\{i, j\}\}$
      $E = E \cup \{\{i, d\}\}$
   end
end

Both Erdős-Renyi random graphs and Watts-Strogatz small-world graphs are homogeneous graphs characterized by low-variance degree distributions. However, many real-world networks, especially large-scale networks, are heterogeneous networks characterized by heavy-tailed power-law degree distributions [4, 13, 74]. In 1999, Barabási and Albert [7] proposed an evolutionary model, known as the BA model, for generating scale-free networks (see Algorithm 2.8). The BA model is based on two principles:

Algorithm 2.8: The Barabási-Albert algorithm for sampling scale-free random graphs.

input : N – the number of nodes
N_0 ($N_0 \ll N$) – the number of nodes in the initial random graph
p – the connection probability for the initial random graph
m ($m \leq N_0$) – the number of links created by each new node
random() – a function returning a pseudo-random number in $[0, 1)$
output: $G = (V, E)$ – an undirected scale-free random graph

G = create a random graph by the algorithm 2.4 with parameters N_0 and p
D = an empty list of nodes

// form D such that a node x appears degree(x) times in D
for $i = 1$ **to** N_0 **do**
  $k = \text{degree}(i)$
  for $j = 1$ **to** k **do**
  | add i to D
  end
end

for $i = N_0 + 1$ **to** N **do**
  $V = V \cup \{i\}$

  dst = an empty set of nodes
  $j = 0$
  while $j < m$ **do**
    $r = 1 + \lfloor \text{random}() \cdot \text{size}(D) \rfloor$
    $o = r$-th element of D
    if $o \notin$ dst **then**
      dst = dst $\cup \{o\}$
      $j = j + 1$
    end
  end

  foreach $d \in$ dst **do**
    $E = E \cup \{\{i, d\}\}$
    add d to D
    add i to D
  end
end

1. *Network growth*. Starting from a small random network, a new node is created in each iteration of the algorithm. Each new node establishes m connections with previously created nodes, where m is one of the parameters of the BA model.
2. *Preferential attachment*. The probability that a new node n connects to an existing node i depends on the degree of i, i.e.

$$P(n \leftrightarrow i) = \frac{\text{degree}(i)}{\sum_{j \in S} \text{degree}(j)} \qquad (2.55)$$

where S denotes the set of "old" nodes, i.e. nodes created in the previous growth steps.

It can be observed that the preferential attachment probability is based on the "rich get richer" principle that is also known as the principle of cumulative advantage or the Matthew effect. Large networks generated by the BA model satisfy a power-law with $\gamma \approx 3$ ($\gamma \to 3$ as $N \to \infty$, where N is the size of the network). In order to show that network growth and preferential attachment are necessary ingredients to produce a network with a power-law degree distribution, Barabási and Albert examined a model that keeps the growing character of the network, but uses an uniform attachment probability (each node has an equal probability to establish a connection with a newly added node). This model results in networks whose degree distribution decays exponentially following $P(k) = C \exp(-\beta k)$. Karpivsky et al. [57] showed that only linear preferential attachment leads to networks having power-law degree distributions.

Various generalizations of the BA model were proposed in order to obtain tunable power-law scaling exponent [32], tunable clustering coefficient [51], tunable assortativity index [47] and scale-free networks with a significant community structure [62, 80]. The model can be generalized in a straightforward manner for directed networks [16]. Amaral et al. [6] investigated effects of node aging (a node cannot establish connections after a fixed number of iterations since its creation) and node capacity (each node has a fixed maximum of connections to other nodes) to the structure of networks whose evolution is governed by the preferential attachment principle. The authors showed that both time and capacity constraints when incorporated into the BA model lead to truncated power-law degree distributions.

Another class of evolutionary scale-free network models are models based on copying mechanisms [54, 56, 58, 60]. The intuition behind copying-based models is related to topic-oriented citation practices that occur in the WWW or scientific publishing. For example, if a paper A references another paper B on the same topic then it is likely that A will also reference a paper referenced by B. Two essential steps are performed in copying-based models when a new node A is created:

- A connects to a "prototype" node that is chosen uniformly at random.
- The rest of links incident with A are created according to the following rule: A connects to a randomly selected node with a probability p, while with the probability $(1 - p)$ a link between A and a randomly selected neighbor of the prototype node is created.

The principle of establishing connections to neighbors of copying-prototype nodes is actually equivalent to the principle of preferential attachment. Namely, the probability that a node B is one of neighbors of the prototype node is proportional to the degree of B because the prototype node is selected uniformly at random, i.e. highly connected nodes have a higher chance to be neighbors of a randomly selected node compared to loosely connected nodes.

References

1. Abbe, E.: Community detection and stochastic block models: recent developments (2017). arXiv:1703.10146
2. Adamic, L.A., Adar, E.: Friends and neighbors on the web. Soc. Netw. **25**(3), 211–230 (2003). https://doi.org/10.1016/S0378-8733(03)00009-1
3. Alba, R.D.: A graph theoretic definition of a sociometric clique. J Math Sociol. **3**(1), 113–126 (1973). https://doi.org/10.1080/0022250X.1973.9989826
4. Albert, R., Barabási, A.L.: Statistical mechanics of complex networks. Rev. Mod. Phys. **74**(1), 47–97 (2002). https://doi.org/10.1103/RevModPhys.74.47
5. Alvarez-Hamelin, J.I., Dall'Asta, L., Barrat, A., Vespignani, A.: K-core decomposition of internet graphs: hierarchies, self-similarity and measurement biases. Netw. Heterog. Media **3**(2), 371–393 (2008). https://doi.org/10.3934/nhm.2008.3.371
6. Amaral, L.A.N., Scala, A., Barthlmy, M., Stanley, H.E.: Classes of small-world networks. Proc. Natl. Acad. Sci. **97**(21), 11149–11152 (2000). https://doi.org/10.1073/pnas.200327197
7. Barabasi, A.L., Albert, R.: Emergence of scaling in random networks. Science **286**(5439), 509–512 (1999). https://doi.org/10.1126/science.286.5439.509
8. Batagelj, V., Zaveršnik, M.: Fast algorithms for determining (generalized) core groups in social networks. Adv. Data Anal. Classif. **5**(2), 129–145 (2011). https://doi.org/10.1007/s11634-010-0079-y
9. Bavelas, A.: Communication patterns in task-oriented groups. J. Acoust. Soc. Am. **22**(6), 725–730 (1950). https://doi.org/10.1121/1.1906679
10. Beauchamp, M.A.: An improved index of centrality. Behav. Sci. **10**(2), 161–163 (1965). https://doi.org/10.1002/bs.3830100205
11. Bender, E.A., Canfield, E.: The asymptotic number of labeled graphs with given degree sequences. J Comb. Theory Ser. A **24**(3), 296–307 (1978). https://doi.org/10.1016/0097-3165(78)90059-6
12. Blondel, V.D., Guillaume, J.L., Lambiotte, R., Lefebvre, E.: Fast unfolding of communities in large networks. J. Stat. Mech Theory Exp. **2008**(10), P10008 (2008)
13. Boccaletti, S., Latora, V., Moreno, Y., Chavez, M., Hwang, D.: Complex networks: structure and dynamics. Phys. Rep. **424**(45), 175–308 (2006). https://doi.org/10.1016/j.physrep.2005.10.009
14. Boldi, P., Vigna, S.: Axioms for centrality. Internet Math. **10**(3–4), 222–262 (2014). https://doi.org/10.1080/15427951.2013.865686
15. Boldi, P., Rosa, M., Vigna, S.: Robustness of social and web graphs to node removal. Soc. Netw. Anal. Min. **3**(4), 829–842 (2013). https://doi.org/10.1007/s13278-013-0096-x
16. Bollobás, B., Borgs, C., Chayes, J., Riordan, O.: Directed scale-free graphs. In: Proceedings of the Fourteenth Annual ACM-SIAM Symposium on Discrete Algorithms, SODA '03, Society for Industrial and Applied Mathematics. pp. 132–139, Philadelphia, PA, USA (2003). https://doi.org/10.1145/644108.644133
17. Bollobás, B., Riordan, O.M.: Mathematical results on scale-free random graphs, pp. 1–34. Wiley-VCH Verlag GmbH & Co. KGaA (2005). https://doi.org/10.1002/3527602755.ch1
18. Bollobás, B.: Random Graphs. Cambridge University Press (2001)
19. Bonacich, P.: Factoring and weighting approaches to status scores and clique identification. J. Math. Sociol. **2**(1), 113–120 (1972). https://doi.org/10.1080/0022250X.1972.9989806
20. Bonacich, P.: Power and centrality: a family of measures. Am. J. Sociol. **92**(5), 1170–1182 (1987). https://doi.org/10.2307/2780000
21. Borgatti, S.P., Everett, M.G.: Models of core/periphery structures. Soc. Netw. **21**(4), 375–395 (2000). https://doi.org/10.1016/S0378-8733(99)00019-2
22. Brandes, U.: A faster algorithm for betweenness centrality. J. Math. Sociol. **25**(2), 163–177 (2001). https://doi.org/10.1080/0022250X.2001.9990249
23. Brandes, U.: On variants of shortest-path betweenness centrality and their generic computation. Soc. Netw. **30**(2), 136–145 (2008). https://doi.org/10.1016/j.socnet.2007.11.001

24. Brandes, U., Delling, D., Gaertler, M., Gorke, R., Hoefer, M., Nikoloski, Z., Wagner, D.: On modularity clustering. IEEE Trans. Knowl. Data Eng **20**(2), 172–188 (2008). https://doi.org/10.1109/TKDE.2007.190689
25. Brin, S., Page, L.: The anatomy of a large-scale hypertextual Web search engine. Comput. Netw. ISDN Syst **30**(1–7), 107–117 (1998)
26. Bron, C., Kerbosch, J.: Algorithm 457: finding all cliques of an undirected graph. Commun. ACM **16**(9), 575–577 (1973). https://doi.org/10.1145/362342.362367
27. Clauset, A., Newman, M.E.J., Moore, C.: Finding community structure in very large networks. Phys. Rev. E **70**, 066111 (2004). https://doi.org/10.1103/PhysRevE.70.066111
28. Clauset, A., Shalizi, C., Newman, M.: Power-law distributions in empirical data. SIAM Rev. **51**(4), 661–703 (2009). https://doi.org/10.1137/070710111
29. Costa,.LdF, Oliveira, O., Travieso, G., Rodrigues, F.A., Villas Boas, P., Antiqueira, L., Viana, M.P., Correa Rocha, L.: Analyzing and modeling real-world phenomena with complex networks: a survey of applications. Adv. Phys. **60**(3), 329–412 (2011). https://doi.org/10.1080/00018732.2011.572452
30. Dijkstra, E.W.: A note on two problems in connexion with graphs. Numerische Mathematik **1**(1), 269–271 (1959). https://doi.org/10.1007/BF01386390
31. Dongen, S.V.: Graph clustering via a discrete uncoupling process. SIAM J. Matrix Anal. Appl. **30**(1), 121–141 (2008). https://doi.org/10.1137/040608635
32. Dorogovtsev, S.N., Mendes, J.F.F., Samukhin, A.N.: Structure of growing networks with preferential linking. Phys. Rev. Lett. **85**, 4633–4636 (2000). https://doi.org/10.1103/PhysRevLett.85.4633
33. Eppstein, D., Strash, D.: Listing all maximal cliques in large sparse real-world graphs, pp. 364–375. Springer, Berlin (2011). https://doi.org/10.1007/978-3-642-20662-7_31
34. Erdős, P., Rényi, A.: On random graphs. I. Publ. Math. Debr. **6**, 290–297 (1959)
35. Erdős, P., Rényi, A.: On the evolution of random graphs. Publ. Math. Inst. Hung. Acad. Sci. **5**, 17–61 (1960)
36. Flake, G.W., Lawrence, S., Giles, C.L., Coetzee, F.M.: Self-organization and identification of web communities. Computer **35**(3), 66–71 (2002). https://doi.org/10.1109/2.989932
37. Fortunato, S., Latora, V., Marchiori, M.: Method to find community structures based on information centrality. Phys. Rev. E **70**, 056104 (2004). https://doi.org/10.1103/PhysRevE.70.056104
38. Fortunato, S.: Community detection in graphs. Phys. Rep. **486**(35), 75–174 (2010). https://doi.org/10.1016/j.physrep.2009.11.002
39. Fortunato, S., Barthlemy, M.: Resolution limit in community detection. Proc. Natl. Acad. Sci. **104**(1), 36–41 (2007). https://doi.org/10.1073/pnas.0605965104
40. Freeman, L.C.: A set of measures of centrality based on betweenness. Sociometry **40**, 35–41 (1977)
41. Freeman, L.C., Borgatti, S.P., White, D.R.: Centrality in valued graphs: a measure of betweenness based on network flow. Soc. Netw. **13**(2), 141–154 (1991). https://doi.org/10.1016/0378-8733(91)90017-N
42. Garlaschelli, D., Loffredo, M.: Patterns of link reciprocity in directed networks. Phys. Rev. Lett. **93**, 268701 (2004). https://doi.org/10.1103/PhysRevLett.93.268701
43. Gilbert, E.N.: Random graphs. Ann. Math. Stat. **30**(4), 1141–1144 (1959)
44. Girvan, M., Newman, M.E.J.: Community structure in social and biological networks. Proc. Natl. Acad. Sci. **99**(12), 7821–7826 (2002). https://doi.org/10.1073/pnas.122653799
45. Goldstein, M.L., Morris, S.A., Yen, G.G.: Problems with fitting to the power-law distribution. Eur. Phys. J. B - Condens. Matter Complex Syst. **41**(2), 255–258 (2004). https://doi.org/10.1140/epjb/e2004-00316-5
46. Good, B.H., de Montjoye, Y.A., Clauset, A.: Performance of modularity maximization in practical contexts. Phys. Rev. E **81**, 046106 (2010). https://doi.org/10.1103/PhysRevE.81.046106
47. Guo, Q., Zhou, T., Liu, J.G., Bai, W.J., Wang, B.H., Zhao, M.: Growing scale-free small-world networks with tunable assortative coefficient. Phys. A Stat. Mech. Appl. **371**(2), 814–822 (2006). https://doi.org/10.1016/j.physa.2006.03.055

48. Hasan, M.A., Zaki, M.J.: A survey of link prediction in social networks, pp. 243–275. Springer, Boston (2011). https://doi.org/10.1007/978-1-4419-8462-3_9
49. Holland, P.W., Laskey, K.B., Leinhardt, S.: Stochastic blockmodels: first steps. Soc. Netw. 5(2), 109–137 (1983). https://doi.org/10.1016/0378-8733(83)90021-7
50. Holme, P., Kim, B.J., Yoon, C.N., Han, S.K.: Attack vulnerability of complex networks. Phys. Rev. E 65, 056109 (2002). https://doi.org/10.1103/PhysRevE.65.056109
51. Holme, P., Kim, B.J.: Growing scale-free networks with tunable clustering. Phys. Rev. E 65, 026107 (2002). https://doi.org/10.1103/PhysRevE.65.026107
52. Jeh, G., Widom, J.: Simrank: A measure of structural-context similarity. In: Proceedings of the Eighth ACM SIGKDD International Conference on Knowledge Discovery and Data Mining, KDD '02. pp. 538–543, ACM, New York, USA (2002). https://doi.org/10.1145/775047.775126
53. Khaouid, W., Barsky, M., Srinivasan, V., Thomo, A.: K-core decomposition of large networks on a single PC. Proc. VLDB Endow. 9(1), 13–23 (2015). https://doi.org/10.14778/2850469.2850471
54. Kleinberg, J.M., Kumar, R., Raghavan, P., Rajagopalan, S., Tomkins, A.: The web as a graph: measurements, models, and methods. In: Asano, T., Imai, H., Lee, D., Nakano, S.i., Tokuyama, T. (eds.) Computing and Combinatorics. Lecture Notes in Computer Science, vol. 1627, pp. 1–17
55. Kleinberg, J.M.: Authoritative sources in a hyperlinked environment. J. ACM 46(5), 604–632 (1999). https://doi.org/10.1145/324133.324140
56. Krapivsky, P.L., Redner, S.: Network growth by copying. Phys. Rev. E 71, 036118 (2005). https://doi.org/10.1103/PhysRevE.71.036118
57. Krapivsky, P.L., Redner, S., Leyvraz, F.: Connectivity of growing random networks. Phys. Rev. Lett. 85, 4629–4632 (2000). https://doi.org/10.1103/PhysRevLett.85.4629
58. Kumar, R., Raghavan, P., Rajagopalan, S., Sivakumar, D., Tomkins, A., Upfal, E.: Stochastic models for the Web graph. In: Proceedings of the 41st Annual Symposium on Foundations of Computer Science, FOCS '00, IEEE Computer Society. pp. 57–65, Washington, USA (2000), https://doi.org/10.1109/SFCS.2000.892065
59. Leicht, E.A., Newman, M.E.J.: Community structure in directed networks. Phys. Rev. Lett. 100, 118703 (2008). https://doi.org/10.1103/PhysRevLett.100.118703
60. Leskovec, J., Kleinberg, J., Faloutsos, C.: Graph evolution: Densification and shrinking diameters. ACM Trans. Knowl. Discov. Data 1(1) (2007). https://doi.org/10.1145/1217299.1217301
61. Leskovec, J., Lang, K.J., Dasgupta, A., Mahoney, M.W.: Community structure in large networks: natural cluster sizes and the absence of large well-defined clusters. Internet Math. 6(1), 29–123 (2009). https://doi.org/10.1080/15427951.2009.10129177
62. Li, C., Maini, P.K.: An evolving network model with community structure. J. Phys. A Math. Gen. 38(45), 9741 (2005). https://doi.org/10.1088/0305-4470/38/45/002
63. Liben-Nowell, D., Kleinberg, J.: The link prediction problem for social networks. In: Proceedings of the Twelfth International Conference on Information and Knowledge Management, CIKM '03. pp. 556–559, ACM, New York, USA (2003). https://doi.org/10.1145/956863.956972
64. Lichtenwalter, R.N., Lussier, J.T., Chawla, N.V.: New perspectives and methods in link prediction. In: Proceedings of the 16th ACM SIGKDD International Conference on Knowledge Discovery and Data Mining, KDD '10. pp. 243–252, ACM, New York, USA, (2010). https://doi.org/10.1145/1835804.1835837
65. Lü, L., Zhou, T.: Link prediction in complex networks: a survey. Phys. A Stat. Mech. Appl. 390(6), 1150–1170 (2011). https://doi.org/10.1016/j.physa.2010.11.027
66. Martínez, V., Berzal, F., Cubero, J.C.: A survey of link prediction in complex networks. ACM Comput. Surv. 49(4), 69:1–69:33 (2016). https://doi.org/10.1145/3012704
67. Mokken, R.J.: Cliques, clubs and clans. Qual. Quant. 13(2), 161–173 (1979). https://doi.org/10.1007/BF00139635
68. Nepusz, T., Petróczi, A., Négyessy, L., Bazsó, F.: Fuzzy communities and the concept of bridgeness in complex networks. Phys. Rev. E 77, 016,107 (2008). https://doi.org/10.1103/PhysRevE.77.016107

69. Newman, M.E.J., Girvan, M.: Finding and evaluating community structure in networks. Phys. Rev. E **69**(2), 026113 (2004). https://doi.org/10.1103/PhysRevE.69.026113
70. Newman, M.E.J.: Analysis of weighted networks. Phys. Rev. E **70**, 056131 (2004). https://doi.org/10.1103/PhysRevE.70.056131
71. Newman, M.E.J.: Assortative mixing in networks. Phys. Rev. Lett. **89**, 208701 (2002). https://doi.org/10.1103/PhysRevLett.89.208701
72. Newman, M.E.J.: Fast algorithm for detecting community structure in networks. Phys. Rev. E **69**, 066133 (2004). https://doi.org/10.1103/PhysRevE.69.066133
73. Newman, M.E.J.: Mixing patterns in networks. Phys. Rev. E **67**, 026126 (2003). https://doi.org/10.1103/PhysRevE.67.026126
74. Newman, M.E.J.: The structure and function of complex networks. SIAM Rev. **45**, 167–256 (2003). https://doi.org/10.1137/S003614450342480
75. Newman, M.E.J.: Power laws, pareto distributions and Zipf's law. Contemp. Phys. **46**(5), 323–351 (2005). https://doi.org/10.1080/00107510500052444
76. Newman, M.: Networks: An Introduction. Oxford University Press Inc, New York (2010)
77. Nicosia, V., Mangioni, G., Carchiolo, V., Malgeri, M.: Extending the definition of modularity to directed graphs with overlapping communities. J. Stat. Mech. Theory Exp. **2009**(03), P03024 (2009)
78. Pastor-Satorras, R., Vázquez, A., Vespignani, A.: Dynamical and correlation properties of the internet. Phys. Rev. Lett. **87**, 258701 (2001). https://doi.org/10.1103/PhysRevLett.87.258701
79. Perra, N., Fortunato, S.: Spectral centrality measures in complex networks. Phys. Rev. E **78**, 036107 (2008). https://doi.org/10.1103/PhysRevE.78.036107
80. Pollner, P., Palla, G., Vicsek, T.: Preferential attachment of communities: the same principle, but a higher level. EPL (Europhys. Lett.) **73**(3), 478 (2006). https://doi.org/10.1209/epl/i2005-10414-6
81. Pons, P., Latapy, M.: Computing communities in large networks using random walks. J. Graph Algorithms Appl. **10**(2), 191–218 (2006). https://doi.org/10.7155/jgaa.00124
82. Radicchi, F., Castellano, C., Cecconi, F., Loreto, V., Parisi, D.: Defining and identifying communities in networks. Proc. Natl. Acad. Sci. **101**(9), 2658–2663 (2004). https://doi.org/10.1073/pnas.0400054101
83. Raghavan, U.N., Albert, R., Kumara, S.: Near linear time algorithm to detect community structures in large-scale networks. Phys. Rev. E **76**, 036106 (2007). https://doi.org/10.1103/PhysRevE.76.036106
84. Rombach, M.P., Porter, M.A., Fowler, J.H., Mucha, P.J.: Core-periphery structure in networks. SIAM J. Appl. Math. **74**(1), 167–190 (2014). https://doi.org/10.1137/120881683
85. Rosvall, M., Bergstrom, C.T.: Maps of information flow reveal community structure in complex networks. Proc. Natl. Acad. Sci. U.S.A. **105**, 1118–1123 (2007). https://doi.org/10.1073/pnas.0706851105
86. Seidman, S.B.: Network structure and minimum degree. Soc. Netw. **5**(3), 269–287 (1983). https://doi.org/10.1016/0378-8733(83)90028-X
87. Seidman, S.B., Foster, B.L.: A graphtheoretic generalization of the clique concept. J. Math. Sociol. **6**(1), 139–154 (1978). https://doi.org/10.1080/0022250X.1978.9989883
88. Shen, H., Cheng, X., Cai, K., Hu, M.B.: Detect overlapping and hierarchical community structure in networks. Phys. A Stat. Mech. Appl. **388**(8), 1706–1712 (2009). https://doi.org/10.1016/j.physa.2008.12.021
89. Tarjan, R.: Depth-first search and linear graph algorithms. SIAM J. Comput. **1**(2), 146–160 (1972). https://doi.org/10.1137/0201010
90. Watts, D.J., Strogatz, S.H.: Collective dynamics of 'small-world' networks. Nature **393**(6684), 409–10 (1998). https://doi.org/10.1038/30918
91. Xie, J., Kelley, S., Szymanski, B.K.: Overlapping community detection in networks: the state-of-the-art and comparative study. ACM Comput. Surv. **45**(4), 43:1–43:35 (2013). https://doi.org/10.1145/2501654.2501657

Part II
Software and Ontology Networks: Complex Networks in Source Code

Chapter 3
Analysis of Software Networks

Abstract Modern software systems are characterized not only by a large number of constituent software entities (e.g. functions, modules, classes), but also by complex networks of dependencies among those entities. Analysis of software networks can help software engineers and researchers to understand and quantify software design complexity and evaluate software systems according to software design quality principles. In this chapter, we firstly give a comprehensive overview of previous research works dealing with analysis of software networks. Then, we present a novel network-based methodology to analyze software systems. The proposed methodology utilizes the notion of enriched software networks, i.e. software networks whose nodes are augmented with metric vectors containing both software metrics and metrics used in complex network analysis. The methodology is empirically validated on enriched software networks that represent large-scale Java software systems at different levels of abstraction.

Modern software systems are complex artifacts. The complexity of a large-scale software system stems not only from a large number of software entities (packages, modules, classes, interfaces, functions, variables, and so on) defined in its source code, but also from dependencies among those entities. More specifically, we can distinguish two layers of complexity in software systems: the internal (algorithmic) complexity of software entities (e.g. the complexity of control flow within an entity) and the structural (design) complexity of dependencies among software entities. Dependency structures present in a software system can be modeled by various types of *software networks* which provide different granularity views to the architecture of the system. Consequently, analysis of software networks can help software engineers and researchers to understand and quantify the design complexity of software systems which in turn may be beneficial for software development and maintenance activities.

Analysis of software networks can also reveal characteristics of software systems with respect to recognized software design quality principles. "Low coupling, high

© Springer International Publishing AG, part of Springer Nature 2019
M. Savić et al., *Complex Networks in Software, Knowledge,
and Social Systems*, Intelligent Systems Reference Library 148,
https://doi.org/10.1007/978-3-319-91196-0_3

cohesion" is one of the basic software design principles promoting effective modularization of software systems [95]. This principle states that dependencies between software entities should be as minimal as possible (low coupling) keeping at the same time strong dependencies among constituent elements of a software entity (high cohesion). High coupling and low cohesion are considered as indicators of poor software quality since highly coupled/poorly cohesive software entities may cause various difficulties in software comprehension, testing and maintenance [8, 9, 21, 32]. Additionally, existing empirical studies suggest that highly coupled software entities tend to be more fault prone [3, 29, 31, 62, 78]. If a software system is structured according to the principle of low coupling and high cohesion then the corresponding software networks will have a modular, clustered structure (high cohesion) and will not contain nodes having a high degree (low coupling).

Internal reuse (afferent coupling) and internal aggregation (efferent coupling) jointly form the structural coupling of a software entity within a software system. Internal reuse is considered a good software engineering practice since it reduces duplicated code inside the system. However, there are two potential negative consequences of extensive internal reuse [9]:

1. *High criticality.* Changes in a highly reused software entity may cause changes in a large number of other entities that directly or indirectly depend on it. Additionally, defects in a highly reused software entity are more likely to propagate to other parts of the system.
2. *Low testability.* Defects in highly reused software entities may not be detected when testing software entities in isolation. This happens when defects need to propagate to other parts of the system and cause failures there in order to be observed.

High efferent coupling may negatively impact the following quality attributes of software systems: understandability, error-proneness, and external reusability [9]. In order to fully comprehend a software entity that aggregates (depends on) a large number of other entities one has also to examine and understand all aggregated software entities. Additionally, the probability that some of aggregated entities contains faults or that is incorrectly reused increases with the number of aggregated entities. Low understandability and high error-proneness cause low maintainability and higher maintenance efforts and costs. Finally, if an entity depends on a large number of other entities then its external reuse in other software projects will be more difficult. The degree of the internal reuse of an entity can be quantified by its in-degree in the corresponding software network. On the other hand, the out-degree an entity corresponds to the degree of its efferent coupling. Therefore, analysis of in-coming and out-going links in software networks can reveal characteristics of afferent and efferent coupling of software entities.

Another strongly advised software design principle says that cyclic dependencies among software entities should be avoided [24, 49, 55]. If two or more software entities are mutually dependent then none of them cannot be tested in isolation. This means that the presence of large cyclic dependencies negatively impacts software testability. As emphasized by Parnas [55], large cyclic dependencies may lead to software systems in which "nothing works until everything works". Large cyclic dependencies may also negatively impact software comprehension. Well structured software systems should have clearly defined, hierarchically ordered layers. Large cyclic dependencies have potential to span across several layers making them mutually dependent, fuzzy and non-hierarchical (i.e. topologically unsortable). Mutually dependent entities form a strongly connected component in the corresponding software network. Therefore, analysis of strongly connected components in software networks can be instrumented (1) to evaluate the design quality of software systems with respect to the principle of avoiding cyclic dependencies, and (2) to understand characteristics of mutually dependent software entities.

The first empirical studies of software systems based on complex network analysis methods, techniques and models were published in 2003, four years after the scale-free and small-world phenomena were initially observed in real-world complex networks from other domains. A comprehensive overview of previous research works investigating the structure of software networks is given in Sect. 3.2. The next section, Sect. 3.3, covers studies analyzing the evolution of software networks.

As the main research contribution of this chapter, we propose a novel methodology to analyze structural complexity of large-scale software systems (Sect. 3.4). In contrast to previous research works dealing with analysis of software networks, our methodology is based on *enriched software networks* – software networks whose nodes are augmented with metric vectors that contain both software metrics and domain-independent metrics used in complex network analysis. To demonstrate the usefulness of the proposed methodology we applied it to enriched software networks extracted from source code of five open-source, widely used, large-scale software systems written in Java. The examined systems, extraction of investigated networks and characteristics of attached metric vectors are explained in Sect. 3.5. We analyzed three types of enriched software networks appearing at different levels of abstraction: package collaboration networks, class collaboration networks and method collaboration networks. The obtained results are presented and discussed in Sect. 3.6. The last section summarizes our main findings and indicates possible further research directions.

3.1 Preliminaries and Definitions

Software networks are directed graphs of static dependencies between source code entities. High-level programming languages provide different mechanisms to define different types of software entities at different levels of abstraction in order to support modularity and reuse of source code. In general, to each entity is assigned a name that is used to reference entity by other entities defined in other parts of the source code. Thus, software networks can be viewed as networks connecting identifiers introduced in the source code.

Most programs written in a procedural programming language consists of procedures (also called subroutines or functions) which collaborate using the call-return mechanism provided by the language. In object-oriented software systems, software entities known as methods collaborate using the same mechanism. Call-return relationships between procedures determine homogeneous software networks that are known as *static call graphs*.

Definition 3.1 (*Static call graph (SCG)*) A static call graph encompasses all procedures defined in a software system. Two procedures A and B are connected by a directed link $A \rightarrow B$ if A explicitly calls B.

Static call graphs for object-oriented (OO) software systems are also known as *method collaboration networks* [33]. It is important to observe that procedure calls through a reflection mechanism, if it is present in a language, do not form static (structural, compile-time) dependencies between procedures, but run-time dependencies.

Classes and interfaces are fundamental constructs in object-oriented programming. A class collaboration network describes dependencies between classes and interfaces defined in an OO software system.

Definition 3.2 (*Class collaboration network (CCN)*) A class collaboration network encompasses all class-level entities (classes and interfaces) defined in an object-oriented software system. Two nodes A and B contained in the CCN are connected by a directed link $A \rightarrow B$ if the class or interface represented by node A references the class or interface represented by node B.

Class-level entity A can reference another class-level entity B in many different ways. A and B are directly connected in the class collaboration network if any of the following holds:

- A extends B or implements B in the case that B is an interface,
- A declares a member variable (field, class attribute) whose type is B,
- A instantiates objects whose type is B,
- A contains a method that declares a local variable whose type is B,
- A contains a method that has a parameter whose type is B,
- A contains a method whose return type is B,
- A contains a method that accesses a member variable declared in B, or
- A contains a method that calls a method declared in B.

If $A \rightarrow B$ then we also say that (1) classes A and B are coupled, (2) class A internally aggregates class B, and (3) class B is internally used by class A. If there is another class C defined in the system such that $C \rightarrow B$ then we say that class B is internally reused since it is used by more than one class. Class collaboration networks can be viewed as simplified class diagrams that preserve only the existence of relations between classes, and discard other types of information about nodes (classes) and links (OO relations). Additionally, homogeneous software networks that represent different forms of class coupling, such as inheritance trees or aggregation networks, can be isolated from class collaboration networks [94].

Object-oriented programming languages usually provide mechanisms to group related class-level entities into packages, namespaces or modules. Package-level entities in OO software systems form so-called *package collaboration networks*.

Definition 3.3 (*Package collaboration network (PCN)*) A package collaboration network encompasses all packages (modules, namespaces) present in an object-oriented software system. Two packages PA and PB are connected by a directed link $PA \to PB$ if package PA contains a class or interface that references at least one class or interface from PB.

3.2 Structure of Software Networks

Previous research works dealing with analysis of software systems using complex network methods and techniques were mainly focused on investigating whether software networks exhibit scale-free and small-world properties. The first empirical evidence of scale-free and small-world properties in software systems was reported by Valverde et al. [84]. The authors analyzed class collaboration networks of JDK (Java Development Kit) and UbiSoft ProRally (a computer game) as undirected graphs. The analysis of the JDK network revealed that the two largest connected components exhibit the small-world property in the Watts-Strogatz sense: their characteristic path lengths are close to characteristic path lengths of comparable random graphs and at the same time their clustering coefficients are significantly larger than clustering coefficients of comparable random graphs. Secondly, the authors reported that degree distributions of those two components follow power-laws which means that they possess the scale-free property. Similar results were obtained for the class collaboration network of UbiSoft ProRally. In their subsequent study [85], the authors investigated the same networks, but this time as directed graphs. They also examined statistical properties of 18 more class collaboration networks mostly associated to software systems written in C++. The main conclusion of the study is that analyzed software networks display the same pattern of node connectivity manifested by power-laws in their degree distributions.

Myers [51] examined software networks associated to six software systems: class collaboration networks of three object-oriented software systems written in C++ (VTK, AbiWord, DM) and static call graphs of three procedural software systems written in C (Linux kernel, MySQL, XMMS). Connected component analysis revealed that all analyzed networks have a giant weakly connected component encompassing a vast majority of nodes (ranging from 86 to 99% of the total number of nodes). On the other side, a small fraction of nodes in analyzed networks belong to non-trivial strongly connected components (approximately 13% in AbiWord and less than 4% for the rest of analyzed systems). Myers analyzed complementary cumulative in-degree and out-degree distributions of investigated networks indicating that they have power-law scaling regions. Consequently, he concluded that investigated systems exhibit the scale-free property. The analysis of degree correlations showed that the networks exhibit a weak disassortative mixing. Additionally, Myers observed that in-degree weakly correlates to out-degree which means that highly reused classes/functions tend to internally aggregate a small number of other classes/functions and vice versa. However, each of investigated systems contains a small fraction of entities that have high in-degree and high out-degree. Myers emphasized that such entities could be problematic concerning their maintenance due to significant internal complexity associated with aggregating the behavior of many other entities and significant external responsibility. Moreover, he provided his personal experience with one such entity in one of the analyzed systems (DM) that was difficult to maintain due to conflicting roles the entity had within the system. Finally, Myers proposed a simple model of software network evolution based on two refactoring techniques: the decomposition of large entities into smaller ones and removal of duplicated code. The experimental evaluation of the model showed that it is capable to reproduce empirically observed heavy-tailed in-degree and out-degree distributions.

De Moura et al. [50] investigated properties of complex networks extracted from four open source software projects written in C/C++ (Linux, XFree86, Mozilla and Gimp). The software networks representing analyzed systems were constructed in the following way:

- each C/C++ header file present in a system is represented by a node in the network, and
- two nodes are connected by an undirected link if the corresponding header files are both included in the same source file.

The authors reported that examined complex networks of header files display scale-free and small-world properties. The complex network of header files contained in the Linux kernel was also studied by Sun et al. [80]. In contrast to De Moura et al., Sun et al. constructed a weighted network where link weights indicate the similarity between connected header files. The weight of an undirected link connecting two header files A and B is computed according to the following formula

$$\text{weight}(A \leftrightarrow B) = \sum_{s \in S} \frac{1}{n_s} \qquad (3.1)$$

where S is the set of source files that include both A and B, and n_s represents the number of header files included in a source file s. The main idea is of the weighting scheme is that the number of source files including both A and B determines the similarity between A and B. The weighting scheme also takes into account the number of included header files in order to penalize source files that include a large number of header files. The results of the analysis revealed that the network exhibits the small-world phenomenon and contains a giant connected component encompassing more than 90% of all nodes. The authors also studied distributions of node degrees, node strengths (the sum of weights of links incident with a node) and link weights indicating that all of them follow power-laws.

The study by Louridas et al. [45] presents analysis of a dataset of nineteen software networks that includes Java class collaboration networks and networks of dependencies among Perl packages, libraries in open-source Unix distributions, Windows DLLs, FreeBSD ports, Tex and Metafont modules, and Ruby libraries. The authors found that in-degree and out-degree distributions of investigated networks can be approximated by power laws implying that the scale-free property in software systems appears in diverse systems and languages. The scale-free property was also reported for class collaboration networks of grid middleware applications [96], agent-oriented applications [79], inter-package dependency networks for various operating system distributions [38, 39, 47], run-time object collaboration networks [57], dynamic method call graphs [11, 60], sorting comparison networks [93] and distributions of widely used software metrics [36].

Puppin and Silvestri [58] investigated a complex network of Java class dependencies emerging from unrelated Java software projects. The authors constructed a class collaboration network encompassing approximately 50000 Java classes found on the Internet and plotted its in-degree distribution. The obtained plot suggests that the in-degree distribution follows a power-law in the tail with the scaling exponent roughly equal to one. The authors exploited the presented evidence to formulate a new ranking strategy for Java software components. More specifically, they proposed a mechanism for ranking Java classes similar to the Google page rank algorithm and presented a prototype of a search engine for Java classes. The initial evaluation showed that the page rank centrality metric applied to the class collaboration network is more effective in ranking Java classes than the classic TF-IDF ranking scheme.

The study by Chatzigeorgiou et al. [14] showed that class collaboration networks of three Java software systems (JUnit, JHotDraw and JRefactory) do not have hub-like cores, i.e. highly coupled classes in those systems do not tend to be directly connected among themselves. The existence of a hub-like core in a complex network is typically checked by computing the assortativity index, but the authors employed an alternative approach based on the S metric proposed by Li et al. [44]. For a graph $G = (V, E)$ having degree sequence $D = \{d_1, d_2, \ldots, d_n\}$ the S metric is defined as

$$S = \sum_{(i,j) \in V} d_i d_j \qquad (3.2)$$

and its normalized variant is given by

$$S' = \frac{S - S_{min}}{S_{max} - S_{min}} \qquad (3.3)$$

where S_{max} and S_{min} are the maximal and the minimal value of S considering all graphs having degree sequence D. S' values close to 1 indicate the presence of a hub-like core in G. For studied software networks S' values are significantly lower than 1 implying that they do not exhibit an assortative mixing pattern. The authors also demonstrated how various graph algorithms such as HITS and spectral graph clustering can be exploited in OO software engineering to identify "God" classes

(classes that "do everything" and violate the principle of uniform distribution of responsibilities), cluster software entities and detect design patterns.

Degree distributions of software networks reflecting specific forms of class coupling (inheritance, aggregation, interface, parameter type and return type coupling) were analyzed by Wheeldon and Counsell [94]. Examined networks were extracted from three Java software systems: Java Development Kit, Apache Ant and Tomcat. The authors tested degree distributions against power-law by performing linear regression on log-log plots that were formed by grouping data values into buckets of exponentially increasing size. The main conclusion of the study is that class collaboration networks restricted to a particular class coupling type are also scale-free networks. Baxter et al. [4] extended the study by Wheeldon and Counsell to a larger corpus of software networks associated to 56 Java software systems. In contrast to all previously mentioned studies, the authors considered power-law, log-normal and stretched exponential distributions to model empirically observed in-degree and out-degree distributions of class collaboration networks restricted to a particular coupling type. The best fits were obtained using the weighted least square fitting technique. This technique estimates parameters of a theoretical distribution by minimizing the following quantity:

$$Q = \sum_{i=1}^{k} \frac{1}{h_i} (h_i - f(\alpha, x_i))^2 \qquad (3.4)$$

where k is the number of data points, h_i is the empirically observed frequency of x_i (the number of nodes whose in-degree/out-degree is equal to x_i), α represents the parameters of the distribution, and $f(\alpha, x_i)$ is the value of the theoretical distribution at x_i. The value of Q itself determines the quality of a fit, i.e. a smaller value indicates a better fit since Q quantifies the difference between empirically observed and theoretically predicted frequencies. The obtained results show that out-degree distributions do not tend to have good power-law fits. On the other hand, power-laws are plausible theoretical models for in-degree distributions. The authors explained this difference by the fact that software developers are inherently aware of out-going class dependencies which is not the case for in-coming class dependencies. Namely, a developer authoring or changing a class can control its out-going dependencies. On the other side, he/she can not control how the class will be internally reused by other developers. The authors generalized the previous observation to the hypothesis that any software metric quantifying something that software developers are inherently aware of will tend to have a "truncated" distribution curve that does not follow a power-law.

Concas et al. [19] performed a comprehensive statistical analysis of an implementation of the Smalltalk object-oriented system. The authors analyzed distributions of the following quantities: the number of methods in a class, the number of attributes in a class, the number of subclasses of a class, the number of calls to a method, the size of methods in terms of LOC (lines of code), the size of classes in terms of LOC, and the in-degree and out-degree of a class in the class collaboration network. Complementary cumulative distributions of examined quantities were tested against power-law and log-normal distributions. The parameters of theoretical distributions were determined using maximum likelihood estimators, while the Pearson's χ^2 test was used to assess the quality of fits. The authors found that the in-degree distributions exhibit power-laws in their tails, while the out-degree distributions exhibit log-normal behavior. In their subsequent study [18], the authors examined statistical properties of metrics used in social network analysis (SNA) computed on two software networks. The authors constructed so called *compilation unit networks* for two software systems written in Java (Eclipse and NetBeans). A compilation unit network reflects dependencies between Java files present in a software project.[1] The following SNA metrics were considered in the empirical evaluation: total node degree, the number of links in an ego network, the number of node pairs that are not directly connected in an ego network, the number of weak components in an ego network when the ego node is removed, reach efficiency (the fraction of nodes within the two step distance from a node), closeness centrality and information centrality. The main conclusion of the study is that empirically observed distributions of SNA metrics, distributions of metrics from the Chidamber–Kemerer (CK) suite adapted for compilation units, and in-degree and out-degree distributions show power-law behavior in their tails. The authors also studied correlations between examined metrics and the number of bugs and reported that SNA metrics and metrics from the CK suite are moderately correlated to the number of defects found in a compilation unit.

Melton and Tempero [49] analyzed characteristics of strongly connected components (SCCs) in class collaboration networks of 78 Java applications. The study suggests that large cyclic class dependencies frequently occur in Java software systems: about 85% of the applications from their experimental corpus have a SCC encompassing more than 10 classes, SCCs larger than 100 classes occur in approximately 45% of examined applications and around 10% of analyzed systems have extremely large SCCs containing more than 1000 classes. The authors also observed that the number of classes involved in cyclic dependencies tends to grow during software evolution. Finally, they noticed that evolutionary dips in the size of SCCs for several applications correspond to major refactoring efforts. Laval et al. [40] observed that large strongly connected components are also present in package collaboration networks of Java applications. The study by Oyetoyan et al. [53] suggests that software entities contained in SCCs tend to be more defect-prone than entities not involved in cyclic dependencies. Al-Mutawa et al. [1] analyzed the shape of SCCs in 107 Java programs. The authors observed that SCCs are in most cases irregular structures and that symmetric SCCs like chains, circles, cliques and stars occur very rarely. Additionally, SCCs exhibiting symmetric shapes tend to be very small. The authors also found that most cyclic dependencies are package local, i.e. they are present between class-level entities belonging to the same package.

Ichii et al. [34] analyzed class collaboration networks associated to four open source software written in Java (Ant, JBoss, JDK and Eclipse). The authors reported

---

[1]It should be noticed that compilation unit networks are considerably different from class collaboration networks since one compilation unit (a Java file) may contain definitions of more than one class and/or interface including also multiple inner classes/interfaces.

that the in-degree distributions almost ideally follow power-laws. On the other hand, the out-degree distributions do not exhibit power-law behavior through the whole range of out-degree values – the distributions have a "peak" for small out-degree values followed by a power-law behavior in tails. The authors also investigated correlations between in-/out-degree and four software metrics (LOC, two variants of Chidamber–Kemerer's WMC and the Chidamber-Kemerer's LCOM) showing that out-degree exhibits a high correlation with metrics of internal complexity (LOC and the variants of WMC).

Taube-Schock et al. [82] investigated characteristics of 97 complex networks associated to software systems written in Java. Examined networks were constructed to encompass not only architectural elements as nodes but also statements contained in methods. The authors analyzed degree distributions of the networks concluding that all of them are heavy-tailed. They emphasized that previous studies "have not considered software systems at the level of statements and variables, limiting the generality of the findings". In contrast to software entities representing architectural elements, statements can not be referenced and they typically reference a low number of methods through method calls that are part of the statement. Since the number of statements in any large-scale software system is significantly higher than the number of architectural elements, software networks including statements as nodes will have a vast majority of low degree nodes. Consequently, the average degree will be low and a small number of nodes representing architectural elements will tend to have a significantly higher degree compared to the average degree. Therefore, heavy-tailed degree distributions will be practically caused by the existence of nodes representing statements, not by dependencies between architectural elements. The authors also computed degree distributions considering only inter-module links (links that constitute coupling among classes) concluding that they are also heavy-tailed for all examined systems.

The study by Šubelj and Bajec [88] investigated whether class collaboration networks of Java software systems exhibit community structures. The authors applied three community detection techniques (the Girvan-Newman algorithm, greedy modularity optimization and label propagation) to class collaboration networks of eight software systems (JUnit, JavaMail, Flamingo, Jung, Colt and three JDK namespaces). The quality of obtained partitions was assessed using the Girvan-Newman modularity measure. The results of the analysis showed that examined software networks possess a strong community structure. However, the comparison between identified communities and existing software packages revealed that identified communities are considerably different than actual packages suggesting that the package structure of analyzed Java systems can be significantly improved towards higher modularization. Strong community structures were also reported for class collaboration networks of Netbeans subsystems [20], the class collaboration network of the Eclipse IDE [43], the static call graph of the Linux kernel [26], method collaboration networks of 10 large-scale Java software systems [59] and backward slice graphs of open source software systems written in C [30].

Turnu et al. [83] analyzed the entropy of degree distributions of Eclipse and Netbeans class collaboration networks in various releases. The main finding of the study is that the entropy of degree distributions exhibits a good correlation with the total number of bugs within a system.

Chong and Lee [16] investigated the structure of class collaboration networks associated to 40 Java open source software systems. The authors used weighted directed graphs to model dependencies between classes where link weights denote the strength of communicational cohesion between classes. More specifically, the weight of a directed link $A \rightarrow B$ is computed according to the following formula:

$$\text{weight}(A \rightarrow B) = \alpha T(A \rightarrow B) + \beta(1 - \text{Comp}(B)) \tag{3.5}$$

where $T(A \rightarrow B)$ is a relative link weight that depends on the type of coupling between A and B, $\text{Comp}(B)$ reflects the internal complexity of class B (measured as the normalized product of cyclomatic complexity and cohesion), and α and β are free parameters in the interval $[0, 1]$ reflecting the importance of corresponding factors. The authors studied distributions of node in-degree, out-degree, weighted degree and betweenness centrality concluding that examined distributions are heavy-tailed. They also investigated correlations between global network-based metrics and SQALE (Software Quality Assessment based on Lifecycle Expectations) ratings. The main finding of the study is that the average weighted degree and shortest path length of SQALE B-rated and SQALE C-rated software systems are significantly higher than those of SQALE A-rated software systems.

3.3 Evolution of Software Networks

Several researchers also investigated the evolution of software networks. Hyland-Wood et al. [33] examined package, class and method collaboration networks of two Java open source software systems (Kowari Metastore and JRDF) considering fifteen-month periods of their development. For each investigated network, the authors constructed an evolutionary sequence of monthly snapshots. The analysis of in-degree and out-degree distributions of monthly snapshots indicated that they follow truncated power-laws. Consequently, the authors concluded that the scale-free property in software systems retains across different levels of abstraction and over the course of software development life cycle. The studies by Jenkins and Kirk [35] and Li et al. [42] also suggest that the scale-free property in class collaboration networks is persistent during software evolution.

The study by Vasa et al. [86] examined the evolution of 12 class collaboration networks extracted from Java software systems. The obtained results suggest that classes having a high in-degree tend to change more frequently compared to classes exhibiting a low efferent coupling.

Zheng et al. [100] examined the evolution of the network reflecting dependencies between packages in the Gento distribution of the Linux operating system. The authors investigated how new nodes integrate into the network concluding that new nodes tend to connect to old nodes with the probability that depends not only on the degree of old nodes but also on their age.

Wen et. al [92] analyzed an evolutionary sequence of fifty monthly snapshots of the static call graph of the Apache HTTP server. The authors observed a densification law [41] in the evolution of the network: the number of links grows super linearly with respect to the number of nodes satisfying the power-law of the form $E(t) \sim N(t)^{1.18}$, where $E(t)$ and $N(t)$ are the number of links and the number of nodes at time t, respectively. This can be considered as a bad phenomenon from the software engineering perspective because it means that the average function coupling increases during software evolution. Similarly to Baxter et al. [4], Wen et. al used the weighted least square fitting technique to examine in-degree and out-degree distributions of each evolutionary snapshot considering power-law, log-normal and stretched exponential distributions as theoretical models. The results of their degree distribution analysis showed that the power-law model provides the best fits for empirical in-degree distributions. On the other hand, the log-normal model is better for out-degree distributions.

Pan et al. [54] studied the evolution of package, class and method collaboration networks of Azareus (a BitTorrent client written in Java). The authors observed an increase in the average coupling at all three levels of abstraction during the evolution of the investigated system. They also reported that degree distributions of evolutionary network snapshots either follow power-law or power-law with exponential cutoff. Evolutionary sequences of package and class collaboration networks exhibit a weak disassortative mixing patterns. On the other hand, the method collaboration network shows a persistent weak assortativity during its evolution. The authors employed the greedy modularity optimization algorithm to identify communities in all networks and examined the evolution of the Girvan-Newman modularity at all three levels of granularity. The obtained results indicate that modularity fluctuates many times over sequential software releases tending towards lower values.

Densification laws [41] in the evolution of class collaboration networks were reported by Vasa et al. [87] and Chatzigeorgiou and Melas [13]. Chatzigeorgiou and Melas [13] examined the evolution of Weka and JFreeChart class collaboration networks. The authors observed that both networks exhibit shrinking diameters over time. They also investigated how new classes integrate into examined networks concluding that the evolution of both networks is guided by the preferential attachment principle. Additionally, obtained results indicate that preferential attachment becomes more intense as investigated systems evolve.

Wang et al. [89, 90] analyzed the evolution of static call graphs corresponding to 223 consecutive versions of the Linux kernel (from version 1.1.0 to 2.4.35). The authors constructed an evolutionary sequence of static call graphs for each Linux kernel module. Then, for each network in the evolutionary sequence, the authors formed a set of nodes called TDNS (top degree node set) that contains 5% nodes with the highest in-degree and 5% nodes with the highest out-degree. Using formed TDNS sets, the authors computed so-called connecting probabilities, i.e. the probability that a node introduced in the next evolutionary snapshot establishes a connection with a node from the TDNS. Static call graphs of five kernel modules exhibit relatively large average connecting probabilities indicating that they exhibit a strong preferential attachment growth. The authors also proposed a method for finding major structural changes in the evolution of software networks that is based on the slope of the average path length change.

Bhattacharya et al. [5] investigated the evolution of static call graphs of eleven open source software systems written in C/C++ (Firefox, Blender, VLC, MySQL, Samba, Bind, Sendmail, openSSH, SQLite, Vsftpd). The authors observed that the average degree increases in time for all investigated networks except for the MySQL static call graph. The number of nodes in strongly connected components of three networks (Samba, MySQL and Blender) also increases in time. Both of previous two observations can be considered as bad phenomena affecting software quality negatively. All examined networks exhibit disassortative mixing patterns. The authors also demonstrated that the page rank metric is a good fault predictor and that it can be used to identify critical functions (functions that are likely to exhibit high severity bugs).

Zanetti and Schweitzer [98] investigated the application of the Girvan-Newman modularity measure (Q) to assess package cohesion in Java software systems. Namely, a Java class collaboration network can be partitioned according to the package structure (the grouping of classes into packages) of class-level entities present in the system. Then, the Q measure can be applied to the obtained partitioning providing a global estimate of software cohesion at the package level. The authors examined the evolution of package cohesion in 28 Java software systems noticing that in a majority of examined systems Q decreases during software evolution. In their subsequent work [99], the authors proposed an automated strategy to remodularize Java software systems based on the move refactoring (moving classes between packages without changing anything else) which maximizes the Q metric.

The study by Fortuna et al. [23] showed that a strong community structure can be found in the complex network describing dependencies between packages in the Debian distribution of the Linux operating system. The authors analyzed the evolution of the community structure observing that the number of communities and the modularity measure increase as the network evolves. Paymal et al. [56] investigated the community structure in class collaboration networks extracted from six consecutive versions of JHotDraw software using the greedy modularity optimization technique. The authors observed that two largest communities contain more than the half of all nodes in each version and that those two communities have continuous and stable growth during software evolution. In one of our previous works [71], we examined the evolution of communities in the Apache Ant class collaboration network relying on four different community detection techniques. The obtained results show that the Walktrap community detection method exhibits the best performance considering the quality of community partitions, evolutionary stability and comparison with actual class groupings into packages. We also noticed that two community detection methods (label propagation and greedy modularity optimization) are extremely sensitive to small evolutionary changes in the structure of the network.

Chaikalis and Chatzigeorgiou [12] studied the evolution of 10 class collaboration networks associated to Java software systems. The authors analyzed how new nodes integrate into examined networks and investigated characteristics of new links between old nodes concluding that the preferential attachment mechanism is strongly present in the evolution of the examined software systems. In one of our previous works [68], we also examined the evolution of one class collaboration network representing Apache Ant. Similarly to Chaikalis and Chatzigeorgiou, we observed that the preferential attachment principle can explain how new Ant classes reference existing Ant classes. Chaikalis and Chatzigeorgiou additionally observed that a majority of newly created links among old nodes are emanating from recently created nodes and that they mostly connect nodes at short distances. Relying on obtained empirical findings, the authors proposed a prediction model for software structure evolution based on the preferential attachment principle, past evolutionary trends and common software engineering principles (e.g. base classes should not reference derived classes).

3.4 Analysis of Enriched Software Networks

Our network-based methodology to analyze software systems is based on the notion of *enriched software networks*. The nodes in an enriched software network are augmented with metrics reflecting quality attributes of corresponding software entities. Characteristics of software entities can be quantitatively expressed not only by software metrics, but also by domain-independent metrics used in complex network analysis. More specifically, network-based centrality metrics can be exploited to estimate functional importance of software entities within a software system considering their position and role in the corresponding software network [46, 75, 76, 97]. Our methodology relies on four such metrics: betweenness centrality [25], HITS hub and authority scores [37] and page rank [10]. If a software entity has a high betweenness centrality in a software network then it can be considered functionally important due to its brokerage role in data and control flow between components of the system. High page ranks indicates functionally important entities that tend to be directly coupled to other functionally important entities. On the other hand, HITS scores enable us to distinguish between two types of entities having a high importance within the system:

- central entities which tend to use other central entities will have high HITS hub scores, while
- central entities which tend to be used by other central entities will have high HITS authority scores.

Usually the first step in complex network analysis is the identification of connected components and characterization of network structure at the macro level. Software networks are directed graphs and they contain both weakly connected and strongly connected components. Weakly connected components can be identified using classic graph traversal algorithms. The breadth-first search (BFS) algorithm is used in our analysis of enriched software networks to identify weakly connected components.

The size of the largest weakly connected component in a software network reflecting dependencies among software modules or classes reflects the overall cohesiveness of the corresponding software project. To investigate whether analyzed networks possess the small-world property, we measure the characteristic path length and clustering coefficient and compare them to the same quantities of a comparable directed random graph. The assortativity index introduced by Newman [52] is used to measure the extent to which highly coupled software entities tend to be directly connected among themselves.

3.4.1 Metric-Based Comparison Test

The metric-based comparison test is the central element of our methodology for analysis of enriched software networks. We proposed an initial version of this test in [70] where it was used to determine properties of highly coupled nodes in enriched class collaboration networks. Here, we generalize the initial proposal to any type of enriched software network and an arbitrary node classifier which divides the set of nodes into two non-overlapping subsets. In the context of our methodological framework, this test is employed to determine characteristics of software entities involved in cyclic dependencies, software entities exhibiting high coupling and software entities causing low cohesion. In other words, it enables investigation and deeper understanding of phenomena that are considered as indicators of poorly designed software systems.

The description of the metric-based comparison test is given in Algorithm 3.1. The metric-based comparison test relies on the Mann-Whitney U (MWU) test [48] and probabilities of superiority [22]. The metric-based comparison test at the input takes a binary classifier (BC) which divides the nodes of an enriched software network into two groups of nodes C_1 and C_2 according to some criterion related to network structure. The test at the output gives five sets of metric descriptors:

1. S_1 – a set of metrics such that for each metric m in S_1 a statistically significant difference between the values of m for software entities contained in C_1 and the values of m for software entities contained in C_2 is not present. In other words, if m belongs to S_1 then there are no substantial differences between C_1 and C_2 regarding the aspect quantified by m.
2. S_2 – a set of metrics such that for each metric m in the set there is a substantial difference between C_1 and C_2 regarding the aspect quantified by m in the sense that the nodes from C_1 strongly tend to have higher values of m compared to the nodes from C_2.
3. S_3 – a set of metrics having characteristics opposite to metrics contained in S_2. If a metric m belongs to S_3 then the nodes from C_2 strongly tend to have higher values of m compared to the nodes contained in C_1.
4. S_4 – a set of metrics such that for each metric m in S_4 the nodes from C_1 tend to have higher values of m compared to the nodes from C_2, but this tendency, although statistically significant, is not too strong.

5. S_5 – a set of metrics having characteristics opposite to metrics contained in S_4. A metric m belongs to this set if it is not present in any of the previously described sets of metrics.

The MWU test is a non-parametric statistical procedure that can be employed to check whether numerical values in one set tend to stochastically dominate over numerical values in some other, independent set [48]. The test does not require a normal distribution of values and it is applicable to sets having different size. Let G_1 and G_2 denote two independent sets of numerical values. The MWU test checks the null hypothesis that the values in G_1 do not tend to be either greater or smaller than the values in G_2. Consequently, the alternative hypothesis is that values in one set tend to be greater than the values in the other set. We use g_1 and g_2 to denote the size of G_1 and G_2, respectively. The MWU test arranges values from both sets into a single ranked sequence R. The minimal value in R has rank 1 and the maximal value has rank $g_1 + g_2$. If there is a group of equal values then the average rank is assigned to each member of the group.[2] Let U_1 be the number of times a value from G_1 precedes a value from G_2 in R:

$$U_1 = g_1 g_2 + \frac{g_1(g_1 + 1)}{2} - S_1 \qquad (3.6)$$

where S_1 is the sum of ranks of values from G_1. If all values in G_1 are smaller than any value in G_2 then we have

$$S_1 = 1 + 2 + \cdots + g_1 = \frac{g_1(g_1 + 1)}{2} \qquad (3.7)$$

which implies $U_1 = g_1 g_2$ (each value from G_1 precedes each value from G_2). On the other hand, if all values in G_1 are larger than any value in G_2 then the following equality holds

$$S_1 = (g_2 + 1) + (g_2 + 2) + \cdots + (g_2 + g_1) = g_1 g_2 + \frac{g_1(g_1 + 1)}{2} \qquad (3.8)$$

which means that $U_1 = 0$ (each value from G_1 succeeds each value from G_2). Similarly, we can obtain U_2 which is the number of times a value from G_2 precedes a value from G_1:

$$U_2 = g_1 g_2 + \frac{g_2(g_2 + 1)}{2} - S_2 \qquad (3.9)$$

where S_2 is the sum of ranks of values from G_2. If the null hypothesis is true then the values of both U_1 and U_2 should be about the half of the total number of pairs:

$$U_1 \approx U_2 \approx \frac{g_1 g_2}{2} \qquad (3.10)$$

---

[2] Other methods for resolving ties such as the smallest rank, the largest rank or a randomly selected rank can be employed as well.

Algorithm 3.1: The metric-based comparison test

---

input : $G = (V, E)$ – an enriched software network

M – the set of metrics associated to nodes in G

BC – a binary classifier dividing V into two non-overlapping subsets

output: S_1, S_2, S_3, S_4, S_5 ($S_{i=1...5} \subseteq M, \bigcup_{i=1}^{5} S_i = M$)

/* initialize S sets */
for $i := 1$ **to** 5 **do**
  | $S_i := \emptyset$
end

/* divide V into C_1 and C_2 according to the binary classifier BC */
$C_1 := \emptyset$
$C_2 := \emptyset$
foreach $v \in V$ **do**
  | **if** $BC(v) =$ positive **then** $C_1 := C_1 \bigcup \{v\}$
  | **else** $C_2 := C_2 \bigcup \{v\}$
end

/* decide to which S set each metric belongs */
foreach $m \in M$ **do**
  | $G_1 := \emptyset$ /* the set containing values of metric m for nodes in C_1 */
  | $G_2 := \emptyset$ /* the set containing values of metric m for nodes in C_2 */
  | **foreach** $v \in C_1$ **do**
  | | $G_1 := G_1 \bigcup \{m(v)\}$
  | **end**
  | **foreach** $v \in C_2$ **do**
  | | $G_2 := G_2 \bigcup \{m(v)\}$
  | **end**
  |
  | $p :=$ apply the Mann-Whitney U test to G_1 and G_2
  |
  | /* compute probabilities of superiority */
  | $ps_1 := 0$
  | $ps_2 := 0$
  | **foreach** $x \in G_1$ **do**
  | | **foreach** $y \in G_2$ **do**
  | | | **if** $x > y$ **then** $ps_1 := ps_1 + 1$
  | | | **else if** $y > x$ **then** $ps_2 := ps_2 + 1$
  | | **end**
  | **end**
  | $PS_1 := ps_1/(|G_1| \cdot |G_2|)$
  | $PS_2 := ps_2/(|G_1| \cdot |G_2|)$
  |
  | **if** $p \geq 0.05$ **then** $S_1 := S_1 \bigcup \{m\}$
  | **else if** $PS_1 \geq 0.75$ **then** $S_2 := S_2 \bigcup \{m\}$
  | **else if** $PS_2 \geq 0.75$ **then** $S_3 := S_3 \bigcup \{m\}$
  | **else if** $PS_1 \geq PS_2$ **then** $S_4 := S_4 \bigcup \{m\}$
  | **else** $S_5 := S_5 \bigcup \{m\}$
end

Let U be the minimum of U_1 and U_2. Mann and Whitney showed that the limit distribution of U is normal regardless of how g_1 and g_2 approach infinity. Consequently, if the standardized value of an empirically observed U is far from zero (the center of the standard normal distribution $N(0, 1)$) then the null hypothesis of the MWU test can be rejected. The metric-based comparison test rejects the null hypothesis of the MWU test if the p-value of U is less than 0.05.

The effect size of the MWU test can be quantified by the so-called probability of superiority [22]. The probability of superiority is the probability that a randomly selected value from the first group is equal to or higher than a randomly selected value from the second group. The metric-based comparison test relies on two probabilities of superiority:

1. $PS_1 = P(X > Y)$ – the probability that a randomly selected value from G_1 (denoted by X) is strictly higher than a randomly selected value from G_2 (denoted by Y), and
2. $PS_2 = P(X < Y)$ – the probability that a randomly selected value from G_2 is strictly higher than a randomly selected value from G_1.

It is important to observe that $PS_1 + PS_2$ is not necessarily equal to 1. Namely, the probability that a randomly selected value from G_1 is equal to a randomly selected value from G_2, denoted by $P(X = Y)$, can be non-zero, i.e.

$$P(X = Y) = 1 - (PS_1 + PS_2) \tag{3.11}$$

PS_1 and PS_2 can be computed in a straightforward manner by comparing each value from G_1 to each value from G_2 (as shown in Algorithm 3.1). If the null hypothesis of the MWU test is rejected and $PS_1 > 0.5$ then we can conclude that values from G_1 tend to be larger than values from G_2. The same applies for $PS_2 > 0.5$ in the opposite case. A higher value of PS_k ($k = 1$ or $k = 2$) implies a stronger stochastic dominance. The metric-based comparison test uses the threshold $T = 0.75$ to mark a strong stochastic domination. Namely, if the null hypothesis of the MWU test is rejected regarding a metric m and $PS_k \geq 0.75$ then we can conclude that software entities belonging to the group k systematically tend to have higher values of m than software entities from the other group.

3.4.2 Analysis of Strongly Connected Components and Cyclic Dependencies

We use the Tarjan algorithm [81] to identify strongly connected components in investigated software networks. Two software entities are either directly or indirectly cyclically dependent if they belong to the same strongly connected component. The Tarjan algorithm also identifies trivial strongly connected components encompassing exactly one node, but we ignore such components since they do not represent

software entities involved in cyclic dependencies. To determine distinctive characteristic of software entities involved in cyclic dependencies, we apply the metric-based
comparison test with the following binary classifier: $BC(n)$ is positive for a node
n if n belongs to a (non-trivial) strongly connected component, otherwise $BC(n)$ is
negative.

The tendency of software entities to form direct mutual dependencies is quantified
using the link reciprocity measure which is the conditional probability $P(B \rightarrow A | A \rightarrow B)$, i.e. the probability that a directed graph which contains a link connecting
A to B ($A \rightarrow B$) also contains a reciprocated link connecting B to A ($B \rightarrow A$). For
each examined software network we also compute the normalized link reciprocity
measure with respect to a comparable random graph [27]. Moreover, we introduce
a new reciprocity measure called path reciprocity that takes into account both direct
and indirect dependencies among software entities.

Definition 3.4 (*Path reciprocity*) Let G be a directed graph, and let $A \rightsquigarrow B$ denote
that there is a directed path from a node A to a node B. The path reciprocity of G is
the conditional probability that $B \rightsquigarrow A$ when $A \rightsquigarrow B$.

In software networks $A \rightsquigarrow B$ means that software entity A either directly or
indirectly depends on software entity B. If $A \rightsquigarrow B$ and $B \rightsquigarrow A$ then A and B are
in a cyclic dependency. The link reciprocity of a software network quantifies the
degree of direct cyclic dependencies present in the corresponding software system,
while its path reciprocity reflects the degree of presence of both direct and indirect cyclic dependencies. The procedure for computing path reciprocity is shown in
Algorithm 3.2. The main idea of the algorithm is that the number of reciprocated
dependencies can be computed from strongly connected components: the number of
reciprocated dependencies in a strongly connected component S containing k nodes
is equal to $k(k-1)$. Then, the total number of reciprocated dependencies is the sum
of reciprocated dependencies over all strongly connected components since strongly
connected components are independent subgraphs. On the other hand, the total number of dependencies can be counted from the sets of reachable nodes considering
each node in the network. Let A denote an arbitrary node and let $R(A)$ be its reachable set (the set containing all nodes B such that $A \rightsquigarrow B$). $R(A)$ can be determined
by executing BFS (or DFS) starting from A. Then, the total number of dependencies is equal to the sum of the cardinality of R-sets over all nodes in the network.
Finally, path reciprocity is the number of reciprocated dependencies divided by the
total number of dependencies.

To analyze the complexity of cyclic dependencies present in a software system
we examine the complexity of strongly connected components of corresponding
software networks. The number of links in a strongly connected component indicates
its complexity. For example, if we have two strongly connected components of the
same size then the one with a higher number of links can be considered more complex.
In general case, a strongly connected component S_1 is more complex than some other
strongly connected component S_2 if

Algorithm 3.2: The algorithm for computing path reciprocity

---

input : $G = (V, E)$ – a directed graph
output: PR – path reciprocity

$r := 0$ /* the number of reciprocated dependencies */
$S :=$ determine strongly connected components of G using the Tarjan algorithm
foreach $s \in S$ **do**
| $r := r + |s|(|s| - 1)$
end

$d := 0$ /* the total number of dependencies */
foreach $v \in V$ **do**
| $rs := \{v\}$ /* the set of nodes reachable from v including v */
| $q := [v]$ /* the queue of nodes */
| **while** $q \neq []$ **do**
| | $n :=$ remove the first element from q
| | **foreach** $w \in V : (n, w) \in E \wedge w \notin rs$ **do**
| | | $q := q + [w]$
| | | $rs := rs \bigcup \{w\}$
| | **end**
| **end**
| $d := d + |rs| - 1$ /* -1 because rs also contains v */
end

$PR := r/d$

$$\frac{L(S_1)}{N(S_1)} > \frac{L(S_2)}{N(S_2)} \tag{3.12}$$

where $L(S)$ and $N(S)$ are the number of links and nodes in a strongly connected component S, respectively. The minimal number of links in any strongly connected component S is equal to the number nodes contained in S. Therefore, $L(S)/N(S)$ actually measures the extent to which S deviates from being a pure cycle containing $N(S)$ mutually reachable nodes. A stronger deviation from the pure cycle indicates a more dense (and consequently more complex) strongly connected component regardless of its size. On the other hand, $L(S)/N(S)$ is equal to the average intra-component degree of S, i.e. the average number of links incident with a node within S whose both ends are in S.

3.4.3   Analysis of Coupling Among Software Entities

The total degree of a node in a software network reflects the degree of the coupling of the corresponding software entity with other entities present in the software system. The coupling of all entities present in the system can be summarized by the total degree distribution of the network. The degree distribution of a network is a probability function P where $P(k)$ is the probability that the degree of a randomly selected node is equal to k. Software networks are directed graphs. Thus, we can distinguish three types of degree distributions:

- The in-degree distribution $P_{in}(k)$ which is the probability that the in-degree of a randomly selected node is equal to k. This distribution summarizes the afferent coupling of software entities present in the system since the in-degree measure in software networks reflects the degree of afferent coupling of a software entity.
- The out-degree distribution $P_{out}(k)$ which is the probability that the out-degree of a randomly selected node is equal to k. The out-degree distribution summarizes the efferent coupling of software entities present in the system.
- The total degree distribution (or just degree distribution) $P_{tot}(k)$ which is the probability that the total degree (total coupling, the sum of in-degree and out-degree) of a randomly selected node is equal to k.

Knowing statistical properties of degree distributions associated to software networks can be useful not only from a theoretical standpoint, but may be also beneficial to software engineering practitioners. Statistical properties of degree distributions of software networks can motivate theoretical principles behind predictive models of software structure evolution. On the other hand, software engineering practitioners can rely on parameters of fitted theoretical models to evaluate software design quality with respect to the principle of low coupling. For example, if the degree distribution of a software network follows a power-law (as frequently stated in the literature) then the scaling exponent of the fitted power-law can be effectively used as an indicator of software design quality – a smaller power-law scaling exponent implies a more skewed distribution of coupling among software entities which further implies a stronger deviation from the principle of low coupling.

In our analysis of enriched software networks, we compute and examine the following descriptive statistics of degree distributions that are directly related to the high coupling phenomenon:

1. Mean degree (μ) – the average number of in-coming/out-going/total links incident with a node. In directed graphs we have that the average in-degree is equal to the average out-degree, while the average total degree is exactly two times higher than the average in-/out-degree.
2. Standard deviation (σ) which measures the amount of variation from the mean degree.
3. The coefficient of variation (c_v) which is a normalized measure of dispersion defined as the ratio of the standard deviation and the mean degree. The coefficient of variation of an exponential distribution is equal to 1. Thus, distributions whose c_v is smaller than 1 are considered low-variance distributions, while distributions exhibiting $c_v > 1$ are considered high-variance distributions.
4. Skewness (G_1) which is the third standardized moment of the distribution and quantifies its asymmetry. A distribution is perfectly symmetric if its skewness is equal to 0 (e.g. normal distributions). A negative skewness indicates that the left tail of the distribution is longer than the right tail, while a positive skewness indicates exactly the opposite. The skewness of an exponential distribution is

equal to 2. Consequently, the skewness of an empirical degree distribution higher than 2 suggests that the distribution is heavy-tailed (i.e. its tail is not exponentially bounded).

Highly coupled software entities have a high total degree in software networks. We call such nodes *hubs*. The main characteristic of Erdős-Renyi random graphs is that their degree distributions can be approximated by Poisson distributions. This further implies that a vast majority of nodes have in-/out-/total degree close to the average degree and that extreme degree values are absent. The coefficient of variation and skewness of a Poisson degree distribution are both equal to $1/\sqrt{\mu}$ where μ is the average node degree. Consequently, the coefficient of variation and skewness of an empirical degree distribution significantly higher than $1/\sqrt{\mu}$ suggest that the corresponding software network strongly deviates from random graphs and contains hubs.

The presence of power-laws in degree distributions of software networks was reported in several studies reviewed in Sect. 3.2. However, the main characteristic of a large majority of those studies is that empirical degree distributions were tested only against power-laws, and usually by applying linear regression on log-log plots which is a biased methodology to confirm the presence of power-laws [17]. Only in a few studies ([4, 19, 69, 92]) non-power-law distributions were additionally considered as theoretical models. In contrast to those studies, our analysis of enriched software networks relies on a robust statistical test introduced by Clauset et al. [17] to examine whether degree distributions follow power-laws. This test additionally includes the log likelihood test to compare the best power-law fit to the best fits of alternative distributions. The test determines parameters of theoretical models relying on maximum likelihood estimators and minimization of the weighted Kolmogorov-Smirnov (KS) statistic. The quality of individual fits is measured as the probability that the weighted KS statistic for empirical data is lower than the same quantity for synthetic data generated by theoretical models. We consider three theoretical models for degree distributions of enriched software networks: power-law (as the null model), exponential and log normal distributions. Empirical power-law degree distributions in evolving complex networks are commonly explained by the linear preferential attachment principle of the Barabaśi-Albert scale-free model [2]. Consequently, to confirm a power-law scaling behavior we have to eliminate alternative attachment principles. Exponential degree distributions characterize complex networks whose evolution is governed by the uniform attachment principle [2]. On the other hand, log-normal degree distribution arise from the nearly linear preferential attachment principle [65].

Knowing differences between highly and loosely coupled software entities can help us to gain a more deeper understanding of high coupling in some software system. This in turn can be beneficial for a variety of software development and maintenance activities. For example, if highly coupled software entities tend to have complex control-flow then software developers should check whether they went through proper white-box testing before internally reusing them. The metric-based comparison test can be applied to enriched software networks in order to determine distinctive characteristics of highly coupled software entities. To apply the test it is necessary to define a binary classifier which separates hubs from non-hubs. In [70] we proposed a hub classifier for class collaboration networks that can be adapted to any software network.

Definition 3.5 (*Hub, highly coupled software entity*) Let V denote the set of nodes in an enriched software network. Let H be a *minimal* subset of V satisfying the following condition:

$$\sum_{h \in H} k(h) > \sum_{o \in V \setminus H} k(o) \tag{3.13}$$

where $k(x)$ represents the total degree of a node x. A software entity is considered highly coupled if it belongs to H.

The main idea of our hub classifier is that a software entity can be considered highly coupled if it belongs to the minimal set of entities whose total coupling is higher than the total coupling of the rest of entities present in the system. It is not hard to see that the H set can be computed in a straightforward manner by employing an adequate greedy algorithm.

Highly coupled software entities may also be significantly different among themselves. We can distinguish three general types of hubs:

1. Hubs whose coupling is dominantly caused by internal reuse. Such software entities exhibit significantly higher afferent than efferent coupling, i.e. their in degree is close to their total degree in a software network.
2. Hubs whose coupling is dominantly caused by aggregation of other entities present in the system. In this case we have that efferent coupling is close to total coupling and significantly higher than afferent coupling.
3. Hubs whose coupling is significantly caused by both reuse and aggregation. Those are entities having both high in-degree and high out-degree in a software network.

In the case that the coupling of hubs tend to be dominantly caused by internal reuse then the presence of high coupling can indicate only negative aspects of extensive internal reuse, not negative aspects of extensive internal aggregation, and vice versa. To determine whether high class coupling was dominantly caused by extensive internal reuse or extensive internal aggregation we proposed the metric named *afferent-efferent coupling balance* [70]. Here, we generalize that metric to any type of software entity.

Definition 3.6 (*Afferent-Efferent Coupling Balance*) The afferent-efferent coupling balance of an enriched software network, denoted by C_k, is the average ratio of in-degree to total degree of nodes whose total degree is higher than or equal to k, i.e.

$$C_k = \frac{\sum_{i \in H_k} k_{in}(i)/k(i)}{|H_k|} = 1 - \frac{\sum_{i \in H_k} k_{out}(i)/k(i)}{|H_k|} \tag{3.14}$$

where H_k denotes the subset of nodes whose total degree is higher than or equal to k, while $k_{in}(i)$, $k_{out}(i)$ and $k(i)$ are in-degree, out-degree and total degree of a node i, respectively.

C_k takes values in the range [0, 1] since in-degree is always smaller than or equal to total degree. Let $D(k)$ denote the set of software entities whose total degree in an arbitrary enriched software network is equal to or higher than k. If $C_k = 1$ then the out-degree of all software entities in D_k is equal to zero, which means that their coupling was entirely caused by internal reuse. On the opposite side, $C_k = 0$ implies that the coupling of all entities in $D(k)$ was entirely caused by aggregation of other software entities.

Definition 3.7 (*Afferent-Efferent Coupling Balance Plot*) An afferent-efferent coupling balance plot is a graph showing C_k for $k \in [m, M]$ where m and M are the minimal and maximal total degree of highly coupled software entities, respectively.

The afferent-efferent coupling balance plot shows the change of C_k for highly coupled software entities. If C_k tends to have a high value which increases with k then we can conclude that internal reuse is the main cause of high coupling. In this case, software maintainers should take appropriate actions to prevent or reduce negative aspects of extensive internal reuse if highly coupled entities tend to cause problems during software evolution. On the opposite side, software maintainers should be aware of negative aspects of extensive internal aggregation, and act accordingly, if C_k tends to have a low value which decreases with k. In the worst case, if C_k takes mid-range values and changes independently from k, software maintainers should act in a way to prevent possible negative consequences of both extensive reuse and extensive aggregation.

3.4.4 Graph Clustering Evaluation Metrics as Software Cohesion Metrics

Let S be an arbitrary compound software entity, i.e. software entity that can define and contain other architectural software entities. For example, S can be a package, module, interface or class. Let E denote the set of constituent elements of S (e.g. variables and methods in the case that S is a class, classes and interfaces when S is a package). Let N be a software network showing dependencies between all software entities present in the system that are at the same level of abstraction as entities in E. If S conforms to the principle of low coupling and high cohesion then entities in E inevitably form a cluster (highly cohesive sub-graph loosely coupled to the rest of the network) in N. For example, if S is a package exhibiting low coupling and high cohesion then classes from S form a cluster in the class collaboration network of the system. Similarly, if S is a class exhibiting low coupling and high cohesion then class variables and methods defined in S form a cluster in an *extended* method collaboration network, i.e. a network encompassing all class variables and methods in the system that shows call dependencies between methods and access dependencies between methods and variables. Consequently, to examine whether S conforms to the principle of low coupling and high cohesion we can check whether its constituent elements form a cluster in N.

Following the previously described reasoning, in one of our previous works [66], we proposed graph clustering evaluation (GCE) metrics to measure cohesion of software entities. In the same paper we analyzed commonly used class cohesion metrics concluding that they estimate cohesiveness of classes relying only on internal dependencies (dependencies between constituent elements of a class) and ignore external dependencies (dependencies between methods and/or variables that are not located in the same class). On the other hand, external dependencies can be extremely important when measuring cohesiveness of software entities:

1. A software entity that has much more external than internal dependencies hardly can be considered cohesive regardless of the density of internal dependencies.
2. If two software entities have the same density of internal dependencies then the entity with a smaller number of external dependencies can be considered more cohesive.

Let $G = (V, E)$ denote an arbitrary directed graph (V is the set of nodes, while E is the set of links). Let C be a cluster of nodes (an arbitrary proper subset of V, i.e. $C \subset V$), and let c denote an arbitrary node in C ($c \in C$). We can distinguish between two types of links incident with nodes in C:

- *intra cluster links* are links between nodes in C (links that stay in the same cluster), and
- *inter-cluster links* are links connecting nodes in C with nodes that are not in C (links that cross cluster boundaries).

Accordingly, we can split the out-degree of c into two quantities:

1. *intra-cluster out-degree* which is the number of out-going intra-cluster links incident with c, and
2. *inter-cluster out-degree* which is the number of out-going inter-cluster links incident with c.

The basic formulation of the graph clustering problem asks for a division of the set of nodes into two clusters of balanced size such that the size of an edge cut (links connecting nodes from different clusters) is minimized. The main idea of GCE metrics based on the size of the edge cut is that a smaller cut implies a more cohesive cluster. The edge cut of an arbitrary cluster C in a directed graph is composed of inter-cluster links emanating from nodes in C. We use E_C to denote the number links in the edge cut of C, i.e.

$$E_C = |\{(x, y) : (x, y) \in E, x \in C, y \notin C\}| = \sum_{x \in C} \text{inter-cluster out-degree}(x)$$

$$(3.15)$$

Similarly, we can define I_C as the number of intra-cluster links emanating from nodes in C, i.e.

$$I_C = |\{(x, y) : (x, y) \in E, x \in C, y \in C\}| = \sum_{x \in C} \text{intra-cluster out-degree}(x)$$

(3.16)

Then, basic GCE metrics based on edge cuts (conductance, expansion and cut-ratio) can be defined as follows.

Definition 3.8 (*Conductance*) The conductance of C is the size of the edge cut normalized by the total number of links emanating from nodes contained in C:

$$\text{Conductance}(C) = \frac{E_C}{E_C + I_C}$$

(3.17)

Definition 3.9 (*Expansion*) The expansion of C is the size of the edge cut divided by the total number of nodes in C:

$$\text{Expansion}(C) = \frac{E_C}{|C|}$$

(3.18)

Definition 3.10 (*Cut-ratio*) The cut-ratio of C is the size of the edge cut normalized by the size of the maximal possible edge cut that C can form with the rest of nodes:

$$\text{Cut-ratio}(C) = \frac{E_C}{|C|(|V| - |C|)}$$

(3.19)

The ratio between the inter-cluster out-degree and total out-degree of a node inspired another class of GCE metrics that are known as out-degree fraction (ODF) graph clustering evaluation metrics. The out-degree fraction of nodes without outgoing links is equal to zero. The main idea behind ODF metrics is that a smaller node ODF implies that the node is more densely connected with nodes belonging to the same cluster than with the rest of nodes.

Definition 3.11 (*Average-ODF*) The average-ODF of C is the average out-degree fraction of nodes in C:

$$\text{Average-ODF}(C) = \frac{1}{|C|} \sum_{c \in C} \frac{|\{(c, d) : (c, d) \in E, d \notin C\}|}{k(c)}$$

(3.20)

where $k(c)$ denotes the out-degree of c.

Definition 3.12 (*Maximum-ODF*) The maximum-ODF of C is the maximal out-degree fraction considering all nodes in C:

$$\text{Maximum-ODF}(C) = \max_{c \in C} \frac{|\{(c, d) : (c, d) \in E, d \notin C\}|}{k(c)}$$

(3.21)

Definition 3.13 (*Flake-ODF*) The Flake-ODF of C is the fraction of nodes in C whose intra-cluster out-degree is higher than inter-cluster out-degree (nodes without out-going links are not counted):

$$\text{Flake-ODF}(C) = \frac{|\{x : x \in C,\ k^{intra}(c) > k^{inter}(c)\}|}{|C|} \tag{3.22}$$

where $k^{intra}(c)$ and $k^{inter}(c)$ denote the intra-cluster out-degree and inter-cluster out-degree of c, respectively.

It should be emphasized that all previously defined GCE metrics except Flake-ODF quantify the lack of cluster cohesion: higher values of conductance, expansion, cut-ratio, average-ODF and maximum-ODF imply less cohesive clusters. Flake-ODF reflects cluster cohesion, i.e. higher values of Flake-ODF indicate more cohesive clusters. GCE metrics are domain-independent metrics. Consequently, we can apply them to any subset of nodes C in a software network. If C are constituent elements of a software entity S then GCE metrics applied to C reflect the degree of cohesiveness of S. It can be noticed that GCE metrics do not ignore external dependencies. To the contrary, they rely on dependencies to external entities in order to determine the extent to which an entity is isolated from the rest of the system. In other words, GCE metrics are based on the following principle: an entity can be considered highly cohesive if its constituent elements are better connected among themselves than with elements defined outside the entity.

In [66] we performed theoretical analysis of GCE metrics using the theoretical framework of software cohesion measurement proposed by Briand et al. [8]. Briand et al. formulated the following properties which a software cohesion metric should satisfy:

1. **Nonnegativity**. A (lack of) cohesion metric M cannot take a negative value.
2. **Normalization**. M takes values in the interval $[0, max]$ where max is the maximum value.
3. **Null value**. M should be equal to zero if R_c is empty, where R_c denotes the set of dependencies among constituent elements of an entity c to which M is applied. In the case that M is a lack of cohesion metric then M should be zero if R_c is maximal possible (i.e. constituent elements of the entity form a dependency clique).
4. **Maximum value**. If R_c is maximal then M should take the maximum value. In the case that M measures the lack of cohesion then it should take the maximum value if $R_c = \emptyset$.
5. **Monotonicity**. Let e be an arbitrary software entity. Let e' be another software entity such that $R_e \subseteq R_{e'}$, i.e. some dependencies have to be introduced in e in order to obtain e'. Then, the following inequalities must hold

$$C(e) \leq C(e') \tag{3.23}$$
$$L(e) \geq L(e') \tag{3.24}$$

Table 3.1 Properties of GCE metrics. P_1 – nonnegativity, P_2 – normalization, P_3 – null value, P_4 – maximum value, P_5 – monotonicity, P_6 – merge property

Metric	P_1	P_2	P_3	P_4	P_5	P_6
Conductance	Yes	Yes	No	Yes	Yes	Yes
Expansion	Yes	No	No	No	Yes	Yes
Cut-ratio	Yes	Yes	No	No	Yes	Yes
Maximum-ODF	Yes	Yes	No	Yes	Yes	Yes
Average-ODF	Yes	Yes	No	Yes	Yes	Yes
Flake-ODF	Yes	Yes	Yes	No	Yes	Yes

where C and L denote a cohesion and a lack of cohesion metric, respectively. In other words, this property states that addition of new internal dependencies should not decrease (resp. increase) the value of C (resp. L).

6. **Merge property**. Let e_1 and e_2 be two unrelated (independent) software entities. This means that e_1 does not reference e_2 and vice versa. Let e be a software entity that is the union of e_1 and e_2. Then the following inequalities must hold

$$C(e) \leq max\{C(e_1), C(e_2)\} \tag{3.25}$$

$$L(e) \geq min\{L(e_1), L(e_2)\} \tag{3.26}$$

This property says that merging two independent software entities must not increase (resp. decrease) the value of a cohesion (resp. lack of cohesion) metric.

Table 3.1 summarizes the results of our theoretical analysis of GCE metrics with respect to the Briand et al. properties (more details can be found in [66]). It can be observed that GCE metrics reflecting lack of cohesion do not satisfy the null value property, while GCE metrics reflecting cohesion does not satisfy the maximum value property. However, we believe that this is not a disadvantage of GCE metrics – GCE metrics violate the null/maximum value property directly because they take external dependencies into account. Additionally, it is very unlikely to observe fully connected software modules in practice (e.g. all classes in a package reference each other; all methods in a class call other and access all attributes defined in the class). Secondly, in such cases GCE metrics favorite loosely coupled software modules emphasizing the principle of low coupling.

3.4.5 Analysis of Cohesion of Software Entities

In our methodological framework GCE metrics are employed to analyze cohesiveness of software entities. GCE metrics are domain-independent measures of cohesion and they can be applied to any type of software network. This means that GCE metrics can be employed to measure cohesiveness of different types of software entities. Another advantage of GCE metrics compared to traditional software cohesion metrics is that GCE metrics are closely related to the Radicchi et al. definitions of clusters in complex networks [61].

Definition 3.14 (*Radicchi strong cluster*) Let C denote a subset of nodes in a software network and let C' be the set of those nodes from C that have out-going links. C is a Radicchi strong cluster if the intra-cluster out-degree of each node in C' is higher than its inter-cluster out-degree, i.e.

$$C \text{ is a Radicchi strong cluster} \iff (\forall c : c \in C \land k(c) > 0) \ k^{intra}(c) > k^{inter}(c)$$

where $k(x)$, $k^{intra}(x)$ and $k^{inter}(x)$ denote the out-degree, intra-cluster out-degree and inter-cluster out-degree of a node x.

Definition 3.15 (*Radicchi weak cluster*) C is a Radicchi weak cluster if the total number of intra-cluster links emanating from nodes in C is higher than the total number of inter-clusters links emanating from nodes in C, i.e.

$$C \text{ is a Radicchi weak cluster} \iff \sum_{c \in C} k^{intra}(c) > \sum_{c \in C} k^{inter}(c)$$

Each Radicchi strong cluster is also Radicchi weak, but conversely it is not necessarily true. If constituent elements of a software entity S form a Radicchi strong cluster in the appropriate software network then the Flake-ODF of S is equal to 1. If S is a Radicchi weak cluster then its conductance is lower than 0.5. Therefore, conductance and Flake-ODF enable the classification of software entities according to the Radicchi definitions of clusters. A software entity whose constituent elements form a Radicchi strong cluster exhibits a strong degree of cohesion. If constituent elements of an entity form a Raddichi weak cluster then its cohesion can be considered satisfactory, otherwise the entity is poorly cohesive.

We will consider a hypothetical class collaboration network shown in Fig. 3.1 to illustrate GCE metrics and the Radicchi definitions of clusters. The network represents a simple object-oriented software system which consists of two packages P and Q both of them containing three classes. It can be observed that the inter-cluster out-degree of class F is higher than its intra-cluster out-degree: this class references one class from its own package and two classes from package P. Therefore, package Q is not a Radicchi strong cluster. This package is neither a Radicchi weak cluster since the number of intra-cluster links emanating from nodes in Q is not higher than the number of inter-cluster links emanating from the same nodes. It can be observed that the system presented in Fig. 3.1 can be refactored in order improve the overall degree of cohesion: if we move class F to package P then both packages would be Radicchi strong.

The presence of low cohesion in a software system can be explained by the existence of software entities whose inter-cluster out-degree is higher than intra-cluster out-degree. We call such entities E entities (E because they have more external

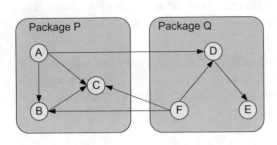

	P	Q
Intra-cluster links	3	2
Inter-cluster links	1	2
Radicchi strong	yes	no
Radicchi weak	yes	no
Conductance	0.25	0.5
Expansion	0.33	0.66
Cut-ratio	0.11	0.22
Maximum-ODF	0.33	0.66
Average-ODF	0.11	0.22
Flake-ODF	1.0	0.66

Fig. 3.1 The class collaboration network of a simple software system and cohesiveness of its packages

than internal dependencies). Class F in Fig. 3.1 is an example of an E-class and we can see that this class directly causes a low cohesion of package Q. In our case studies we will use the metric-based comparison test to determine distinctive characteristics of E-classes and E-methods in order to gain a more deeper understanding of the low cohesion phenomena at different levels of abstraction.

3.5   Experimental Dataset

Using the SNEIPL tool [67, 73, 74], we formed a dataset of enriched software networks associated to five open-source, widely used, large-scale software systems written in Java. The dataset includes enriched package, class and method collaboration networks of the following software projects:

1. Tomcat (version 7.0.29) – a HTTP web server that implements the official Java Servlet and JavaServer Pages specifications.
2. Lucene (version 3.6.0) – a search engine library providing full text indexing and searching functionalities.
3. Ant (version 1.9.2) – a tool for automating software build processes.
4. Xerces (version 2.11.0) – a library for parsing, validating, serializing and manipulating XML files.
5. JFreeChart (version 1.0.17) – a framework for creating and displaying charts.

The basic characteristics of enriched software networks from the experimental dataset are shown in Table 3.2.

SNEIPL enriches nodes of extracted software networks with domain-independent metrics used in complex network analysis, as well as with domain-dependent (software) metrics. The complete list of metrics attached to nodes of extracted enriched software networks is shown in Table 3.3. The metrics listed in the first part of the table

Table 3.2 The basic characteristics of enriched software networks from our experimental dataset. PCN, CCN and MCN denote package collaboration networks, class collaboration networks and method collaboration networks, respectively. N and L are the number of nodes and the number of links in the corresponding network, respectively

Software system	Version	LOC	PCN		CCN		MCN	
			N	L	N	L	N	L
Tomcat	7.0.29	329924	97	493	1494	6841	15311	20823
Lucene	3.6.0	111763	16	57	789	3544	5943	7114
Ant	1.9.2	219094	71	319	1175	5521	10341	14783
Xerces	2.11.0	216902	40	211	876	4775	9171	12051
JFreeChart	1.0.17	226623	37	228	624	3218	8759	12195

are domain-independent metrics of coupling, cohesion and functional importance. Those metrics are computed for any type of software network extracted by SNEIPL. The second part of the table shows domain-dependent metrics.

SNEIPL isolates package, class and method collaboration networks from an architectural graph-based representation of software systems known as general dependency network [74]. The general dependency network (GDN) of a system encompasses software entities appearing at different levels of abstraction and two types of dependencies among them: horizontal dependencies between entities from the same level of abstraction and vertical dependencies between entities from different levels of abstraction. Vertical dependencies show the hierarchical decomposition of the system, i.e. a vertical dependency $A \rightarrow B$ means that entity A defines and contains entity B. SNEIPL extracts GDNs from a language-independent, enriched Concrete Syntax Tree (eCST) representation of source code [63, 64, 72] that is formed by ANTLR-generated source code parsers.

SNEIPL computes domain-independent metrics of coupling and functional importance directly from extracted software networks. GCE metrics at the package level are computed from class collaboration networks partitioned into clusters according to the package organization of a system (classes belonging to the same package are put in the same cluster according to the vertical dependencies between packages and classes in the GDN). To compute GCE metrics at the class level SNEIPL relies on extended method collaboration networks (networks encompassing methods and class variables that show call dependencies between methods and access dependencies between methods and variables) partitioned into clusters according to the class organization of the system (variables and methods belonging to the same class are in the same cluster; again vertical GDN dependencies are exploited to partition networks into clusters).

Regarding domain-dependent metrics, LOC and cyclomatic complexity for methods and classes are computed from the eCST representation. The rest of metrics of internal complexity are computed from the GDN representation relying on existing vertical dependencies. At the class level, SNEIPL also computes coupling and inheritance metrics from the Chidamber–Kemerer metrics suite [15] – coupling between

Table 3.3 The list of metrics attached to nodes of enriched software networks extracted by the SNEIPL tool

Abbr.	Metric	Network	Metric type
IN	In-degree	All	Coupling
OUT	Out-degree	All	Coupling
TOT	Total degree	All	Coupling
BET	Betweenness centrality	All	Importance
PR	Page rank	All	Importance
HITSA	HITS authority score	All	Importance
HITSH	HITS hub score	All	Importance
COND	Conductance	All	Cohesion
EXP	Expansion	All	Cohesion
CUTR	Cut-ratio	All	Cohesion
AVGODF	Average out-degree fraction	All	Cohesion
MAXODF	Maximum out-degree fraction	All	Cohesion
FODF	Flake out-degree fraction	All	Cohesion
LOC	Lines of code	CCN MCN	Complexity
CC	Cyclomatic complexity	CCN MCN	Complexity
NUMA	The number of attributes in a class	CCN	Complexity
NUMM	The number of methods in a class	CCN	Complexity
NUME	The number of entities within a package	PCN	Complexity
ABST	Package abstractness	PCN	Complexity
CBO	Coupling between objects	CCN	Coupling
NOC	The number of children	CCN	Inheritance
DIT	The depth in inheritance tree	CCN	Inheritance
LCOM	Lack of cohesion in methods	CCN	Cohesion
HM	Hitz-Montanzeri LCOM	CCN	Cohesion
HM2	Hitz-Montanzeri LCOM2	CCN	Cohesion
TCC	Tight class cohesion	CCN	Cohesion
LCC	Loose class cohesion	CCN	Cohesion

objects (CBO), the number of children (NOC) and the depth in inheritance tree (DIT). The CBO for a class A is the total number of other classes that are directly coupled to A. If A does not form cyclic dependencies with classes to which it is directly coupled then CBO(A) is equal to the total degree of A in the class collaboration network. Besides class collaboration networks, SNEIPL also forms class dependency graphs restricted to particular forms of class coupling. One of such graphs is inheritance network which shows inheritance relationships between classes present in a system. Inheritance networks are used to compute and enrich nodes of class collaboration networks with inheritance metrics (NOC and DIT).

SNEIPL also enriches nodes of class collaboration networks with widely used class cohesion metrics: LCOM from the Chidamber–Kemerer suite [15], two LCOM metrics proposed by Hitz and Montanzeri [32], and two cohesion metrics proposed by Bieman and Kang [6] (TCC – tight class cohesion and LCC – loose class cohesion). LCOM by Chidamber and Kemerer is based on data-coupling between methods: two methods of a class A are data-coupled if they both access to at least one class variable declared in A. Then, the LCOM of A is the number of non-data-coupled methods of A (denoted by P) reduced by the number of data-coupled methods (denoted by Q) if $P > Q$, or zero otherwise. As the name of the metric indicates, LCOM measures the lack of class cohesion. Let $G(C)$ denote a directed graph that shows data-coupling dependencies between methods defined in a class C. Let $G'(C)$ be an extension of $G(C)$ with method call dependencies and let $G''(C)$ denote an extension of $G'(C)$ with indirect method call dependencies. The first LCOM metrics proposed by Hitz and Montanzeri represents the number of weakly connected components in $G(C)$. The second one is the number of weakly connected components in $G'(C)$. The TCC metric proposed by Bieman and Kang is equivalent to the density of $G'(C)$, while LCC is the density of $G''(C)$. SNEIPL computes all previously mentioned class cohesion metrics from subgraphs of extended method collaboration networks induced by methods and class variables defined in a class.

3.6 Results and Discussion

We firstly computed and examined the basic structural characteristics of enriched software networks from our experimental dataset: the number of nodes in the largest weakly connected component, average path length, clustering coefficient and assortativity index. Additionally, we computed the average path length and clustering coefficient of comparable Erdős-Renyi random graphs (directed random graphs having the same number of nodes and links as our networks). Obtained values are presented in Table 3.4. Each examined network has a giant weakly connected component encompassing a vast majority of nodes. Giant connected components in package and class collaboration networks are substantially larger than giant connected components in method collaboration networks. This can be explained by a larger number of isolated nodes in method collaboration networks. However, it should be emphasized that an isolated node in a method collaboration network does not necessarily imply that the

Table 3.4 The basic structural characteristics of enriched software networks from our experimental dataset. LWCC – the percentage of nodes contained in the largest weakly connected component, APL – the average path length, APL_r – the average path length of a comparable random graph, CC – the clustering coefficient, CC_r – the clustering coefficient of a comparable random graph, A – the Newman's assortativity index. PCN, CCN and MCN stand for package, class and method collaboration networks, respectively

Network	Software	LWCC [%]	APL	APL_r	CC	CC_r	A
PCN	Tomcat	94.85	1.82	2	0.319	0.0526	–0.08
	Lucene	100	1.7	1.59	0.5896	0.1917	–0.34
	Ant	100	2.1	2.08	0.7018	0.0553	–0.48
	Xerces	100	2.28	1.64	0.346	0.1218	–0.15
	JFreeChart	94.59	2.14	1.49	0.5178	0.1605	–0.3
CCN	Tomcat	92.37	3.3	3.26	0.2393	0.0033	–0.1
	Lucene	99.62	3.75	3.09	0.2365	0.0055	–0.05
	Ant	99.4	4.68	3.18	0.2392	0.004	–0.07
	Xerces	99.77	3.04	2.88	0.2548	0.006	–0.07
	JFreeChart	97.92	2.75	2.74	0.2149	0.0085	–0.11
MCN	Tomcat	59.89	1.89	6.37	0.0258	0.0002	–0.03
	Lucene	61.18	1.79	6.35	0.028	0.0005	–0.06
	Ant	65.89	4.56	6.16	0.0309	0.0003	–0.03
	Xerces	58.23	2.28	5.99	0.0271	0.0004	–0.2
	JFreeChart	62.59	1.67	5.97	0.0141	0.0004	–0.18

corresponding method is unused. Firstly, Java methods can be dynamically invoked through the Java reflection API. Secondly, isolated nodes in method collaboration networks may represent call-back methods (one of the most frequent examples is `toString` method present in almost every Java class) or methods without internal dependencies that are directly exposed to users in the case of libraries.

The average path lengths of analyzed package and class collaboration networks are close to the predictions made by the Erdős-Renyi random graph model. Therefore, analyzed networks can be considered small-world networks. Method collaboration networks exhibit the ultra small-world property, i.e. their characteristic path lengths are considerably smaller than the predictions by the Erdős-Renyi model. Moreover, examined software networks exhibit the (ultra) small-world property in the Watts-Strogatz sense [91]: their clustering coefficients are considerably higher than clustering coefficients of comparable random graphs (at least two times higher for package collaboration networks, 25 times higher for class collaboration networks, and 35 times higher in the case of method collaboration networks). Obtained values of the assortativity index indicate that analyzed networks exhibit weak to moderate disassortativity ($A < 0$). This means that highly coupled software entities do not tend to be directly connected among themselves.

The existence of giant weakly connected components (GWCCs) in examined class collaboration networks (CCNs) is not surprising. The existence of a GWCC in a CCN means that a vast majority of classes defined in the corresponding system work together in order to realize desired functionalities. The existence of two or more large weakly connected components in a CCN, none of them being giant, would imply large non-interacting software components (sets of classes) realizing unrelated functionalities. In such situations it is easier for software developers, testers and maintainers to have separate software projects for each weakly connected component of the CCN. This means that the size of the largest weakly connected component (WCC) can be used to estimate the overall cohesiveness of a software project. Relatively small values of the largest WCC (values below 50%) would definitely indicate one of the following two scenarios:

- The project is in an early phase of the development that is conducted in a bottom-up manner, i.e. the software product is not growing from a central core of classes, but from several independent parts that are not yet assembled together.
- The project is providing a set of unrelated functionalities, none of them being the main feature of the project.

In the case that a software project has a good degree of overall cohesiveness (has a WCC whose size is above 90%) then small size WCCs (including isolated nodes) correspond to:

- software components that are in an early phase of development and not yet functionally integrated into the product, or
- deprecated components that are not removed from the source code distribution.

The existence of GWCCs in examined software networks is not also surprising from the theoretical point of view. Let $G(N, p)$ denote an Erdős-Renyi random graph having N nodes where p is the probability that two nodes are connected. The topological properties of $G(N, p)$ vary as a function of p, displaying a phase transition at the critical probability $p_c = 1/N$ that corresponds to the critical average degree $k_c = 1$ [7]. If the average degree of a large random graph exceeds k_c then it will have a GWCC with a very high probability (this probability tends to one as N tends to infinity). The average degree of all examined software networks is higher than k_c which means that the simplest model of complex networks can explain empirically observed GWCCs.

3.6.1 Strongly Connected Components and Cyclic Dependencies

We identified strongly connected components (SCCs) in enriched package, class and method collaboration networks from our dataset using the Tarjan algorithm. The basic structural characteristics of SCCs are summarized in Table 3.5. It can be observed that there are different patterns of strong connectivity in software networks

Table 3.5 The basic structural characteristics of strongly connected components: $\#SCC$ – the number of strongly connected components, $LSCC$ – the percentage of nodes in the largest strongly connected component, $N(SCC)$ – the percentage of nodes contained in all strongly connected components, R – link reciprocity, R_n – normalized link reciprocity with respect to a comparable Erdős-Renyi random graph, R^p – path reciprocity. PCN, CCN and MCN stand for package, class and method collaboration networks, respectively

Network type	Software	$\#SCC$	$LSCC$ [%]	$N(SCC)$ [%]	R	R_n	R^p
PCN	Tomcat	10	13.4021	49.4845	0.191	0.145	0.162
	Lucene	1	56.25	56.25	0.386	0.195	0.511
	Ant	3	40.8451	46.4789	0.276	0.226	0.372
	Xerces	4	62.5	77.5	0.199	0.074	0.615
	JFreeChart	1	75.6757	75.6757	0.325	0.185	0.794
CCN	Tomcat	56	12.72	35.74	0.078	0.075	0.179
	Lucene	40	17.87	35.23	0.08	0.075	0.162
	Ant	27	24.34	35.06	0.046	0.042	0.237
	Xerces	32	13.81	32.76	0.078	0.072	0.118
	JFreeChart	19	7.05	17.63	0.032	0.024	0.048
MCN	Tomcat	37	0.2743	0.9732	0.004	0.003	0.021
	Lucene	12	0.0673	0.4543	0.004	0.004	0.001
	Ant	23	0.5996	1.1024	0.003	0.003	0.009
	Xerces	14	0.338	0.9377	0.002	0.002	0.013
	JFreeChart	5	0.0913	0.2398	0.002	0.002	0.002

at different levels of abstraction. All package collaboration networks, except the package collaboration network of Tomcat, contain a small number of large SCCs. Moreover, package collaboration networks of Lucene, Xerces and JFreeChart contain a giant SCC encompassing more than 50% of packages present in those systems. The size of the largest SCC and the number of nodes contained in all SCCs of the Tomcat package collaboration network indicate that this network significantly deviates from being an acyclic graph. For all package networks we have high values of both link and path reciprocity, much higher than for software networks at lower levels of abstraction. Therefore, we can conclude that examined software systems exhibit significant and large cyclic dependencies at the package level. Secondly, for all systems except Tomcat we have that $R^p \gg R$ at the package level which means that indirect cyclic package dependencies occur more frequently than direct cyclic package dependencies.

In contrast to package collaboration networks, giant strongly connected components are absent at the class level, but class collaboration networks contain a relatively large number of SCCs. The majority of SCCs in class collaboration networks are small size components (components that encompass two or three classes), but those networks also contain SCCs of non-negligible size:

- Tomcat has 38 small size SCCs (67.8% of the total number of SCCs present in the network). The largest SCC encompasses 190 class-level entities (Java classes and interfaces).
- Lucene has 27 small size SCCs (67.5%). The largest SCC contains 141 class-level entities.
- Ant has 14 small size SCCs (51.85%). The largest SCC contains 286 class-level entities.
- Xerces has 22 small size SCCs (68.7%). The largest SCC contains 121 class-level entities.
- JFreeChart has 13 small size SCCs (68.4%). The largest SCC contains 44 class-level entities.

The size of the largest SCC (ranging from 7 to 25% of the total number of nodes) and the total number of nodes involved in cyclic dependencies (ranging from 17 to 36% of the total number of nodes) indicate that investigated class collaboration networks significantly deviate from acyclic graphs, but not as strong as package collaboration networks. The reciprocity of links in class collaboration networks is relatively small ($0.03 < R \leq 0.08$) but still higher than expected by a random chance ($R_n > 0$). It can be seen that the path reciprocity of class collaboration networks is significantly higher than the link reciprocity. This means that cyclic class dependencies are mostly indirect. Additionally, class collaboration networks of Tomcat, Ant and Lucene show considerably high path reciprocity. For example, the path reciprocity of the Ant class collaboration network is equal to 0.237 which means that nearly quarter of all class dependencies (both direct and indirect) present in this system are cyclic dependencies.

Cyclic dependencies are almost absent at the method level. For each examined method collaboration network, we have that (1) the number of nodes contained in SCCs is less than 1.2% of the total number of nodes, and (2) the size of the largest SCC is less than 0.7% of the total number of nodes. Method collaboration networks also exhibit considerably small link and path reciprocity. The absence of large SCCs is not surprising since large cyclic dependencies in method collaboration networks would imply large chains of mutually recursive methods.

From the software engineering perspective, large cyclic dependencies among software entities are undesirable and considered as design anti-patterns. Although various software methodologies advise to avoid cyclic dependencies, our results show that large cycles are present in the package and class structure of investigated software systems. Additionally, our results are consistent with the findings of Melton and Tempero [49] who showed that large cyclic dependencies are common for classes present in Java software systems, as well as with the findings of Laval et al. [40] who reported large cyclic dependencies at the package level in 4 Java software systems.

One of distinctive characteristics of class collaboration networks compared to other types of software networks is that they contain a large number of strongly connected components exhibiting a large variability in size. As emphasized in Sect. 3.4, the complexity of a SCC S can be expressed by its average intra-component degree. The average intra-component degree of S is equal to $L(S)/N(S)$, where $L(S)$ and $N(S)$ are the number of links and nodes in S, respectively. We computed the Spearman

Table 3.6 The densification of strongly connected components in class collaboration networks. $\rho(N(S), \frac{L(S)}{N(S)})$ – the Spearman's rank correlation between the size and average intra-component degree of SCCs, α – the scaling exponent of a densification power-law $L(S) \approx N(S)^{\alpha}$

Software system	$\rho(N(S), \frac{L(S)}{N(S)})$	α
Tomcat	0.975	1.29
Lucene	0.965	1.27
Ant	0.982	1.27
Xerces	0.977	1.33
JFreeChart	0.825	1.32

correlation between the size and average intra-component degree of SCCs contained in examined class collaboration networks. Obtained values are given in Table 3.6. It can be observed that there is a strong Sperman correlation (always higher than 0.8) between two previously mentioned structural quantities. This means that SCCs strongly tend to *densify* with size: larger SCCs deviate to a larger extent from being a pure cycle of classes compared to smaller SCCs. From the software engineering point of view, this result indicates that classes contained in large SCCs are extremely hard to comprehend, test and refactor since the density of cyclic dependencies increases with the number of mutually dependent classes.

A strong correlation between the size of SCCs and their average intra-component degrees suggests that the number of links within a SCC grows super-linearly with the number of nodes:

$$L(S) \approx N(S)^{\alpha}, \alpha > 1 \tag{3.27}$$

Higher values of the scaling exponent α indicate a higher degree of link densification within strongly connected components (or, equivalently, a higher degree of densification of cyclic dependencies). For each examined system, we made a log-log plot in which each point represents one strongly connected component (Fig. 3.2). The x coordinate of a point corresponds to the number of nodes, while the y coordinate corresponds to the number of links in the SCC represented by the point. Then, we fitted a power-law curve of the form x^{α} using the chi-square minimization technique. It can be seen that the relationship between the number of nodes and the number or links within SCCs can be very well approximated by power-laws (the coefficient of determination is higher than 0.9 in all cases). Obtained power-law exponents (given on the plots but also in Table 3.6) indicate the degree of densification of SCCs in analyzed systems and we can observe that Xerces exhibits the strongest densification of cyclic class dependencies.

The scaling exponent of an empirically observed densification law for strongly connected components in class collaboration networks can be instrumented as an indicator of software design quality. Smaller scaling exponents imply lower complexity of strongly connected components and better software designs with respect to the "avoid cyclic dependencies" principle. In an ideal case, when nearly all strongly

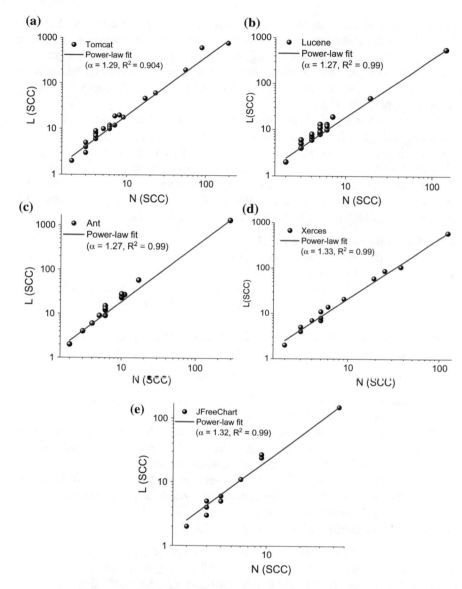

Fig. 3.2 The densification of strongly connected components in **a** Tomcat, **b** Lucene, **c** Ant, **d** Xerces and **e** JFreeChart class collaboration networks

connected components are pure circles, α is close to 1. On the other hand, the maximal value of the scaling exponent ($\alpha = 2$) corresponds to the case when nearly all strongly connected components are cliques of mutually dependent classes. The highest α is obtained for Xerces which means that a randomly selected SCC of Xerces is more denser and, consequently, more harder to comprehend than comparable SCCs from other analyzed systems.

In order to determine characteristics of strongly connected components in enriched class and package collaboration networks, we applied the metric-based comparison test to two groups of nodes: nodes belonging to SCCs constitute the first group (denoted by C_1), while nodes not involved in cyclic dependencies belong to the second group (denoted by C_2). The obtained results are presented in Table 3.7 for class collaboration networks and in Table 3.8 for package collaboration networks. For each metric M, both tables show the following quantities:

- $\overline{C_1}$: the average value of M for nodes in C_1.
- $\overline{C_2}$: the average value of M for nodes in C_2.
- U: the test statistic of the Mann-Whitney U (MWU) test.
- p: the p-value of U. The null hypothesis of the MWU test (no statistically significant differences between C_1 and C_2 regarding metric M) is rejected if the p-value is less than 0.05.
- PS_1: the probability of superiority of C_1 over C_2 with respect to M.
- PS_2: the probability of superiority of C_2 over C_1 with respect to M.

For example, the first row in Table 3.7 shows that:

1. the average LOC (the number of lines of code) of Tomcat classes involved in cyclic dependencies is 289.17,
2. the average LOC of Tomcat classes not involved in cyclic dependencies is 161.06,
3. there is a statistically significant difference in LOC between those two groups of classes (the null hypothesis of the MWU test is rejected),
4. the probability that the LOC a randomly selected class involved in cyclic dependencies is strictly higher than the LOC of a randomly selected class not involved in cyclic dependencies is 0.56, and
5. the inverse probability of superiority is 0.43.

The null hypothesis of the MWU test is accepted for the NUMA metric (the number of attributes in a class) in the case of Tomcat, And and JFreeChart class collaboration networks. This means that classes involved in cyclic dependencies in mentioned software systems do not tend to declare significantly more class attributes compared to the rest of classes and vice versa. The absence of statistically significant differences in inheritance metrics can be observed for some of the analyzed enriched class collaboration networks:

- The null hypothesis of the MWU test is accepted for the NOC metric in the case of Tomcat and Lucene: highly extensible classes (classes exhibiting a high reuse through inheritance) in those two systems do not tend to be involved in cyclic dependencies more than lowly extensible classes and vice versa.

Table 3.7 The results of the metric-based comparison test for strongly connected components in enriched class collaboration networks. The column "null hyp." denotes whether the null hypothesis of the MWU test is accepted. Bold probabilities of superiority indicate considerable differences between classes involved in cyclic dependencies and classes not involved in cyclic dependencies with respect to the corresponding metric

Software	Metric	$\overline{C_1}$	$\overline{C_2}$	U	p	Null hyp.	PS_1	PS_2
Tomcat	LOC	289.17	161.06	291224	$<10^{-4}$	Rejected	0.56	0.43
	CC	27.32	17.24	281184	0.002	Rejected	0.48	0.38
	NUMA	5.21	4.77	270537	0.075	**Accepted**	0.47	0.41
	NUMM	14.26	8	308247	$<10^{-4}$	Rejected	0.57	0.36
	IN	6.38	3.57	339696	$<10^{-4}$	Rejected	0.56	0.24
	OUT	7.29	3.06	386846	$<10^{-4}$	Rejected	0.72	0.21
	CBO	12.68	6.64	368293	$<10^{-4}$	Rejected	0.68	0.25
	NOC	0.46	0.22	262066	0.472	**Accepted**	0.09	0.06
	DIT	0.6	0.42	297607	$<10^{-4}$	Rejected	0.34	0.18
	BET	1847.29	124.63	386286	$<10^{-4}$	Rejected	0.68	0.17
	PR	0.000896	0.000543	341903	$<10^{-4}$	Rejected	0.66	0.33
	HITSA	0.008266	0.005221	337456	$<10^{-4}$	Rejected	0.66	0.34
	HITSH	0.017819	0.009386	375515	$<10^{-4}$	Rejected	0.73	0.27
	COND	0.39	0.25	323221	$<10^{-4}$	Rejected	0.56	0.30
	FODF	0.61	0.74	317090	$<10^{-4}$	Rejected	0.27	0.51
Lucene	LOC	207.19	94.66	94019	$<10^{-4}$	Rejected	0.66	0.33
	CC	21.18	15.37	96546	$<10^{-4}$	Rejected	0.64	0.27
	NUMA	3.42	2.8	88067	$<10^{-4}$	Rejected	0.56	0.32
	NUMM	10.11	6.1	87275	$<10^{-4}$	Rejected	0.58	0.35
	IN	6.03	3.65	90841	$<10^{-4}$	Rejected	0.55	0.27
	OUT	7.27	2.98	110913	$<10^{-4}$	Rejected	0.74	0.18
	CBO	12.28	6.63	100382	$<10^{-4}$	Rejected	0.68	0.26
	NOC	0.72	0.39	76054	0.1	**Accepted**	0.2	0.13
	DIT	0.97	0.72407	81111	0.0009	Rejected	0.4	0.26
	BET	1522.37	102.47	111895	$<10^{-4}$	Rejected	0.74	0.16
	PR	0.001415	0.001187	90250	$<10^{-4}$	Rejected	0.63	0.36
	HITSA	0.017006	0.008393	96260	$<10^{-4}$	Rejected	0.68	0.32
	HITSH	0.03222	0.013092	107163	$<10^{-4}$	Rejected	**0.75**	0.25
	COND	0.40	0.20	101945	$<10^{-4}$	Rejected	0.68	0.25
	FODF	0.61	0.81	98626	$<10^{-4}$	Rejected	0.23	0.61
Ant	LOC	194.4	162.2	176965	0.00036	Rejected	0.56	0.43
	CC	15.69	14.29	174507	0.0018	Rejected	0.51	0.39
	NUMA	4.36	5.01	160886	0.504	**Accepted**	0.42	0.44
	NUMM	9.86	8.23	172325	0.006	Rejected	0.51	0.42
	IN	9.78	1.95	228723	$<10^{-4}$	Rejected	0.64	0.19
	OUT	5.93	4.02	200409	$<10^{-4}$	Rejected	0.59	0.31
	CBO	15.11	5.98	215327	$<10^{-4}$	Rejected	0.65	0.28

Table 3.7 (continued)

	NOC	1.17	0.21	168921	0.0343	Rejected	0.16	0.08
	DIT	1.11	1.14	157439	0.9624	**Accepted**	0.35	0.35
	BET	3255.16	103.95	253004	$<10^{-4}$	Rejected	**0.76**	0.14
	PR	0.001744	0.000369	242063	$<10^{-4}$	Rejected	**0.77**	0.22
	HITSA	0.011703	0.002198	227472	$<10^{-4}$	Rejected	0.72	0.28
	HITSH	0.023104	0.018486	192418	$<10^{-4}$	Rejected	0.61	0.39
	COND	0.32	0.25	178782	$<10^{-4}$	Rejected	0.50	0.37
	FODF	0.69	0.74	174857	$<10^{-4}$	Rejected	0.34	0.45
Xerces	LOC	385.88	155.29	105178	$<10^{-4}$	Rejected	0.62	0.37
	CC	53.89	14.47	101519	$<10^{-4}$	Rejected	0.51	0.31
	NUMA	8.16	4.66	100067	$<10^{-4}$	Rejected	0.53	0.34
	NUMM	14.96	8.29	108053	$<10^{-4}$	Rejected	0.61	0.33
	IN	6.49	4.94	108948	$<10^{-4}$	Rejected	0.56	0.27
	OUT	9.19	3.62	125479	$<10^{-4}$	Rejected	0.7	0.21
	CBO	14.39	8.57	113871	$<10^{-4}$	Rejected	0.64	0.29
	NOC	0.74	0.24	92605	0.0214	Rejected	0.19	0.09
	DIT	1.56	0.75	104528	$<10^{-4}$	Rejected	0.44	0.2
	BET	1888.39	108.65	131778	$<10^{-4}$	Rejected	0.74	0.18
	PR	0.001643	0.000897	116151	$<10^{-4}$	Rejected	0.69	0.31
	HITSA	0.011387	0.012732	101701	0.000001	Rejected	0.6	0.4
	HITSH	0.02032	0.01189	106463	$<10^{-4}$	Rejected	0.63	0.37
	COND	0.44	0.18	121018	$<10^{-4}$	Rejected	0.64	0.2
	FODF	0.55	0.82	119662	$<10^{-4}$	Rejected	0.19	0.6
JFreeChart	LOC	548.27	245.49	30856	0.13	**Accepted**	0.54	0.45
	CC	36.28	15.69	29545	0.46	**Accepted**	0.46	0.42
	NUMA	8.06	3.88	30721	0.15	**Accepted**	0.48	0.39
	NUMM	25.6	11.56	31787	0.04	Rejected	0.54	0.41
	IN	10.23	4.07	44351	$<10^{-4}$	Rejected	**0.75**	0.18
	OUT	7.48	4.66	35014	$<10^{-4}$	Rejected	0.57	0.33
	CBO	16.79	8.73	37484	$<10^{-4}$	Rejected	0.63	0.31
	NOC	1.04	0.33	32150	0.02	Rejected	0.22	0.08
	DIT	0.55	0.92	31164	0.09	**Accepted**	0.26	0.36
	BET	924.85	38.88	45216	$<10^{-4}$	Rejected	**0.76**	0.15
	PR	0.002716	0.001364	45590	$<10^{-4}$	Rejected	**0.81**	0.19
	HITSA	0.02862	0.009961	41477	$<10^{-4}$	Rejected	0.73	0.27
	HITSH	0.02655	0.02206	30701	0.16	**Accepted**	0.54	0.46
	COND	0.30	0.29	29005	0.67	**Accepted**	0.44	0.41
	FODF	0.68	0.67	28642	0.83	**Accepted**	0.40	0.42

- The null hypothesis of the MWU test is accepted for the DIT metric in the case of Ant and JFreeChart. This means that highly specialized classes present in those two software systems do not tend to be involved in cyclic dependencies more than highly abstract classes and vice versa.

The null hypothesis of the MWU test is also accepted for two metrics of internal complexity (LOC and CC) in the case of JFreeChart. It can be seen that the average value of LOC/CC for JFreeChart classes involved in cyclic dependencies is two times higher compared to the average value of LOC/CC for the rest of JFreeChart classes. However, the probability that a class from a SCC has strictly higher internal complexity (estimated either by LOC or CC) than a class not involved in cyclic dependencies is close to the opposite probability of superiority. This means that strongly connected components in the JFreeChart class collaboration network contain a small fraction of outliers (extremely large and complex classes) that increase the average value of LOC/CC.

The null hypothesis of the MWU test is rejected considering all present metrics only in the case of the Xerces class collaboration network, but a significant superiority in metric values of classes from strongly connected components is absent. The largest differences ($PS_1 > 0.7 \wedge PS_2 < 0.25$) are present regarding OUT (out-degree) and BET (betweenness centrality) which means that Xerces classes involved in cyclic dependencies moderately tend to exhibit higher efferent coupling and moderately tend to be the most central classes in the system.

Classes involved in cyclic dependencies tend to exhibit a lower degree of class cohesion (higher COND and lower FODF) compared to classes not involved in cyclic dependencies in all systems except JFreeChart. Observed differences, although statistically significant, are not too drastic. The null hypothesis of the MWU test is accepted for COND and FODF in the case of JFreeChart. This means JFreeChart classes involved in cyclic dependencies possess the same degree of class cohesion as classes not involved in cyclic dependencies.

Drastic differences in metric values of classes belonging to SCCs and classes that are not involved in cyclic dependencies are present in three systems: Ant, JFreeChart and Lucene. Ant and JFreeChart classes involved in cyclic dependencies strongly tend to have significantly higher values of BET and PR (page rank) implying that strongly connected components dominantly contain central and functionally most important classes present in those two systems. Additionally, highly reused classes in JFreeChart strongly tend to be involved in cyclic dependencies. Lucene classes involved in cyclic dependencies strongly tend to have higher values of the HITS hub score. This means that central Lucene classes referencing other central Lucene classes are dominantly located in strongly connected components. The existence of a strongly connected core of functionally most important classes implies that refactorings aimed to eliminate or significantly reduce cyclic class dependencies have to deal with dependencies present in the core of the system. Consequently, such refactoring are extremely hard to perform. Additionally, they may have a big impact to system evolution since reorganization of dependencies between core classes may cause a complete redesign of the system. This is especially the case with JFreeChart whose strongly connected core dominantly contains the most reused classes.

The results of the metric-based comparison test for SCCs in package collaboration networks are presented in Table 3.8. Lucene contains only 16 packages and its package collaboration network is omitted from this analysis due to an extremely small size for a statistical test. In all systems except Xerces, packages involved in cyclic dependencies tend to contain a larger number of class-level entities (in the case of JFreeChart this tendency is very strong). Ant and JFreeChart packages involved in cyclic dependencies also tend to be more abstract compared to packages that are not cyclically dependent. They also exhibit the same degree of package cohesion as packages not involved in cyclic dependencies. On the other hand, Xerces and Tomcat packages involved cyclic dependencies tend to be significantly less cohesive compared to packages not involved in cyclic dependencies.

The results of the metric-based comparison test show that Ant, Xerces and JFreeChart packages involved in cyclic dependencies strongly tend to have higher afferent and total coupling compared to packages not involved in cyclic dependencies. Additionally, packages exhibiting high page rank and HITS authority score are dominantly located in strongly connected components. Therefore, we can conclude that package collaboration networks of Ant, Xerces and JFreeChart have a core-periphery structure with strongly connected cores. The core-periphery structure of JFreeChart is the most apparent. Each JFreeChart package from the strongly connected core has strictly higher afferent coupling, page rank and HITS authority score than any package not participating in cyclic dependencies ($PS_1 = 1$, $PS_2 = 0$). From the software engineering perspective, the presence of a core-periphery structure in a package collaboration network where the core is strongly connected implies that the core of the system does not have a layered (hierarchical) organization. In other words, there is no hierarchy among the most functionally important components of the system. This may have several negative consequences to system maintainability:

- core components are hard to understand and comprehend (due to the absence of a hierarchical ordering),
- core components cannot be tested in isolation,
- faults in core components are likely to propagate through the entire core compromising the stability of the whole system, and
- external reusability of core components is limited (external reuse of one component from a strongly connected core without eliminating cyclic dependencies implies that all core components have to be externally reused).

The package structure of Xerces is extremely interesting. The Xerces package collaboration network has the strongly connected core encompassing packages that are significantly less cohesive than packages located on the periphery of the network. This means that functionally less important Xerces packages are better designed than core Xerces packages.

Table 3.8 The results of the metric-based comparison test for strongly connected components in enriched package collaboration networks

Software	Metric	$\overline{C_1}$	$\overline{C_2}$	U	p	Null hyp.	PS_1	PS_2
Tomcat	NUME	14.33	8.77	1601	0.0021	Rejected	0.66	0.29
	ABST	0.09	0.16	1203	0.8455	**Accepted**	0.33	0.31
	IN	5.58	4.59	1706	0.0001	Rejected	0.69	0.23
	OUT	7.50	2.71	1857	$<10^{-4}$	Rejected	**0.76**	0.18
	TOT	13.08	7.31	1796	$<10^{-4}$	Rejected	0.74	0.22
	BET	43.98	4.42	1770	$<10^{-4}$	Rejected	0.67	0.17
	PR	0.0120	0.0086	1799	$<10^{-4}$	Rejected	**0.76**	0.23
	HITSA	0.0584	0.0518	1484	0.0263	Rejected	0.63	0.37
	HITSH	0.0970	0.0470	1608	0.0018	Rejected	0.68	0.32
	COND	0.62	0.44	1509	0.02	Rejected	0.63	0.35
	FODF	0.33	0.52	1464	0.04	Rejected	0.32	0.56
Ant	NUME	18.48	6.61	903	0.0014	Rejected	0.69	0.25
	ABST	0.11	0.06	814	0.0311	Rejected	0.55	0.25
	IN	8.91	0.66	1118	$<10^{-4}$	Rejected	**0.87**	0.09
	OUT	5.88	3.29	854	0.0089	Rejected	0.62	0.25
	TOT	14.79	3.95	1016	$<10^{-4}$	Rejected	**0.77**	0.15
	BET	74.03	0.26	957	0.0001	Rejected	0.56	0.04
	PR	0.0243	0.0052	1144	$<10^{-4}$	Rejected	**0.91**	0.09
	HITSA	0.0927	0.0102	1130	$<10^{-4}$	Rejected	**0.90**	0.10
	HITSH	0.1189	0.0991	774	0.0901	**Accepted**	0.61	0.37
	COND	0.66	0.64	664	0.67	**Accepted**	0.46	0.52
	FODF	0.22	0.30	637	0.91	**Accepted**	0.40	0.42
Xerces	NUME	17.06	20.44	144	0.8713	**Accepted**	0.51	0.47
	ABST	0.17	0.44	157	0.5598	**Accepted**	0.34	0.47
	IN	6.26	1.89	226	0.0051	Rejected	**0.75**	0.12
	OUT	6.45	1.22	228	0.0042	Rejected	**0.78**	0.14
	TOT	12.71	3.11	251	0.0003	Rejected	**0.87**	0.07
	BET	45.63	0.84	231	0.0030	Rejected	0.71	0.06
	PR	0.0303	0.0066	213	0.0173	Rejected	**0.76**	0.23
	HITSA	0.1255	0.0415	225	0.0056	Rejected	**0.80**	0.19
	HITSH	0.1449	0.0237	249	0.0004	Rejected	**0.89**	0.11
	COND	0.52	0.29	198	0.04	Rejected	0.71	0.29
	FODF	0.46	0.72	205	0.04	Rejected	0.25	0.72

(continued)

Table 3.8 (continued)

Software	Metric	$\overline{C_1}$	$\overline{C_2}$	U	p	Null hyp.	PS_1	PS_2
JFreeChart	NUME	19.93	5.11	222	0.0006	Rejected	**0.87**	0.1
	ABST	0.2	0.04	218	0.0011	Rejected	**0.84**	0.1
	IN	8.14	0	252	$<10^{-4}$	Rejected	**1.00**	0
	OUT	7.04	3.44	181	0.0515	**Accepted**	0.68	0.24
	TOT	15.18	3.44	226	0.0004	Rejected	**0.88**	0.09
	BET	38.71	0	225	0.0005	Rejected	**0.79**	0
	PR	0.0343	0.0042	252	$<10^{-4}$	Rejected	**1.00**	0
	HITSA	0.1535	0.0018	252	$<10^{-4}$	Rejected	**1.00**	0
	HITSH	0.1562	0.0846	185	0.0384	Rejected	**0.73**	0.27
	COND	0.55	0.54	133	0.80	**Accepted**	0.47	0.53
	FODF	0.37	0.47	130	0.89	**Accepted**	0.44	0.47

Table 3.9 lists all packages present in the Lucene package collaboration network and shows their structural characteristics. It can be seen that the most coupled packages are located in the strongly connected component of the network. Lucene packages involved in cyclic dependencies strongly tend to be the most central packages in the network: the afferent coupling, betweenness centrality and HITS authority score of all Lucene packages that are not cyclically dependent, except the root package "org.apache.lucene", are equal to zero. Additionally, those packages have lower values of the page rank metric compared to packages from the strongly connected component. Consequently, it can be concluded that the Lucene package collaboration network also exhibits a core-periphery structure with a strongly connected core.

3.6.2 Degree Distribution Analysis

The basic descriptive characteristics of in-degree, out-degree and total degree distributions of examined enriched software networks are shown in Table 3.10. The table also shows the coefficient of variation and skewness of degree distributions of comparable Erdős-Renyi random graphs. It can be noticed that the maximal value of in-degree, out-degree and total degree in package collaboration networks is less than 100. As a rule of thumb, a distribution with a power-law scaling region should exhibit an approximately linear relationship on a log-log plot over at least two orders of magnitude [77]. According to this rule, power-laws can be excluded for all degree distributions of package collaboration networks and out-degree distributions of class and method collaboration networks. Examined package collaboration networks, although representing systems that can be considered large-scale (each of them contains more than 100000 lines of code), are relatively small (have less than 100 nodes). Evidently, networks having less than 100 nodes cannot exhibit a variability of in-/out-/total

Table 3.9 Characteristics of nodes in the Lucene enriched package collaboration network. "O.A.L" stands for "org.apache.lucene". The column "SCC" denotes whether a package belongs to the strongly connected component of the network

Package	SCC	TOT	IN	OUT	BET	PR	HITSA	HITSH
O.A.L.util	Yes	19	14	5	38.42	0.21	0.59	0.24
O.A.L.index	Yes	17	10	7	31.58	0.19	0.44	0.36
O.A.L.search	Yes	13	6	7	19.58	0.08	0.28	0.4
O.A.L.analysis	Yes	11	6	5	5.25	0.07	0.3	0.33
O.A.L.analysis.tokenattributes	Yes	9	7	2	1.33	0.1	0.36	0.18
O.A.L.store	Yes	8	6	2	2.83	0.11	0.27	0.18
O.A.L.document	Yes	7	4	3	0	0.06	0.25	0.23
O.A.L.search.spans	Yes	5	2	3	0	0.02	0.11	0.22
O.A.L.util.packed	Yes	3	1	2	0	0.04	0.06	0.15
O.A.L.queryParser	No	6	0	6	0	0.01	0	0.38
O.A.L.search.payloads	No	4	0	4	0	0.01	0	0.24
O.A.L.collation	No	3	0	3	0	0.01	0	0.21
O.A.L.search.function	No	3	0	3	0	0.01	0	0.22
O.A.L.analysis.standard	No	3	0	3	0	0.01	0	0.21
O.A.L.util.fst	No	2	0	2	0	0.01	0	0.15
O.A.L	No	1	1	0	0	0.05	0.04	0

Table 3.10 The basic characteristics of degree distributions of enriched software networks from our experimental dataset. μ – mean, σ – standard deviation, ER – the coefficient of variation and skewness of degree distributions of comparable Erdős-Renyi random graphs, c_v – the coefficient of variation, G_1 – skewness, M – maximum value

Software	Network	Distribution	μ	σ	ER	c_v	G_1	M
Tomcat	PCN	Total degree	10.16	9.85	0.31	0.97*	2.22	58
	PCN	In degree	5.08	8.06	0.44	1.59	3.95	58
	PCN	Out degree	5.08	5.26	0.44	1.04	2.03	30
	CCN	Total degree	9.16	15.49	0.33	1.69	8.17	293
	CCN	In degree	4.58	13.41	0.47	2.93	11.76	293
	CCN	Out degree	4.58	6.94	0.47	1.52	3.47	73
	MCN	Total degree	2.72	8.27	0.61	3.04	30.69	429
	MCN	In degree	1.36	7.66	0.86	5.63	38.74	429
	MCN	Out degree	1.36	3.21	0.86	2.36	6.02	64
Lucene	PCN	Total degree	7.12	5.41	0.37	0.76*	1.07*	19
	PCN	In degree	3.56	4.26	0.53	1.2	1.16	14
	PCN	Out degree	3.56	1.97	0.53	0.55*	0.41**	7
	CCN	Total degree	8.99	11.57	0.33	1.29	5.94	175
	CCN	In degree	4.49	9.82	0.47	2.19	7.22	153
	CCN	Out degree	4.49	5.3	0.47	1.18	2.93	46
	MCN	Total degree	2.39	4.01	0.65	1.67	8.09	118
	MCN	In degree	1.2	3.27	0.91	2.73	13.43	117
	MCN	Out degree	1.2	2.39	0.91	1.99	5.28	37
Ant	PCN	Total degree	8.99	14.4	0.33	1.6	3.55	77
	PCN	In degree	4.49	11.66	0.47	2.6	3.87	64
	PCN	Out degree	4.49	3.72	0.47	0.83*	2.75	25
	CCN	Total degree	9.4	24.66	0.33	2.63	15.29	534
	CCN	In degree	4.7	23.5	0.46	5	16.58	533
	CCN	Out degree	4.7	5.45	0.46	1.16	2.38	40
	MCN	Total degree	2.86	9	0.59	3.15	34.66	558
	MCN	In degree	1.43	8.52	0.84	5.96	41.05	558
	MCN	Out degree	1.43	2.95	0.84	2.06	4.83	42
Xerces	PCN	Total degree	10.55	8.38	0.31	0.79*	0.78*	33
	PCN	In degree	5.28	6.31	0.44	1.2	1.67*	25
	PCN	Out degree	5.28	5.22	0.44	0.99*	1*	17
	CCN	Total degree	10.9	14.22	0.3	1.3	2.87	106
	CCN	In degree	5.45	10.96	0.43	2.01	4.22	105
	CCN	Out degree	5.45	8.71	0.43	1.6	3.84	91
	MCN	Total degree	2.63	6.15	0.62	2.34	17.89	245
	MCN	In degree	1.31	5.38	0.87	4.1	25.95	244
	MCN	Out degree	1.31	2.95	0.87	2.24	5.31	52

(continued)

Table 3.10 (continued)

Software	Network	Distribution	μ	σ	ER	c_v	G_1	M
JFreeChart	PCN	Total degree	12.32	10.09	0.28	0.82*	1.02*	43
	PCN	In degree	6.16	7.34	0.4	1.19	1.39*	28
	PCN	Out degree	6.16	5.1	0.4	0.83*	1.25*	22
	CCN	Total degree	10.31	15.93	0.31	1.54	5.79	211
	CCN	In degree	5.16	13.88	0.44	2.69	8.02	211
	CCN	Out degree	5.16	7.44	0.44	1.44	4.41	93
	MCN	Total degree	2.78	10.56	0.6	3.79	47.1	750
	MCN	In degree	1.39	10.09	0.85	7.25	54.04	750
	MCN	Out degree	1.39	3.22	0.85	2.31	6.3	66

Two stars indicate that c_v (resp. G_1) of the corresponding distribution is lower than c_v (resp. G_1) of degree distributions of comparable Erdős-Renyi random graphs. One star marks that c_v (resp. G_1) is lower compared to c_v (resp. G_1) of exponential distributions. PCN – package collaboration network, CCN – class collaboration network, MCN – method collaboration network

degrees that spans two or more orders of magnitude. Nevertheless, we will test their degree distributions against power-laws using statistically robust techniques.

The coefficient of variation, skewness and maximal in-degree, out-degree and total degree in class and method collaboration networks indicate the presence of hubs – classes and methods whose afferent, efferent and total coupling are considerably higher compared to the average values. For example, the total degree of the most coupled class defined in Ant is 56 times higher than the average total degree, the in-degree of the most internally reused Ant class is 112 times higher than the average in-degree, and the out-degree of the Ant class that references the largest number of other Ant classes is 8.5 times higher than the average out-degree. For Ant methods we have that the total degree of the most coupled Ant method is 195 times higher than the average total degree, the most internally reused Ant method has in-degree 390 times higher compared to the average in-degree, and the Ant method calling the highest number of other methods has out-degree 29 times higher than the average out-degree. The coefficient of variation and skewness of degree distributions of Erdős-Renyi random graphs are both equal to $1/\sqrt{\mu}$ where μ denotes the average node degree. For random graphs comparable to examined class and method collaboration networks $1/\sqrt{\mu}$ is always smaller than 1 since $\mu > 1$. The coefficient of variation of degree distributions of analyzed class and method collaboration networks is always higher than 1, while the skewness is significantly higher than 1, indicating a strong deviation from the Erdős-Renyi model of random graphs and the presence of strong node hubness. Additionally, the skewness of degree distributions associated to class and method collaboration networks is considerably higher than the skewness of exponential distributions which is equal to 2. This implies that those distributions are heavy-tailed since their tails are not exponentially bounded.

Lucene, Xerces and JFreeChart do not exhibit strong node hubness at the package level. The coefficient of variation of in-degree, out-degree and total degree distributions of package collaboration networks associated to previously mentioned software systems is either considerably less than or very close to the coefficient of variation of exponential distributions which is equal to 1. The same also holds for the skewness of mentioned distributions. In other words, mentioned distributions are not highly skewed to the right and cannot be considered heavy-tailed. The same properties also hold for out-degree distributions of Tomcat and Ant package collaboration networks. On the other hand, in-degree distributions of Tomcat and Ant package collaboration networks are considerably skewed to the right ($G_1 \gg 2$) and can be considered heavy-tailed. Therefore, we can conclude the following:

1. Lucene, Xerces and JFreeChart do not tend to contain packages exhibiting considerably high afferent and efferent coupling.
2. Ant and Tomcat do not tend to contain packages having considerably high efferent coupling, but those two software systems contain packages with considerably high afferent coupling.

Using the power-law test introduced by Clauset et al. [17] (whose implementation is provided by the poweRlaw R package [28]), we examined whether degree distributions of investigated software systems follow power-laws (and, consequently, exhibit the scale-free property). Additionally, the powerRlaw package provides the likelihood ratio test that can be instrumented to compare the best power-law fit to the best fits of alternative theoretical distributions in the identified power-law scaling region. The results of the power-law test for package collaboration networks are shown in Table 3.11. The null hypothesis of the power-law test (power-law is a plausible statistical model for empirical data) is accepted if obtained p-value is higher than 0.1 [17]. We can see that power-laws are plausible statistical models for a majority of empirical distributions. The bootstrapping step of the power-law test cannot be performed in the case of the Lucene out-degree distribution due to an extremely small power-law scaling region. Lower bounds of power-law scaling regions (denoted by x_m in Table 3.11) are always higher than 1 which means that the power-law behavior appear in tails of the distributions.

The results of the likelihood ratio test which compares the best power-law fit to the best fits of log-normal and exponential models in the power-law scaling region are also shown in Table 3.11. The value of the log likelihood ratio is denoted by R_d where d is an alternative statistical model ("ln" – log-normal, "e" – exponential). A positive and statistically significant R_d ($R_d > 0$, $p(R_d) < 0.1$) indicates that the goodness of the power-law fit is significantly better compared to the goodness of the fit of the alternative distribution. On the other side, a negative and statistically significant R_d ($R_d < 0$, $p(R_d) < 0.1$) implies that the alternative model provides better fit to the tail of the distribution. The obtained values of R_d and $p(R_d)$ indicate that:

• The power-law model is significantly better than the exponential model only for tails of total and in-degree distributions of the Ant package collaboration network. On the other hand, the exponential model provides significantly better fits to tails of in-degree and out-degree distributions of the JFreeChart package collaboration network. In all other cases, the exponential model is equally plausible as the power-law model.

Table 3.11 The results of the power-law test for package collaboration networks. x_m – the lower bound of the power-law scaling region, L – the size of the power-law scaling region, α – the power-law scaling exponent, p – the p-value of the power-law test statistic, the "Acc." column denotes whether the null hypothesis of the power-law test is accepted, R_{ln} – the log likelihood ratio between the best power-law and log-normal fits, $p(R_{ln})$ – the p-value of R_{ln}, R_e – the log likelihood ratio between the best power-law and exponential fits, $p(R_e)$ – the p-value of R_e

Software	Distr.	x_m	L	α	p	Acc.	R_{ln}	$p(R_{ln})$	R_e	$p(R_e)$
Tomcat	Total degree	9	49	2.78	0.06	No				
	In degree	5	53	2.38	0.77	Yes	-0.67	0.49	0.77	0.43
	Out degree	8	22	3.52	0.73	Yes	-0.38	0.69	0.38	0.69
Lucene	Total degree	3	16	2.07	0.32	Yes	-0.88	0.37	-0.85	0.39
	In degree	6	9	4.03	0.61	Yes	-0.14	0.88	0.3	0.76
	Out degree	3	4	3.05	/	/				
Ant	Total degree	4	73	2.48	0.52	Yes	-0.97	0.33	4.12	$<10^{-4}$
	In degree	2	62	1.83	0.78	Yes	-0.52	0.6	2.65	0.008
	Out degree	4	21	3.38	0.91	Yes	-0.2	0.83	1.13	0.25
Xerces	Total degree	13	20	3.84	0.28	Yes	-0.64	0.51	-0.69	0.48
	In degree	1	24	1.6	0.05	No				
	Out degree	7	10	3.23	0.42	Yes	-0.78	0.43	-1.06	0.28
JFreeChart	Total degree	16	27	4.05	0.63	Yes	-0.49	0.61	-0.32	0.74
	In degree	4	24	2.03	0.02	No				
	Out degree	3	19	2.05	0.01	No				

- The power-law model is never preferred over the log-normal model. The log-normal model provides better fits to tails of all degree distributions ($R_{ln} < 0$) and in three cases (in-degree and out-degree distributions of JFreeChart and the in-degree distribution of Xerces) the power-law model can be rejected in favor of the log-normal model due to statistically significant differences.

In other words, there is a moderate support for the power-law behavior in tails of degree distributions of package collaboration networks: power-law fits are plausible in tails, but alternative models are either equally plausible or even provide better fits. Additionally, the power-law scaling region is extremely small for some distributions. Therefore, we also investigated which model provides the best fits considering the whole range of degree values ($x_m = 1$). The results of the likelihood ratio test are given in Table 3.12. The value of the log likelihood ratio is denoted by $R\left(\frac{d_1}{d_2}\right)$, where d_1 and d_2 are two theoretical models ("pw" – power-law, "ln" – log-normal, "e" – exponential). A positive and statistically significant R ($R\left(\frac{d_1}{d_2}\right) > 0$, $p < 0.1$) indicates that d_1 is preferred over d_2, while a negative and statistically significant value of R indicates exactly the opposite [17]. The log-normal model provides better fits than the power-law model in all cases ($R\left(\frac{pw}{ln}\right) < 0$), and in 13 (out of 15) cases the goodness of the log-normal fit is significantly better compared to the goodness of the power-law fit. A similar situation can be observed with the exponential model: it significantly outperforms the power-law model in 12 cases, and the power-law model is preferred only for one distribution (the in-degree distribution of Ant). Therefore, we can conclude that alternative theoretical distributions are more plausible models for degree distributions of package collaboration networks considering the whole range of degree values.

The results of the power-law test for class collaboration networks are summarized in Table 3.13. We can see that power-laws are plausible statistical models for tails of all distributions ($p > 0.1$, $x_m > 1$). The power-law model provides better fits to a majority of distributions in comparison with the exponential model (the total degree distribution of Xerces and out-degree distributions of Tomcat and Ant are only three exceptions). However, the power-law model is never strongly preferred over the log-normal model. To the contrary, the log-normal model provides better fits to all degree distributions ($R_{ln} < 0$) except for the total degree distribution of Ant where a statistically significant difference in the goodness of fit is absent (see Fig. 3.3). The goodness of log-normal fits is significantly better in the case of three distributions: the Tomcat in-degree distribution, the Lucene in-degree distribution and the Xerces in-degree distribution. For those three distributions, the power-law model can be rejected in favor of the log-normal model.

In the same way as for package collaboration networks, we investigated which of considered theoretical models provides the best fits to degree distributions of class collaboration networks through the whole range of degree values. The results of the likelihood ratio test are presented in Table 3.14. The log-normal model provides the best fits to all distributions. Additionally, it significantly outperforms the power-law model in all cases ($R\left(\frac{pw}{ln}\right) < 0$, $p < 0.1$). The same situation can be observed when we compare exponential and power-law models: the exponential model provides significantly better fits than the power-law model in all cases ($R\left(\frac{pw}{e}\right) < 0$, $p < 0.1$). On the other hand, the log-normal model significantly outperforms the exponential model for a vast majority of distributions. In the case of out-degree distributions of Lucene and Ant (two distributions with the lowest coefficient of variation and skewness, see Table 3.10), the log-normal model is slightly better, but statistically

Table 3.12 The results of the likelihood ratio tests for package collaboration networks considering the whole range of values ($x_m = 1$)

Software	Distr.	$R\left(\frac{pw}{ln}\right)$	p	$R\left(\frac{pw}{e}\right)$	p	$R\left(\frac{ln}{e}\right)$	p	Best
Tomcat	Total degree	-6.69	$<10^{-4}$	-7.83	$<10^{-4}$	-0.05	0.95	Exp
	In degree	-3.3	0.0009	-0.96	0.33	1.41	0.16	ln
	Out degree	-4.65	$<10^{-4}$	-5.1	$<10^{-4}$	-1.26	0.2	Exp
Lucene	Total degree	-3.17	0.001	-4.44	$<10^{-4}$	0.34	0.72	ln
	In degree	-1.45	0.14	-2.01	0.04	-1.19	0.23	Exp
	Out degree	-8	$<10^{-4}$	-11.49	$<10^{-4}$	3.87	$<10^{-4}$	ln
Ant	Total degree	-5.25	$<10^{-4}$	-3.19	0.001	5.81	$<10^{-4}$	ln
	In degree	-1.29	0.19	2.31	0.02	3.27	0.001	ln
	Out degree	-5.24	$<10^{-4}$	-8.16	$<10^{-4}$	1.55	0.12	ln
Xerces	Total degree	-3.97	$<10^{-4}$	-4.44	$<10^{-4}$	-2.61	0.009	Exp
	In degree	-1.81	0.06	-0.33	0.73	0.88	0.37	ln
	Out degree	-2.77	0.005	-2.45	0.01	-0.94	0.34	Exp
JFreeChart	Total degree	-4.97	$<10^{-4}$	-6.66	$<10^{-4}$	-0.42	0.67	Exp
	In degree	-2.29	0.02	-2.28	0.02	-1.61	0.1	Exp
	Out degree	-3.7	0.0002	-5.06	$<10^{-4}$	-0.22	0.81	Exp

significant differences in the goodness of fit are absent. Complementary cumulative out-degree distributions of Lucene and Ant class collaboration networks are shown in Fig. 3.4. We can visually observe that out-degree values in tails are higher compared to the exponential fits concluding that those distributions are heavy-tailed. Other distributions associated to class collaboration networks are also heavy-tailed since their tails are not exponentially bounded: they have higher coefficient of variation and skewness than exponential distributions ($c_v(\exp) = 1$, $G_1(\exp) = 2$), and the log-normal fits through the whole range of values are better compared to the exponential fits.

The results of the power-law test for degree distributions of method collaboration networks are shown in Table 3.15. The power-law model is plausible only for 6 distributions of method coupling: in-degree and out-degree distributions of the Tomcat

Table 3.13 The results of the power-law test for class collaboration networks

Software	Distr.	x_m	L	α	p	Acc.	R_{ln}	$p(R_{ln})$	R_e	$p(R_e)$
Tomcat	Total degree	17	276	2.8	0.99	Yes	−0.68	0.49	1.95	0.05
	In degree	4	289	2.07	0.74	Yes	−1.84	0.06	3.32	0.0009
	Out degree	19	54	3.69	0.46	Yes	−1.07	0.28	−0.75	0.45
Lucene	Total degree	10	165	2.68	0.64	Yes	−1.39	0.16	1.24	0.21
	In degree	3	150	1.95	0.17	Yes	−2.76	0.005	1.75	0.08
	Out degree	11	35	3.72	0.84	Yes	−0.29	0.76	1.28	0.19
Ant	Total degree	11	523	2.6	1.00	Yes	0.41	0.68	2.77	0.005
	In degree	7	526	2.08	0.78	Yes	−0.38	0.7	3.36	0.0007
	Out degree	14	26	4.17	0.42	Yes	−0.94	0.34	−0.29	0.76
Xerces	Total degree	49	57	4.65	0.83	Yes	−0.75	0.45	−0.85	0.39
	In degree	6	99	2.1	0.60	Yes	−2.45	0.01	1.18	0.23
	Out degree	27	64	4.29	0.66	Yes	−0.22	0.81	0.53	0.59
JFreeChart	Total degree	14	197	2.65	0.88	Yes	−0.47	0.63	2.38	0.01
	In degree	9	202	2.31	0.94	Yes	−0.4	0.68	2.31	0.02
	Out degree	14	79	3.79	0.14	Yes	−0.07	0.94	1.16	0.24

method collaboration network, the total degree distribution of the Lucene method collaboration network and all degree distributions of the Ant method collaboration network. The power-law scaling can be ruled out with a high confidence for the rest of the distributions. Therefore, it can be concluded that Xerces and JFreeChart do not exhibit the scale-free property at the method level. Focusing on the cases for which the power-law hypothesis is not rejected, we can notice that the power-law model outperforms the exponential model for three distributions. Secondly, statistically significant differences between power-law and log-normal fits are absent: the log-normal model is slightly better in 5 cases, while the power-law model is slightly better only for the total degree distribution of the Lucene method collaboration network. Finally, the log-normal model outperforms other two theoretical models when fitting is performed through the whole range of degree values (see Table 3.16).

Fig. 3.3 The complementary cumulative total degree distribution of the Ant class collaboration network. This is the only degree distribution associated to class collaboration networks from our experimental dataset for which the power-law model provides the best fit in the tail compared to the alternative models. However, the difference between the best power-law fit and the best log-normal fit is not statistically significant

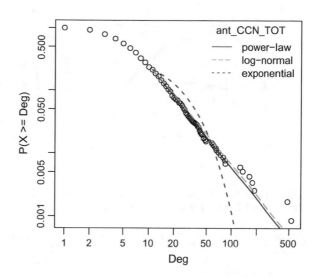

Table 3.14 The results of the likelihood ratio tests for class collaboration networks considering the whole range of values ($x_m = 1$)

Software	Distr.	$R\left(\frac{pw}{\ln}\right)$	p	$R\left(\frac{pw}{e}\right)$	p	$R\left(\frac{\ln}{e}\right)$	p	Best
Tomcat	Total degree	−19.56	$<10^{-4}$	−7.80	$<10^{-4}$	3.81	0.0001	ln
	In degree	−5.04	$<10^{-4}$	5.77	$<10^{-4}$	6.64	$<10^{-4}$	ln
	Out degree	−12.88	$<10^{-4}$	−6.23	$<10^{-4}$	3.78	0.0002	ln
Lucene	Total degree	−19.66	$<10^{-4}$	−14.44	$<10^{-4}$	3.86	0.0001	ln
	In degree	−4.68	$<10^{-4}$	4.70	$<10^{-4}$	6.12	$<10^{-4}$	ln
	Out degree	−11.88	$<10^{-4}$	−10.12	$<10^{-4}$	0.31	0.75	ln
Ant	Total degree	−18.89	$<10^{-4}$	−4.24	$<10^{-4}$	3.26	0.001	ln
	In degree	−3.06	0.002	4.24	$<10^{-4}$	4.61	$<10^{-4}$	ln
	Out degree	−13.63	$<10^{-4}$	−11.27	$<10^{-4}$	−0.88	0.38	ln
Xerces	Total degree	−18.48	$<10^{-4}$	−10.86	$<10^{-4}$	5.82	$<10^{-4}$	ln
	In degree	−5.08	$<10^{-4}$	7.00	$<10^{-4}$	9.15	$<10^{-4}$	ln
	Out degree	−10.19	$<10^{-4}$	−2.06	0.04	5.37	$<10^{-4}$	ln
JFreeChart	Total degree	−16.20	$<10^{-4}$	−9.28	$<10^{-4}$	4.07	$<10^{-4}$	ln
	In degree	−4.51	$<10^{-4}$	3.57	0.0004	4.94	$<10^{-4}$	ln
	Out degree	−8.95	$<10^{-4}$	−3.40	0.0007	2.49	0.013	ln

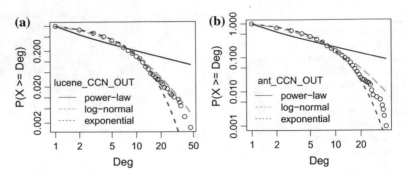

Fig. 3.4 The complementary cumulative out-degree distribution of **a** the Lucene class collaboration network and **b** the Ant class collaboration network

Table 3.15 The results of the power-law test for method collaboration networks

Software	Distr.	x_m	L	α	p	Acc.	R_{ln}	$p(R_{ln})$	R_e	$p(R_e)$
Tomcat	Total degree	9	420	2.93	0.01	No				
	In degree	3	426	2.44	0.29	Yes	−0.67	0.49	5.31	$<10^{-4}$
	Out degree	13	51	3.6	0.55	Yes	−1.26	0.21	0.51	0.61
Lucene	Total degree	16	102	3.84	0.88	Yes	0.19	0.84	1.29	0.19
	In degree	2	115	2.32	0.05	No				
	Out degree	3	34	2.63	0.0004	No				
Ant	Total degree	8	237	2.78	0.13	Yes	−0.14	0.88	3.2	0.001
	In degree	2	242	2.38	0.52	Yes	−0.03	0.97	5.84	$<10^{-4}$
	Out degree	17	35	4.93	0.86	Yes	−0.54	0.58	0.27	0.78
Xerces	Total degree	6	239	2.71	0	No				
	In degree	4	240	2.43	0.02	No				
	Out degree	5	47	2.62	$<10^{-4}$	No				
JFreeChart	Total degree	3	747	2.31	$<10^{-4}$	No				
	In degree	2	748	2.22	$<10^{-4}$	No				
	Out degree	2	64	2.26	$<10^{-4}$	No				

Table 3.16 The results of the likelihood ratio tests for method collaboration networks considering the whole range of values ($x_m = 1$)

Software	Distr.	$R\left(\frac{pw}{ln}\right)$	p	$R\left(\frac{pw}{e}\right)$	p	$R\left(\frac{ln}{e}\right)$	p	Best
Tomcat	Total degree	−24.99	$<10^{-4}$	1.72	0.08	6.59	$<10^{-4}$	ln
	In degree	−5.56	$<10^{-4}$	6.26	$<10^{-4}$	6.73	$<10^{-4}$	ln
	Out degree	−19.28	$<10^{-4}$	−0.07	0.93	11	$<10^{-4}$	ln
Lucene	Total degree	−17.72	$<10^{-4}$	−2.94	0.003	6.21	$<10^{-4}$	ln
	In degree	−6.63	$<10^{-4}$	5.31	$<10^{-4}$	7.04	$<10^{-4}$	ln
	Out degree	−10.33	$<10^{-4}$	1.18	0.23	7.24	$<10^{-4}$	ln
Ant	Total degree	−21.88	$<10^{-4}$	0.69	0.48	5.17	$<10^{-4}$	ln
	In degree	−3.45	0.0005	5.83	$<10^{-4}$	6.16	$<10^{-4}$	ln
	Out degree	−16.76	$<10^{-4}$	−1.22	0.21	9.98	$<10^{-4}$	ln
Xerces	Total degree	−18.78	$<10^{-4}$	1.95	0.05	7.11	$<10^{-4}$	ln
	In degree	−7.48	$<10^{-4}$	5.49	$<10^{-4}$	6.56	$<10^{-4}$	ln
	Out degree	−10.18	$<10^{-4}$	9.08	$<10^{-4}$	13.76	$<10^{-4}$	ln
JFreeChart	Total degree	−17.54	$<10^{-4}$	2.15	0.03	4.58	$<10^{-4}$	ln
	In degree	−5.25	$<10^{-4}$	4.13	$<10^{-4}$	4.46	$<10^{-4}$	ln
	Out degree	−11	$<10^{-4}$	6.93	$<10^{-4}$	12.25	$<10^{-4}$	ln

Summarizing obtained results, it can be concluded that examined software networks are not scale-free for two reasons:

1. The log-normal model provides better fits to empirical degree distributions compared to the power-law model considering the whole range of degree values.
2. The tails of degree distributions of examined package collaboration and class collaboration networks can be modeled by power-laws, but alternative models are either equally plausible or even provide better fits.

From the practical point of view, our findings imply that the scaling exponent of fitted power-laws cannot be instrumented in software engineering practice as a metric of software quality with respect to the principle of low coupling. Our findings also have implications for theoretical models of software evolution: mechanisms generating power-laws cannot be used to guide and formulate founding principles of theoretical models aimed to explain and predict changes in software structure.

One of prominent characteristics of scale-free networks is that they contain hubs – highly connected nodes whose degree is significantly higher compared to the average node degree. The same holds for all examined class and method collaboration networks, as well as for package collaboration networks of Ant and Tomcat. Those networks contain hubs due to heavy-tailed degree distributions that are best explained by the log-normal model. The emergence of hubs and power-law scaling behavior in complex networks can be explained by the preferential attachment principle of the Barabási-Albert model [2]. The preferential attachment principle formalizes one of frequently observed tendencies in the evolution of real-world complex networks: new nodes strongly tend to attach to hubs, thus increasing their hubness. Formally, the probability Π that a new node connects to an old node is directly proportional to the degree of the old node (denoted by k):

$$\Pi(k) \propto k \tag{3.28}$$

On the other hand, log-normal degree distributions arise from nearly-linear preferential attachments [65] of the form

$$\Pi_{nl}(k) \propto \frac{k}{1 + a \ln k} \tag{3.29}$$

Therefore, both log-normal and power-law degree distributions can be explained by some form of preferential attachment with the difference that the preferential attachment probability in scale-free networks is slightly higher compared to networks having log-normal degree distributions.

From the software engineering perspective, heavy-tailed in-degree distributions in class and method collaboration networks imply a broad spectrum of class/method reuse: most classes/methods present in a system are not internally reused (i.e. they are internally used by exactly one class/method), but there is a significant number of classes/methods whose degree of internal reuse is far above the average. The nearly-linear preferential attachment principle, the generating mechanism for log-normal degree distributions, suggests the tendency of increasing internal reuse for already highly reused software entities as software evolves. Software engineering practice encourages internal code reuse and this tendency may seem very desirable. However, modifications in a highly reused software entity may affect a very large number of other entities which directly or indirectly depend on it. In addition, highly reused classes/methods are particularly critical (externally responsible) because defects in them are more likely to propagate to other parts of the system [9]. In cases when defects need to propagate to other parts of the system and cause failures there in order to be detected, they may not be detected when testing highly reused entities in isolation. In other words, highly reused software entities are hard to test when the context of their internal reuse is required to observe an unexpected behavior. Thus, identifying highly reused classes and methods, especially ones which do not realize simple functionalities, and their effective testing or validation before they become extremely reused may help to prevent the following conflict situation: the presence of highly reused, hard-to-modify and hard-to-test entities with an increasing tendency of internal reuse which makes them even more critical and harder to maintain in terms of modifiability and use-context-required testability.

Out-going degree of a node in a software network reflects the degree of internal aggregation of other software entities. The presence of heavy-tailed out-degree distributions in class and method collaboration networks implies a broad spectrum of internal class/method aggregation. As stated by Briand et al. [9], a high degree of internal aggregation may negatively impact the understandability of software entities and make them more prone to faults and errors. Our results show that real-world Java software systems contain classes and methods exhibiting a high degree of internal aggregation. Thus, it is necessary to monitor and control out-going dependencies of such entities during software evolution in order to prevent negative aspects of high efferent coupling.

3.6.3 Analysis of Highly Coupled Software Entities

Software development methodologies advise to keep total class/method coupling as low as possible. Empirically observed heavy-tailed total degree distributions of class and method collaboration networks imply the presence of highly coupled classes/methods in corresponding software systems. From the software engineering perspective, this phenomenon is considered bad because highly coupled software entities can cause difficulties in software maintenance, testing and comprehension. Let C denote the set of nodes of an arbitrary enriched class/method collaboration network. We split C into two subsets H and O ($C = H \cup O$, $H \cap O = \emptyset$) where

1. H contains hubs (highly coupled entities), see Definition 3.5, and
2. O contains entities that are not highly coupled (non-hubs).

Table 3.17 shows the fraction of hubs in our networks and their minimal total coupling. It can be observed that in examined software systems 13–17% classes are highly coupled. The total coupling of highly coupled classes is equal to or higher than 14. The number of highly coupled methods vary from 9 to 14% and such methods are coupled to at least 5 other methods (the average method coupling varies from 2.4 to 2.8 which means that the coupling of highly coupled methods is at least two times higher than the average method coupling).

The metric-based comparison test can be applied to H and O sets in order to determine distinctive characteristics of hubs and get a more deeper understanding of the high coupling phenomenon in real-world software systems. The results of the test for class collaboration networks are shown in Table 3.18. The null hypothesis of the MWU test is accepted for the DIT metric (the depth in inheritance tree) in four out of five systems (all except Lucene). This result implies that highly coupled classes tend to exhibit the same degree of specialization as loosely coupled classes, i.e. highly coupled classes do not tend to be localized in class inheritance hierarchies. In all other aspects (voluminosity, internal complexity, reuse through inheritance, centrality in

Table 3.17 The percentage of hubs ($|H|$) and the minimal total degree of hubs (H_d) in class and method collaboration networks

	Class collaboration network		Method collaboration network	
Software system	$|H|$ [%]	H_d	$|H|$ [%]	H_d
Tomcat	13.79	17	9.87	7
Lucene	17.11	14	14.35	5
Ant	12.85	15	12.11	6
Xerces	13.7	23	9.68	7
JFreeChart	14.74	17	9.41	7

the design structure, class cohesion) differences between hub and non-hub classes are statistically significant in all examined systems: highly coupled classes tend to contain more lines of code, they have more complex control-flow, they tend to be more reused through inheritance, and they tend to be more functionally important compared to loosely coupled classes. Highly coupled Ant classes strongly tend to have higher cyclomatic complexity implying that they are harder to test than loosely coupled classes (the white-box testing requires significantly more test cases for hubs than for non-hubs). Finally, highly coupled classes in all examined systems tend to be significantly less cohesive indicating that, besides negative implications of high class coupling, they may also exhibit negative effects of low class cohesion.

Drastic differences between hubs and non-hubs present in all examined systems are related to the following metrics:

1. LOC. Highly coupled classes tend to be considerably more voluminous than non-hub classes. The average LOC of hubs is at least three times higher than the average LOC of non-hubs. The probability that a randomly selected hub class contains more lines of code than a randomly selected non-hub class is in the range [0.77, 0.86] indicating strong superiority of highly coupled classes regarding LOC.
2. BET. Highly coupled classes tend to have a considerably higher betweenness centrality than loosely coupled classes. The average betweenness centrality of hubs is at least 13 times higher than the average betweenness centrality of loosely coupled classes (in the case of Ant 30 times higher, for Tomcat 28 times higher). This means that hubs tend to occupy central positions in class collaboration networks implying their vital role to the overall functionality of investigated software systems.
3. HITSA. The most functionally important classes that are internally reused by other functionally important classes strongly tend to be highly coupled.

The results of the metric-based comparison test for hubs in method collaboration networks are given in Table 3.19. Statistically significant differences between highly and loosely coupled methods are present in all examined systems regarding all aspects quantified by considered metrics. Highly coupled methods tend to contain more lines of code than loosely coupled methods. This tendency is very strong in the

Table 3.18 The results of the metric-based comparison test for hubs in enriched class collaboration networks. $\overline{C_1}$ and $\overline{C_2}$ denote the average value of the corresponding metric for hub classes and non-hub classes, respectively. The column "null hyp." denotes whether the null hypothesis of the MWU test (no statistically significant differences between hubs and non-hubs) is accepted. PS_1 and PS_2 are the probability of superiority of hubs and non-hubs, respectively. Bold probabilities of superiority indicate substantial differences between hubs and non-hubs with respect to the corresponding metric

Software	Metric	$\overline{C_1}$	$\overline{C_2}$	U	p	Null hyp.	PS_1	PS_2
Tomcat	LOC	699.03	137.29	204480	$<10^{-4}$	Rejected	**0.84**	0.15
	CC	71.47	13.69	176951	$<10^{-4}$	Rejected	0.68	0.22
	NUMA	12.18	3.89	178509	$<10^{-4}$	Rejected	0.7	0.22
	NUMM	31.68	7.21	206308	$<10^{-4}$	Rejected	**0.84**	0.13
	NOC	1.25	0.18	136294	0.006	Rejected	0.18	0.05
	DIT	0.63	0.47	126201	0.35	**Accepted**	0.27	0.23
	BET	4777.76	169.75	208000	$<10^{-4}$	Rejected	**0.83**	0.11
	PR	0.002055	0.000474	179470	$<10^{-4}$	Rejected	0.74	0.26
	HITSA	0.03057	0.00288	197367	$<10^{-4}$	Rejected	**0.81**	0.18
	HITSH	0.03866	0.00869	191786	$<10^{-4}$	Rejected	**0.78**	0.2
	COND	0.37	0.29	144701	$<10^{-4}$	Rejected	0.54	0.34
	FODF	0.65	0.7	142019	0.0001	Rejected	0.34	0.51
Lucene	LOC	332.5	93.03	69044	$<10^{-4}$	Rejected	**0.78**	0.22
	CC	37.11	13.32	62822	$<10^{-4}$	Rejected	0.66	0.25
	NUMA	7.52	2.93	59358	$<10^{-4}$	Rejected	0.62	0.28
	NUMM	15.17	5.91	65950	$<10^{-4}$	Rejected	0.72	0.23
	NOC	1.68	0.26	55865	$<10^{-4}$	Rejected	0.35	0.09
	DIT	0.46	0.88	53421	0.0002	Rejected	0.2	0.4
	BET	2828.94	139.12	71292	$<10^{-4}$	Rejected	**0.76**	0.15
	PR	0.00347	0.000809	70588	$<10^{-4}$	Rejected	**0.79**	0.2
	HITSA	0.04507	0.00451	74288	$<10^{-4}$	Rejected	**0.84**	0.15
	HITSH	0.04022	0.01568	59136	$<10^{-4}$	Rejected	0.66	0.32
	COND	0.35	0.25	53638	$<10^{-4}$	Rejected	0.56	0.34
	FODF	0.64	0.64	52542	0.0004	Rejected	0.32	0.51
Ant	LOC	574.78	114.32	132568	$<10^{-4}$	Rejected	**0.86**	0.14
	CC	49.67	9.63	124300	$<10^{-4}$	Rejected	**0.78**	0.17
	NUMA	12.37	3.66	117880	$<10^{-4}$	Rejected	0.73	0.2
	NUMM	26.08	6.25	133043	$<10^{-4}$	Rejected	**0.85**	0.13
	NOC	2.88	0.2	95863	$<10^{-4}$	Rejected	0.31	0.07
	DIT	1.28	1.11	83411	0.11	**Accepted**	0.39	0.32
	BET	7715.53	249.41	129761	$<10^{-4}$	Rejected	**0.8**	0.12
	PR	0.003835	0.000411	117779	$<10^{-4}$	Rejected	**0.75**	0.23
	HITSA	0.02957	0.00201	123589	$<10^{-4}$	Rejected	**0.79**	0.19
	HITSH	0.04227	0.01685	123316	$<10^{-4}$	Rejected	**0.79**	0.2
	COND	0.37	0.26	98375	$<10^{-4}$	Rejected	0.6	0.33
	FODF	0.66	0.73	96166	$<10^{-4}$	Rejected	0.33	0.57

Table 3.18 (continued)

Xerces	LOC	766.36	145.84	71757	$<10^{-4}$	Rejected	**0.79**	0.21
	CC	100.54	15.78	64222	$<10^{-4}$	Rejected	0.64	0.22
	NUMA	22.35	3.18	66545	$<10^{-4}$	Rejected	0.69	0.22
	NUMM	24.35	8.26	67912	$<10^{-4}$	Rejected	0.73	0.23
	NOC	1.46	0.23	53140	0.003	Rejected	0.26	0.09
	DIT	0.68	1.07	50199	0.06	**Accepted**	0.23	0.33
	BET	3897.61	182.87	72166	$<10^{-4}$	Rejected	**0.76**	0.16
	PR	0.003367	0.000788	67360	$<10^{-4}$	Rejected	0.74	0.25
	HITSA	0.05635	0.0053	77198	$<10^{-4}$	Rejected	**0.85**	0.15
	HITSH	0.05368	0.00846	68238	$<10^{-4}$	Rejected	0.74	0.24
	COND	0.3	0.25	51778	0.01	Rejected	0.48	0.34
	FODF	0.69	0.74	46820	0.01	Rejected	0.32	0.46
JFreeChart	LOC	854.12	202.85	37753	$<10^{-4}$	Rejected	**0.77**	0.23
	CC	59.83	12.32	36467	$<10^{-4}$	Rejected	0.71	0.22
	NUMA	11.97	3.35	33972	$<10^{-4}$	Rejected	0.65	0.27
	NUMM	37.25	10.02	38177	$<10^{-4}$	Rejected	**0.77**	0.21
	NOC	1.77	0.23	31594	$<10^{-4}$	Rejected	0.35	0.06
	DIT	1	0.83	27197	0.08	**Accepted**	0.39	0.27
	BET	931.58	67.69	39678	$<10^{-4}$	Rejected	**0.77**	0.14
	PR	0.004705	0.001066	35461	$<10^{-4}$	Rejected	0.71	0.26
	HITSA	0.05979	0.0052	41045	$<10^{-4}$	Rejected	**0.83**	0.15
	HITSH	0.06227	0.01604	38268	$<10^{-4}$	Rejected	**0.78**	0.21
	COND	0.38	0.27	29625	0.001	Rejected	0.54	0.33
	FODF	0.59	0.68	28092	0.02	Rejected	0.35	0.5

case of Tomcat, Ant and Xerces. They also tend to have more complex control-flow implying that they are harder to comprehend and test than loosely coupled methods. Finally, hub methods tend to exhibit higher centrality in the method collaboration network compared to non-hubs by all considered centrality indices. Therefore, it can be concluded that functionally most important methods tend to be highly coupled. Consequently, changes in highly coupled methods may have bigger impact to the overall system stability and evolution.

Looking back to the basic characteristics of degree distributions of examined enriched class and method collaboration networks (Table 3.10), we can spot that the coefficient of variation, skewness and maximal degree of in-degree distributions are considerably higher compared to the same quantities for out-degree distributions. Tails of in-degree distributions are longer and heavier than tails of out-degree distributions suggesting that highly coupled classes/methods contained in tails of total degree distributions incline to have higher afferent than efferent coupling. For example, the Tomcat class having the highest total degree at the same time has the highest in-degree. This means that the coupling of the most coupled class in Tomcat is entirely

Table 3.19 The results of the metric-based comparison test for hubs in enriched method collaboration networks

Software	Metric	$\overline{C_1}$	$\overline{C_2}$	U	p	Null hyp.	PS_1	PS_2
Tomcat	LOC	41.43	7.83	16981968	$<10^{-4}$	Rejected	**0.79**	0.16
	CC	8.88	1.10	16758874	$<10^{-4}$	Rejected	0.72	0.11
	BET	168.01	10.87	15970716	$<10^{-4}$	Rejected	0.61	0.08
	PR	0.00015	0.00006	15410002	$<10^{-4}$	Rejected	0.69	0.21
	HITSA	0.00319	0.00025	15372074	$<10^{-4}$	Rejected	0.69	0.21
	HITSH	0.01087	0.00087	16930797	$<10^{-4}$	Rejected	**0.75**	0.13
Lucene	LOC	25.73	7.24	3174409	$<10^{-4}$	Rejected	0.70	0.24
	CC	5.68	1.49	3155571	$<10^{-4}$	Rejected	0.60	0.15
	BET	231.58	7.17	3301046	$<10^{-4}$	Rejected	0.59	0.07
	PR	0.00036	0.00014	3447817	$<10^{-4}$	Rejected	**0.75**	0.16
	HITSA	0.00670	0.00035	3517471	$<10^{-4}$	Rejected	**0.76**	0.14
	HITSH	0.00722	0.00138	3261323	$<10^{-4}$	Rejected	0.68	0.18
Ant	LOC	28.18	7.05	9131832	$<10^{-4}$	Rejected	**0.76**	0.15
	CC	5.26	1.00	8801845	$<10^{-4}$	Rejected	0.68	0.13
	BET	2589.15	89.82	8824654	$<10^{-4}$	Rejected	0.63	0.08
	PR	0.00024	0.00008	8445790	$<10^{-4}$	Rejected	0.69	0.21
	HITSA	0.00474	0.00030	8567170	$<10^{-4}$	Rejected	0.70	0.19
	HITSH	0.01309	0.00156	9315483	$<10^{-4}$	Rejected	**0.78**	0.14
Xerces	LOC	52.45	8.45	5737374	$<10^{-4}$	Rejected	**0.75**	0.19
	CC	11.75	1.38	5681649	$<10^{-4}$	Rejected	0.68	0.13
	BET	254.16	13.93	5525703	$<10^{-4}$	Rejected	0.58	0.08
	PR	0.00029	0.00009	5846469	$<10^{-4}$	Rejected	**0.75**	0.16
	HITSA	0.00155	0.00013	5980268	$<10^{-4}$	Rejected	**0.76**	0.14
	HITSH	0.00022	0.00198	5398129	$<10^{-4}$	Rejected	0.66	0.20
JFreeChart	LOC	34.49	7.66	4852403	$<10^{-4}$	Rejected	0.70	0.22
	CC	5.37	0.96	4775212	$<10^{-4}$	Rejected	0.59	0.13
	BET	74.39	2.31	4695939	$<10^{-4}$	Rejected	0.52	0.08
	PR	0.00034	0.00009	4972039	$<10^{-4}$	Rejected	0.69	0.17
	HITSA	0.00295	0.00015	4997100	$<10^{-4}$	Rejected	0.69	0.17
	HITSH	0.00456	0.00337	4499054	$<10^{-4}$	Rejected	0.62	0.24

caused by its internal reuse since this class does not reference any other class defined in Tomcat. The same applies for the most coupled method in Tomcat. In other words, basic statistical quantities describing empirically observed in-degree and out-degree distributions of class/method collaboration networks suggest that there is some kind of disbalance between in-degree and out-degree of highly coupled classes/methods where in-degree (afferent coupling) tends to be significantly higher than out-degree (efferent coupling). Figures 3.5, 3.6, 3.7, 3.8 and 3.9 show afferent-efferent coupling

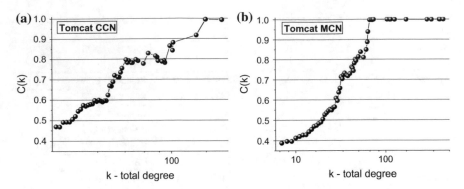

Fig. 3.5 The AEC balance plot for **a** the class and **b** method collaboration network of Tomcat

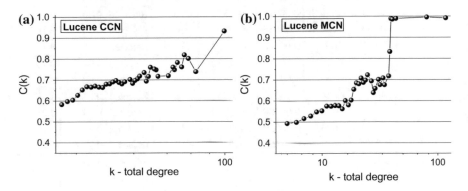

Fig. 3.6 The AEC balance plot for **a** the class and **b** method collaboration network of Lucene

(AEC) balance plots (see Definition 3.7) for class and method collaboration networks of investigated software systems. It can be seen that $C_k \gg 0.5$ for $k > 20$ in all networks implying that high class/method coupling is dominantly caused by extensive internal reuse. Moreover, C_k increases with k implying that the dominance of afferent over efferent coupling increases with total class/method coupling. In all examined method collaboration networks we have that $C_k \approx 1$ for extremely large values of k, i.e. the coupling of the most coupled methods is entirely or almost entirely caused by immense internal reuse.

Highly coupled classes defined in Ant are strongly caused by internal reuse. On average, 92% of the total number of links incident with an Ant class having total degree higher than 50 are in-coming links and only 8% are out-going links. Highly coupled classes in Tomcat, Lucene and JFreeChart are dominantly caused by internal reuse. For those three systems we have that $C_k > 0.7$ for classes having total degree higher than 50 which means that on average 70% of the total coupling is caused by internal reuse (and this percentage increases as class coupling increases). C_k of the Xerces class collaboration network shows substantially different behavior compared to the rest of the networks. It starts to decrease from 0.71 at $k = 50$ to 0.43 at $k = 96$.

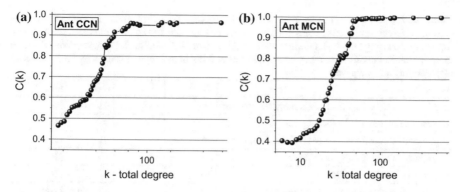

Fig. 3.7 The AEC balance plot for **a** the class and **b** method collaboration network of Ant

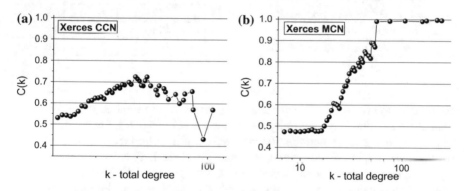

Fig. 3.8 The AEC balance plot for **a** the class and **b** method collaboration network of Xerces

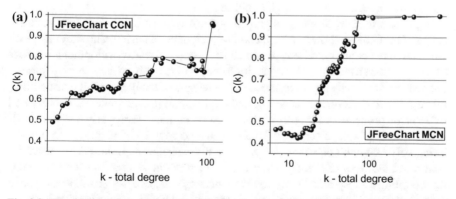

Fig. 3.9 The AEC balance plot for **a** the class and **b** method collaboration network of JFreeChart

Table 3.20 The top ten most coupled classes in Xerces. TOT, IN and OUT denote the degree of total, afferent and efferent class coupling, respectively. The HC column indicates the cause of high class coupling (A - high coupling caused dominantly by internal class aggregation, R - high coupling caused dominantly by internal class reuse, AR - high coupling caused by both extensive aggregation and extensive reuse

Class	TOT	IN	OUT	HC
org.apache.xerces.impl.xs.traversers.XSDHandler	106	15	91	A
org.apache.xerces.xni.XNIException	105	105	0	R
org.apache.xerces.impl.xs.XMLSchemaValidator	96	15	81	A
org.apache.xerces.util.SymbolTable	87	86	1	R
org.apache.xerces.xni.QName	86	86	0	R
org.apache.xerces.dom.CoreDocumentImpl	80	47	33	AR
org.apache.xerces.impl.xs.SchemaGrammar	79	35	44	AR
org.apache.wml.dom.WMLDocumentImpl	76	37	39	AR
org.apache.xerces.impl.Constants	73	72	1	R
org.apache.xerces.impl.XMLEntityManager	68	28	40	AR

Extremely highly coupled Xerces classes (classes whose total coupling is higher than 50) are dominantly caused by internal reuse ($C_{50} = 0.71$), but the dominance of afferent over efferent coupling decreases for higher values of total coupling. The top ten most coupled Xerces classes are shown in Table 3.20. It can be observed that high coupling of two classes from the list is strongly caused by extensive internal aggregation of other Xerces classes, extensive internal reuse is the reason for high class coupling of four classes, while for the rest of classes both reuse and aggregation are significant factors of their high coupling.

High class/method coupling is generally considered as a signal of poor software designs. Our findings show that high coupling originates from immense internal reuse which is, to the contrary, widely accepted as a good software development practice. This seems to be a paradox. However, high coupling strongly caused only by internal reuse can indicate only negative aspects of extensive internal reuse (high criticality), not negative aspects of extensive internal aggregation (lower understandability and higher error-proneness). If highly reused software entities are properly tested (or validated) in early phases of software development and do not tend to change frequently during software evolution then we can actually consider high coupling caused by reuse as an indicator of a good rather than poor software design. In such situations, high coupling corresponds to low code redundancy and good design choices made in an early phase of software development that enabled positive aspects of code reuse. On the other hand, if highly coupled entities tend to cause problems during software evolution, in the sense that highly coupled entities tend to change frequently and force changes in a large number of other entities that depend on them, then it is necessary to control high afferent coupling and keep it as low as possible. Our results also show that highly coupled entities exhibit higher internal complexity than loosely coupled entities. Therefore, developers before reusing highly coupled entities should

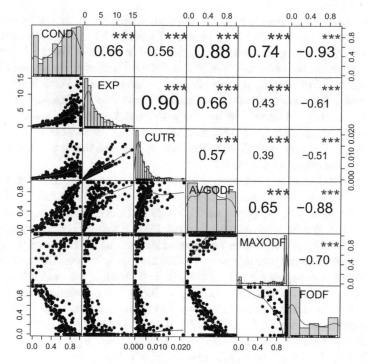

Fig. 3.10 The Spearman correlation matrix for GCE metrics at the package level

be aware of their complex control flow and check whether they went through a proper white-box testing in order to avoid errors caused by faults propagated from reused entities.

3.6.4 Correlations Between Cohesion Metrics

In our methodological framework, we proposed graph clustering evaluation (GCE) metrics to quantify cohesion of packages and classes. As already emphasized, two principals advantages of GCE metrics over commonly used software cohesion metrics is that (1) they do not ignore external dependencies, and (2) they are closely related to the Radicchi definitions of clusters in complex networks. Two GCE metrics (conductance and Flake-ODF) can be directly used to identify Radicchi strong and Radicchi weak software entities (conductance for Radicchi weak entities, Flake-ODF for Raddichi strong entities).

A software entity whose constituent elements do not form a Radicchi strong or weak cluster in the appropriate software network can be considered poorly cohesive. On the other hand, Radicchi strong software entities are exceptionally strongly cohesive, while software entities that are Radicchi weak exhibit a satisfactory degree of cohesion. Nodes in package and class collaboration networks from our experimental dataset are enriched with five GCE metrics (COND, EXP, CUTR, AVGODF and MAXODF) quantifying the lack of cohesion (lower values of those metrics mean more cohesive entities) and one GCE metric (FODF) reflecting the degree of cohesion (a higher FODF implies a more cohesive software entity). Additionally, nodes in our class collaboration networks are enriched with five "traditional" class cohesion measures: LCOM introduced by Chidamber and Kemerer [15], HM and HM2 proposed by Hitz and Montazeri [32], and TCC and LCC introduced by Bieman and Kang [6]. LCOM, HM and HM2 are the lack of cohesion metrics, while TCC and LCC reflect the degree of cohesion.

Figure 3.10 shows the Spearman correlation matrix for GCE metrics at the package level. All packages present in examined software systems are jointly taken into account when computing correlations. Three stars in a cell of the correlation matrix mean that the correlation between corresponding GCE metrics is statistically significant. It can be observed that Spearman correlations among GCE metrics quantifying the lack of cohesion (all metrics except FODF) are mostly strong. The strongest correlation is between expansion and cut-ratio ($\rho = 0.9$), while the weakest correlation is between cut-ratio and maximum-ODF ($\rho = 0.39$). The correlations between conductance and the rest of GCE metrics quantifying the lack of cohesion are medium to strong indicating that EXP, CUTR, AVGODF and MAXODF in most of the cases will be able to separate Radicchi weak from poorly cohesive packages.

FODF is the only of used GCE metrics which does not quantify the lack of cohesion, but the degree of cohesion. Therefore, there are strong negative Spearman correlations between FODF and the rest of GCE metrics. The presence of strong negative correlations suggests that GCE metrics quantifying the lack of cohesion will have a good performance in identifying extremely cohesive packages. Conductance has the strongest negative correlation with FODF implying that this particular GCE metric is able not only to separate Radicchi weak packages from poorly cohesive ones, but also to separate Radicchi weak from Radicchi strong packages.

We also computed correlations between GCE metrics at the class level, as well as correlations between GCE metrics and traditional class cohesion metrics. Figure 3.11 shows the clustered heat map plot of the Spearman correlation matrix. The red color denotes a positive, while blue indicates a negative Spearman correlation. Darker colors imply stronger correlations. The correlations between different GCE metrics at the class level are statistically significant and very strong: the absolute value of the Spearman correlation coefficient varies from 0.85 to 0.99. Therefore, we can conclude that all GCE metrics will be very successful in identifying Radicchi strong, Radicchi weak and poorly cohesive classes. Strong correlations can also be observed among LCOM, HM and HM2, as well as between TCC and LCC. In other words, class cohesion metrics can be clustered into three groups according to the presence of strong correlations:

- GCE metrics are in the first group,
- LCOM, HM and HM2 are in the second group, and
- TCC and LCC are in the third group of strongly correlating class cohesion metrics.

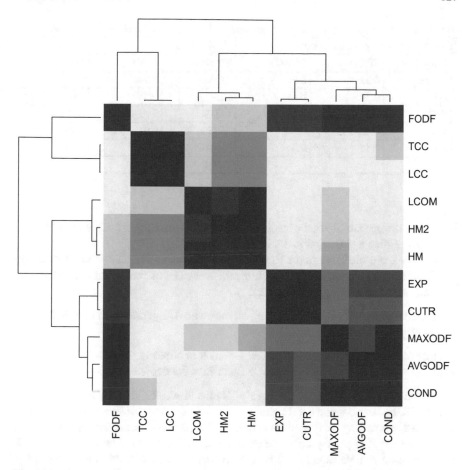

Fig. 3.11 The clustered heat map plot of the Spearman correlation matrix for class cohesion metrics

The correlations between metrics from the second group and metrics from the third group are weak to medium: the strongest correlation is between LCC and HM2 ($\rho = -0.47$) and the weakest correlation is between LCOM and LCC ($\rho = -0.17$). On the other hand, correlations between GCE metrics and metrics from two other groups are in the range from -0.06 to 0.34 which means that they are either absent or weak.

The absence of strong correlations between GCE metrics and widely used class cohesion metrics suggests that traditional class cohesion metrics are unable to distinguish between highly and poorly cohesive classes with respect to the Radicchi notions of clusters in complex networks. To check this we computed the following probabilities of stochastic dominance taking into account all classes present in examined software systems:

Table 3.21 The ability of class cohesion measures to identify Radicchi strong/weak classes

Traditional class cohesion measures			GCE metrics		
Metric (M)	$S(M)$	$W(M)$	Metric (M)	$S(M)$	$W(M)$
LCOM	0.33	0.48	COND	0.02	0
HM	0.34	0.45	EXP	0.03	0.07
HM2	0.34	0.41	CUTR	0.03	0.09
TCC	0.45	0.53	AVGODF	0.01	0.05
LCC	0.45	0.54	MAXODF	0.0006	0.08
			FODF	1	0.97

- $S(M)$ – the probability that a Radicchi strong class has a higher value of a cohesion metric M than a class that is not Radicchi strong, and
- $W(M)$ – the probability that a Radicchi weak class has a higher value of cohesion metric M than a class that is not Radicchi weak (let us recall that each Radicchi strong entity is also Raddichi weak).

If M is able to identify Radicchi strong classes with a high precision then $S(M)$ will be either

- close or equal to 0 in the case that M is a lack of cohesion metric, or
- close or equal to 1 in the case that M quantifies the degree of cohesion.

The same holds for $W(M)$ in the case of Radicchi weak classes. Clearly, we have that $W(\text{COND}) = 0$ since conductance has a well defined threshold with respect to the Radicchi weak property: a class is Radicchi weak if its conductance is lower than 0.5, otherwise it is poorly cohesive. Similarly, $S(\text{FODF}) = 1$ because the FODF measure has a well defined threshold with respect to the Radicchi strong property – the FODF of Radicchi strong classes is equal to 1, while non-Radicchi strong classes have FODF smaller than 1. The obtained values of S and W (Table 3.21) show us that traditional class cohesion metrics are unable to identify Radicchi strong/weak classes with a high precision indicating that GCE metrics are much better choice for measuring class cohesion.

3.6.5 Analysis of Package and Class Cohesion

The number of Radicchi strong and Radicchi weak packages and classes in investigated software systems are shown in Table 3.22. It can be observed that an extremely small number of packages are Raddichi strong clusters in class collaboration networks. The number of Radicchi weak packages varies from 22 to 50% implying that a majority of packages are poorly cohesive in the sense that classes more frequently reference classes from other packages than from their own packages. Therefore, it

Table 3.22 The number of Radicchi strong (RS), Radicchi weak (RW) and poorly cohesive (PC) packages and classes in examined software systems. N denotes the total number of packages/classes

Level	System	N	RS		RW		PC	
Package	Tomcat	97	14	(14.43%)	33	(34.02%)	64	(65.98%)
	Lucene	16	1	(6.25%)	8	(50%)	8	(50%)
	Ant	71	5	(7.04%)	13	(18.31%)	58	(81.69%)
	Xerces	40	7	(17.5%)	18	(45%)	22	(55%)
	JFreeChart	37	4	(10.81%)	13	(35.14%)	24	(64.86%)
Class	Tomcat	1494	732	(49%)	1057	(70.75%)	437	(29.25%)
	Lucene	789	380	(48.16%)	596	(75.54%)	193	(24.46%)
	Ant	1175	543	(46.21%)	894	(76.09%)	281	(23.91%)
	Xerces	876	471	(53.77%)	653	(74.54%)	223	(25.46%)
	JFreeChart	624	269	(43.11%)	430	(68.91%)	194	(31.09%)

can be concluded that investigated software systems exhibit a poor cohesion at the package level.

The study by Šubelj and Bajec [88] showed, and several later studies confirmed, that class collaboration networks possess a strong community structure, i.e. classes can be grouped into cohesive clusters that are loosely coupled among themselves. Šubelj and Bajec also observed that clusters in class collaboration networks identified by various community detection methods weakly correspond to existing packages. Our results provide an explanation for their observation – classes from the same package mostly do not form a cluster in the class collaboration network. Consequently, community detection/graph clustering techniques applied to class collaboration networks are unable to reveal the package structure of corresponding software systems.

Investigated software systems possess considerably better cohesion at the class level. Approximately half of classes are Radicchi strong clusters in extended method collaboration networks which means that they exhibit strong cohesion. It can be observed that 25–30% classes do not satisfy the Radicchi weak clustering criterion. Such classes are candidates for refactorings aimed to improve system modularity and prevent potential negative consequences of low cohesion (lower readability and external reusability, unlocalized propagation of changes and increased maintenance efforts).

A low degree of package cohesiveness can be explained by a large number of classes that reference a higher number of classes from other packages than from their own packages. We use E to denote such classes. The percentage of E classes in examined software system varies from 21% in Lucene (one of two systems that have the lowest fraction of poorly cohesive packages) to 52% in Ant (the system with the highest fraction of poorly cohesive packages). We instrumented the metric-based comparison test to determine distinctive characteristics of E classes regarding their internal complexity, coupling and cohesion. The obtained results are shown in Table 3.23. It can be observed that

Table 3.23 The results of the metric-based comparison test for E classes (classes that reference a higher number of classes from other packages than from their own packages). $\overline{C_1}$ and $\overline{C_2}$ are the average values of the corresponding metric for E and non-E classes, respectively. PS_1 is the probability of superiority of E classes over non-E classes, while PS_2 is the opposite probability of superiority

Software	Metric	$\overline{C_1}$	$\overline{C_2}$	U	p	Null hyp.	PS_1	PS_2
Tomcat	LOC	305.30	150.97	341108	$<10^{-4}$	Rejected	0.66	0.34
	CC	30.39	15.43	357297	$<10^{-4}$	Rejected	0.65	0.26
	IN	3.26	5.33	301783	$<10^{-4}$	Rejected	0.31	0.48
	OUT	7.69	2.81	406964	$<10^{-4}$	Rejected	**0.76**	0.18
	NOC	0.3	0.31	267239	0.2383	**Accepted**	0.1	0.06
	DIT	0.68	0.38	294521	$<10^{-4}$	Rejected	0.32	0.18
	COND	0.48	0.2	388003	$<10^{-4}$	Rejected	0.71	0.2
	EXP	1.27	0.47	397011	$<10^{-4}$	Rejected	0.73	0.19
Lucene	LOC	132.17	134.88	57408	0.029	Rejected	0.55	0.44
	CC	30.75	13.87	55859	0.1117	**Accepted**	0.49	0.41
	IN	4.48	4.5	53114	0.5903	**Accepted**	0.42	0.39
	OUT	4.91	4.38	60404	0.0009	Rejected	0.54	0.37
	NOC	0.51	0.5	52565	0.7430	**Accepted**	0.17	0.16
	DIT	0.89	0.79	53199	0.5680	**Accepted**	0.34	0.31
	COND	0.34	0.25	61468	0.0002	Rejected	0.54	0.35
	FODF	0.67	0.75	58914	0.0058	Rejected	0.32	0.46
Ant	LOC	210.87	132.72	213650	$<10^{-4}$	Rejected	0.62	0.38
	CC	18.38	10.85	221011	$<10^{-4}$	Rejected	0.59	0.31
	IN	3.47	6.04	214127	$<10^{-4}$	Rejected	0.28	0.53
	OUT	6.47	2.77	263448	$<10^{-4}$	Rejected	0.73	0.2
	NOC	0.54	0.56	174392	0.7128	**Accepted**	0.12	0.11
	DIT	1.52	0.72	240600	$<10^{-4}$	Rejected	0.58	0.18
	COND	0.35	0.19	234103	$<10^{-4}$	Rejected	0.61	0.25
	FODF	0.65	0.81	223237	$<10^{-4}$	Rejected	0.24	0.54
Xerces	LOC	354.21	159.84	118498	$<10^{-4}$	Rejected	0.66	0.33
	CC	44.09	17.78	126358	$<10^{-4}$	Rejected	0.65	0.23
	IN	4.78	5.84	95167	0.0852	**Accepted**	0.45	0.38
	OUT	10.33	2.64	140350	$<10^{-4}$	Rejected	**0.75**	0.17
	NOC	0.51	0.34	94754	0.1081	**Accepted**	0.17	0.1
	DIT	0.59	1.26	98934	0.0057	Rejected	0.24	0.35
	COND	0.39	0.19	122494	$<10^{-4}$	Rejected	0.61	0.23
	FODF	0.61	0.8	120096	$<10^{-4}$	Rejected	0.21	0.56

(continued)

Table 3.23 (continued)

Software	Metric	$\overline{C_1}$	$\overline{C_2}$	U	p	Null hyp.	PS_1	PS_2
JFreeChart	LOC	462.62	169.00	70333	$<10^{-4}$	Rejected	0.73	0.27
	CC	30.89	10.15	67375	$<10^{-4}$	Rejected	0.66	0.26
	IN	5.2	5.12	51518	0.1183	**Accepted**	0.38	0.46
	OUT	8.64	2.4	75661	$<10^{-4}$	Rejected	**0.75**	0.17
	NOC	0.64	0.31	53435	0.0156	Rejected	0.18	0.07
	DIT	1.18	0.59	61814	$<10^{-4}$	Rejected	0.48	0.19
	COND	0.41	0.2	68978	$<10^{-4}$	Rejected	0.66	0.23
	FODF	0.55	0.77	66820	$<10^{-4}$	Rejected	0.23	0.62

- E classes tend to contain more lines of code than non-E classes. The average LOC of E classes in Lucene is close to the average LOC of non-E classes, but the probability of superiority of E classes regarding the LOC metric is higher than the opposite probability of superiority. Additionally, the difference in the probabilities of superiority is statistically significant. This means that Lucene contains a small fraction of extremely voluminous non-E classes that increase the average value of LOC.
- E classes in all systems except Lucene tend to have higher cyclomatic complexity than non-E classes.
- E classes do not tend to be more internally reused than non E classes. There are no statistically significant differences in the afferent coupling of E and non-E classes in Lucene, Xerces and JFreeChart. For other two systems, E classes tend to be significantly less internally reused compared to non-E classes.
- Contrary to the afferent coupling, E classes tend to have significantly higher efferent coupling than non-E classes in all examined systems. This tendency is very strong in Tomcat, Xerces and JFreeChart ($PS_1 \geq 0.75$).
- There are no statistically significant differences in the NOC of E classes and non-E classes in all systems except JFreeChart. E classes in JFreeChart tend to be more frequently extended than non-E classes. This can be considered as a bad phenomenon from the software engineering perspective since JFreeChart classes more often extend classes contributing to low cohesion at the package level than classes whose efferent coupling is localized within packages they belong to.
- E classes tend to have higher DIT than non-E classes in all systems except Lucene. This means that non-E classes tend to be more abstract than E classes, i.e. E classes are mostly located at the periphery of class inheritance trees.
- The null hypothesis of the MWU test is rejected in all examined systems with respect to COND and FODF. E classes tend to be exhibit higher COND and lower FODF which implies that they are less cohesive than non-E classes.

Observed trends imply that low package cohesion in investigated software systems is caused dominantly by poorly cohesive classes having high internal complexity, high efferent coupling and high degree of specialization.

Table 3.24 The results of the metric-based comparison test for E methods. $\overline{C}_1$ and $\overline{C}_2$ are the average values of the corresponding metric for E and non-E methods, respectively. PS_1 is the probability of superiority of E methods over non-E methods, while PS_2 is the opposite probability of superiority

Software	Metric	$\overline{C}_1$	$\overline{C}_2$	U	p	Null hyp.	PS_1	PS_2
Tomcat	LOC	31.17	8.27	20791802	$<10^{-4}$	Rejected	**0.78**	0.17
	CC	5.79	1.31	19686796	$<10^{-4}$	Rejected	0.66	0.14
	IN	1.02	1.41	12917021	0.905	**Accepted**	0.33	0.32
	OUT	5.45	0.77	22686908	$<10^{-4}$	Rejected	**0.83**	0.07
Lucene	LOC	21.53	8.65	2168326	$<10^{-4}$	Rejected	0.66	0.25
	CC	4.02	1.88	2073155	$<10^{-4}$	Rejected	0.54	0.19
	IN	0.92	1.23	1549747	0.773	**Accepted**	0.31	0.32
	OUT	4.21	0.88	2592680	$<10^{-4}$	Rejected	**0.78**	0.1
Ant	LOC	19.4	8.07	9538096	$<10^{-4}$	Rejected	0.71	0.2
	CC	3.36	1.23	8924480	$<10^{-4}$	Rejected	0.59	0.18
	IN	1.24	1.46	4667661	0.274	**Accepted**	0.36	0.35
	OUT	4.68	0.92	10921783	$<10^{-4}$	Rejected	**0.83**	0.09
Xerces	LOC	28.92	9.85	7686795	$<10^{-4}$	Rejected	0.67	0.24
	CC	6.04	1.74	7288895.5	$<10^{-4}$	Rejected	0.54	0.18
	IN	0.98	1.37	5462396	0.24	**Accepted**	0.32	0.3
	OUT	3.86	0.86	8617351.5	$<10^{-4}$	Rejected	0.73	0.12
JFreeChart	LOC	21.56	7.94	8161442	$<10^{-4}$	Rejected	0.74	0.2
	CC	3.14	1.02	6719588	$<10^{-4}$	Rejected	0.44	0.17
	IN	0.84	1.5	5557430	0.003	Rejected	0.27	0.32
	OUT	4.4	0.8	9148248	$<10^{-4}$	Rejected	**0.81**	0.08

As for classes, we use E to denote methods having a higher number of external dependencies (method calls and access to variables that are not localized within a class) than internal dependencies (dependencies that are localized within a class). If a software system contains more E methods than non-E methods then its cohesion at the class level will be inevitably poor. Lucene has the lowest fraction of E methods (9.64%), while the largest percentage of E methods is present in JFreeChart (16.54%). The results of the comparison between E and non-E methods regarding their internal complexity and coupling are shown in Table 3.24. E methods tend to have higher LOC and CC than non-E methods. The null hypothesis of the MWU test is accepted for afferent coupling in all systems except JFreeChart. E methods in JFreeChart are significantly less internally reused than non-E methods. Finally, a drastic difference between E and non-E methods can be observed regarding their efferent coupling, i.e. E methods strongly tend to call more methods than non-E methods. Therefore, poorly cohesive classes are dominantly caused by methods having high internal complexity and high efferent coupling.

The results of our analysis of package and class cohesion in 5 real-world, widely-used software systems suggest that high internal complexity of software entities and high efferent coupling are intrinsically linked to low cohesion. Large methods having a complex control flow and calling a large number of other methods cause poorly cohesive classes. Poorly cohesive classes that contain complex methods and reference a large number of other classes cause poor cohesion at the package level. Therefore, our empirical findings suggest that improving cohesion in software systems, besides leading to a more coherent, understandable and extensible modular structure, may also positively influence software quality attributes associated to code complexity (understandability and testability) and efferent coupling (error-proneness and external reusability).

3.7 Conclusions

Real-world software systems are characterized by complex structural dependencies between constituent software entities. In this chapter we proposed a novel, statistically robust, network-based methodology to examine coupling and cohesion in software systems. The main idea of the methodology is to enrich nodes of software networks with both software metrics and metrics used in complex network analysis in order to determine characteristics of highly coupled software entities, software entities involved in cyclic dependencies and software entities causing low cohesion.

Using the proposed methodology, we analyzed enriched package, class and method collaboration networks of five open-source software systems written in Java (Tomcat, Lucene, Ant, Xerces and JFreeChart). The analysis of basic network characteristics revealed that examined networks contain giant weakly connected components, exhibit the small-world property in the Watts-Strogatz sense and show weak to moderate disassortativity. On the other hand, patterns of strong connectivity vary across the levels of abstraction: package collaboration networks tend to contain giant strongly connected components, class collaboration networks contain a large number of strongly connected components displaying a large variability in size, while strongly connected components at the method level are almost absent. The path reciprocity in investigated class and package collaboration networks is significantly higher than the link reciprocity indicating that cyclic package and class dependencies are mostly indirect.

The application of the metric-based comparison test revealed that package collaboration networks exhibit a core-periphery structure with a strongly connected core encompassing the most coupled and functionally the most important packages. The analysis of strongly connected components in class collaboration networks showed that they tend to densify with size. The densification of strongly connected components at the class level can be modeled by a power-law whose scaling exponent can be used as a metric of software design quality with respect to the principle of avoiding cyclic dependencies. Strongly connected components in two investigated systems (Ant and JFreeChart) strongly tend to contain functionally the most important classes implying that refactorings aimed at reducing cyclic class dependencies may have a big impact to the structure and evolution of those two systems.

In contrast to previous studies analyzing degree distributions of software networks, we employed the power-law test introduced by Clauset et al. to examine whether studied networks possess the scale-free property. Additionally, we compared the best power-law fits to the best fits of log-normal and exponential distributions. The obtained results show that high coupling in real software systems cannot be accurately modeled by power-law distributions, further implying that the scaling exponent of fitted power-laws is not a reliable metric of software design quality. The log-normal model provides better fits to empirical distributions compared to power-law suggesting that the nearly-linear preferential attachment principle governs the evolution of software networks.

Due to heavy-tailed degree distributions, examined class and method collaboration networks contain hubs – classes and methods whose afferent, efferent and total coupling are considerably higher compared to the average values. We demonstrated that our methodology enables the identification of key differences between hubs and non-hubs. More specifically, our experimental results indicate that highly coupled classes/methods tend to be more voluminous, more internally complex and more functionally important than loosely coupled classes/methods. Additionally, highly coupled classes tend to be significantly less cohesive than loosely coupled classes and they do not tend to be localized in class inheritance hierarchies. The analysis of highly coupled classes and methods also showed that they tend to have significantly higher afferent than efferent coupling. Moreover, the extent of the domination of afferent over efferent coupling increases with total coupling in all examined systems except Xerces. This result implies that extremely highly coupled classes/methods in real software systems are caused dominantly by their extensive internal reuse. This further means that the presence of high coupling may indicate only negative aspects of extensive internal reuse (high criticality and context-required testability), not negative aspects of extensive internal aggregation (low understandability, high error-proneness and low external reusability).

Our methodology to analyze enriched software networks relies on graph clustering evaluation (GCE) metrics to quantify cohesion of packages and classes. The conducted empirical evaluation of GCE metrics showed that they do not exhibit significant correlations with traditional software cohesion metrics. Moreover, we demonstrated that traditional software cohesion metrics are unable to distinguish between highly and poorly cohesive classes with respect to the Radicchi notions of clusters in complex networks. A majority of packages in examined systems are not Radicchi weak clusters implying that investigated systems exhibit a poor cohesion at the package level. On the other hand, analyzed systems display considerably better cohesion at the class level. The application of the metric-based comparison test revealed that poorly cohesive packages (resp. classes) are dominantly caused by classes (resp. methods) having high internal complexity and high efferent coupling. Additionally, classes causing poor cohesion at the package level tend to be poorly cohesive and possess a high degree of specialization (i.e. they are mostly located at lower levels of class inheritance hierarchies).

Using the methodology proposed in this chapter we investigated the structure of enriched software networks. However, the proposed methodology can also be instrumented to study the evolution of enriched software networks. By applying the metric-based comparison test to evolutionary sequences of enriched software networks we are able to study evolutionary changes in characteristics of strongly connected components, highly coupled software entities and software entities causing low cohesion. Considering the evolution of software networks, software entities can be divided into two classes according to the following three binary classifiers:

1. An entity belongs to the positive class if it is referenced by one or more entities created in a subsequent software release, otherwise it belongs to the negative class.
2. If an entity references newly created entities then it belongs to the positive class, otherwise it belongs to the negative class.
3. An entity that is changed in a subsequent release is associated with the positive class, otherwise it is in the negative class.

The application of the metric-based comparison test with the first two classifiers can reveal how new software entities integrate into a software system, thus giving valuable insights related to the evolution of real-world software systems. On the other hand, the application of the metric-based comparison test with the third classifier can provide valuable insights related to the actual software maintenance practices.

References

1. Al-Mutawa, H.A., Dietrich, J., Marsland, S., McCartin, C.: On the shape of circular dependencies in java programs. In: Proceedings of the 2014 23rd Australian Software Engineering Conference, ASWEC '14, pp. 48–57. IEEE Computer Society, Washington, DC (2014). https://doi.org/10.1109/ASWEC.2014.15

2. Barabasi, A.L., Albert, R.: Emergence of scaling in random networks. Science **286**(5439), 509–512 (1999). https://doi.org/10.1126/science.286.5439.509
3. Basili, V., Briand, L., Melo, W.: A validation of object-oriented design metrics as quality indicators. IEEE Trans Softw Eng **22**(10), 751–761 (1996). https://doi.org/10.1109/32.544352
4. Baxter, G., Frean, M., Noble, J., Rickerby, M., Smith, H., Visser, M., Melton, H., Tempero, E.: Understanding the shape of java software. In: Proceedings of the 21st Annual ACM SIGPLAN Conference on Object-oriented Programming Systems, Languages, and Applications, OOPSLA '06, pp. 397–412. ACM, New York (2006). https://doi.org/10.1145/1167473.1167507
5. Bhattacharya, P., Iliofotou, M., Neamtiu, I., Faloutsos, M.: Graph-based analysis and prediction for software evolution. In: Proceedings of the 34th International Conference on Software Engineering, ICSE '12, pp. 419–429. IEEE Press, Piscataway (2012)
6. Bieman, J.M., Kang, B.K.: Cohesion and reuse in an object-oriented system. In: Proceedings of the 1995 Symposium on Software Reusability, SSR '95, pp. 259–262. ACM, New York (1995)
7. Bollobás, B.: Random Graphs. Cambridge University Press, Cambridge (2001)
8. Briand, L.C., Daly, J.W., Wüst, J.: A unified framework for cohesion measurement in object-oriented systems. Empir. Softw. Eng. **3**(1), 65–117 (1998). https://doi.org/10.1023/A:1009783721306
9. Briand, L.C., Daly, J.W., Wüst, J.K.: A unified framework for coupling measurement in object-oriented systems. IEEE Trans. Softw. Eng. **25**(1), 91–121 (1999)
10. Brin, S., Page, L.: The anatomy of a large-scale hypertextual Web search engine. Comput. Netw. ISDN Syst. **30**(1–7), 107–117 (1998). https://doi.org/10.1016/S0169-7552(98)00110-X
11. Cai, K.Y., Yin, B.B.: Software execution processes as an evolving complex network. Inf. Sci. **179**(12), 1903–1928 (2009). https://doi.org/10.1016/j.ins.2009.01.011
12. Chaikalis, T., Chatzigeorgiou, A.: Forecasting Java software evolution trends employing network models. IEEE Trans. Softw. Eng. **41**(6), 582–602 (2015). https://doi.org/10.1109/TSE.2014.2381249
13. Chatzigeorgiou, A., Melas, G.: Trends in object-oriented software evolution: investigating network properties. In: Proceedings of the 34th International Conference on Software Engineering, ICSE '12, pp. 1309–1312. IEEE Press, Piscataway (2012). https://doi.org/10.1109/ICSE.2012.6227092
14. Chatzigeorgiou, A., Tsantalis, N., Stephanides, G.: Application of graph theory to OO software engineering. In: Proceedings of the 2006 International Workshop on Workshop on Interdisciplinary Software Engineering Research, WISER '06, pp. 29–36. ACM, New York (2006). https://doi.org/10.1145/1137661.1137669
15. Chidamber, S.R., Kemerer, C.F.: A metrics suite for object oriented design. IEEE Trans. Softw. Eng. **20**(6), 476–493 (1994). https://doi.org/10.1109/32.295895
16. Chong, C.Y., Lee, S.P.: Analyzing maintainability and reliability of object-oriented software using weighted complex network. J. Syst. Softw. **110**(C), 28–53 (2015). https://doi.org/10.1016/j.jss.2015.08.014
17. Clauset, A., Shalizi, C., Newman, M.: Power-law distributions in empirical data. SIAM Rev. **51**(4), 661–703 (2009). https://doi.org/10.1137/070710111
18. Concas, G., Marchesi, M., Murgia, A., Tonelli, R.: An empirical study of social networks metrics in object-oriented software. Adv. Softw. Eng. **2010**, 4:1–4:21 (2010). https://doi.org/10.1155/2010/729826
19. Concas, G., Marchesi, M., Pinna, S., Serra, N.: Power-laws in a large object-oriented software system. IEEE Trans. Softw. Eng. **33**(10), 687–708 (2007). https://doi.org/10.1109/TSE.2007.1019
20. Concas, G., Monni, C., Orr, M., Tonelli, R.: A study of the community structure of a complex software network. In: 2013 4th International Workshop on Emerging Trends in Software Metrics (WETSoM), pp. 14–20 (2013). https://doi.org/10.1109/WETSoM.2013.6619331
21. Eder, J., Kappel, G., Schrefl, M.: Coupling and cohesion in object-oriented systems, Technical report. University of Klagenfurt (1992)

22. Erceg-Hurn, D.M., Mirosevich, V.M.: Modern robust statistical methods: an easy way to maximize the accuracy and power of your research. Am. Psychol. **63**(7), 591–601 (2008). https://doi.org/10.1037/0003-066X.63.7.591
23. Fortuna, M.A., Bonachela, J.A., Levin, S.A.: Evolution of a modular software network. Proc. Natl. Acad. Sci. **108**(50), 19985–19989 (2011). https://doi.org/10.1073/pnas.1115960108
24. Fowler, M.: Reducing coupling. IEEE Softw. **18**(4), 102–104 (2001). https://doi.org/10.1109/MS.2001.936226
25. Freeman, L.C.: A set of measures of centrality based on betweenness. Sociometry **40**, 35–41 (1977). https://doi.org/10.2307/3033543
26. Gao, Y., Zheng, Z., Qin, F.: Analysis of Linux kernel as a complex network. Chaos Solitons Fractals **69**, 246–252 (2014). https://doi.org/10.1016/j.chaos.2014.10.008
27. Garlaschelli, D., Loffredo, M.: Patterns of link reciprocity in directed networks. Phys. Rev. Lett. **93**, 268,701 (2004). https://doi.org/10.1103/PhysRevLett.93.268701
28. Gillespie, C.: Fitting heavy tailed distributions: the power law package. J. Stat. Softw. **64**(2), 1–16 (2015). https://doi.org/10.18637/jss.v064.i02
29. Gyimothy, T., Ferenc, R., Siket, I.: Empirical validation of object-oriented metrics on open source software for fault prediction. IEEE Trans. Softw. Eng. **31**(10), 897–910 (2005). https://doi.org/10.1109/TSE.2005.112
30. Hamilton, J., Danicic, S.: Dependence communities in source code. In: 28th IEEE International Conference on Software Maintenance (ICSM), pp. 579–582 (2012). https://doi.org/10.1109/ICSM.2012.6405325
31. Harrison, R., Counsell, S., Nithi, R.: Coupling metrics for object-oriented design. In: Proceedings of Fifth International Software Metrics Symposium (Metrics 1998), pp. 150–157 (1998). https://doi.org/10.1109/METRIC.1998.731240
32. Hitz, M., Montazeri, B.: Measuring coupling and cohesion in object-oriented systems. In: Proceedings of the International Symposium on Applied Corporate Computing, pp. 25–27 (1995)
33. Hylland-Wood, D., Carrington, D., Kaplan, S.: Scale-free nature of Java software package, class and method collaboration graphs. Technical Report. TR-MS1286, MIND Laboratory, University of Maryland, College Park (2006)
34. Ichii, M., Matsushita, M., Inoue, K.: An exploration of power-law in use-relation of Java software systems. In: Proceedings of the 19th Australian Conference on Software Engineering, ASWEC '08, pp. 422–431. IEEE Computer Society, Washington, DC (2008). https://doi.org/10.1109/ASWEC.2008.4483231
35. Jenkins, S., Kirk, S.R.: Software architecture graphs as complex networks: a novel partitioning scheme to measure stability and evolution. Inf. Sci. **177**, 2587–2601 (2007). https://doi.org/10.1016/j.ins.2007.01.021
36. Jing, L., Keqing, H., Yutao, M., Rong, P.: Scale free in software metrics. In: 30th Annual International Computer Software and Applications Conference (COMPSAC'06), vol. 1, pp. 229–235 (2006). https://doi.org/10.1109/COMPSAC.2006.75
37. Kleinberg, J.M.: Authoritative sources in a hyperlinked environment. J. ACM **46**(5), 604–632 (1999). https://doi.org/10.1145/324133.324140
38. Kohring, G.A.: Complex dependencies in large software systems. Adv. Complex Syst. **12**(06), 565–581 (2009). https://doi.org/10.1142/S0219525909002362
39. Labelle, N., Wallingford, E.: Inter-package dependency networks in open-source software. In: Proceedings of the 6th International Conference on Complex Systems (ICCS), paper no. 226 (2006)
40. Laval, J., Falleri, J., Vismara, P., Ducasse, S.: Efficient retrieval and ranking of undesired package cycles in large software systems. J. Object Technol. **11**(1), 1–24 (2012). https://doi.org/10.5381/jot.2012.11.1.a4
41. Leskovec, J., Kleinberg, J., Faloutsos, C.: Graph evolution: Densification and shrinking diameters. ACM Trans. Knowl. Discov. Data (TKDD) **1**(1) (2007). https://doi.org/10.1145/1217299.1217301

42. Li, H., Huang, B., Lu, J.: Dynamical evolution analysis of the object-oriented software systems. In: 2008 IEEE Congress on Evolutionary Computation (IEEE World Congress on Computational Intelligence), pp. 3030–3035 (2008). https://doi.org/10.1109/CEC.2008.4631207
43. Li, H., Zhao, H., Cai, W., Xu, J.Q., Ai, J.: A modular attachment mechanism for software network evolution. Phys. A Stat. Mech. Appl. **392**(9), 2025–2037 (2013). https://doi.org/10.1016/j.physa.2013.01.035
44. Li, L., Alderson, D., Doyle, J.C., Willinger, W.: Towards a theory of scale-free graphs: definition, properties, and implications. Int. Math. **2**(4), 431–523 (2005). https://doi.org/10.1080/15427951.2005.10129111
45. Louridas, P., Spinellis, D., Vlachos, V.: Power laws in software. ACM Trans. Softw. Eng. Methodol. **18**(1), 2:1–2:26 (2008). https://doi.org/10.1145/1391984.1391986
46. Ma, Y.T., He, K.Q., Li, B., Liu, J., Zhou, X.Y.: A hybrid set of complexity metrics for large-scale object-oriented software systems. J. Comput. Sci. Technol. **25**(6), 1184–1201 (2010). https://doi.org/10.1007/s11390-010-9398-x
47. Maillart, T., Sornette, D., Spaeth, S., von Krogh, G.: Empirical tests of Zipf's law mechanism in open source Linux distribution. Phys. Rev. Lett. **101**, 218,701 (2008). https://doi.org/10.1103/PhysRevLett.101.218701
48. Mann, H.B., Whitney, D.R.: On a test of whether one of two random variables is stochastically larger than the other. Ann. Math. Stat. **18**(1), 50–60 (1947). https://doi.org/10.2307/2236101
49. Melton, H., Tempero, E.: An empirical study of cycles among classes in Java. Empir. Softw. Eng. **12**(4), 389–415 (2007). https://doi.org/10.1007/s10664-006-9033-1
50. de Moura, A.P.S., Lai, Y.C., Motter, A.E.: Signatures of small-world and scale-free properties in large computer programs. Phys. Rev. E **68**(1), 017,102 (2003). https://doi.org/10.1103/PhysRevE.68.017102
51. Myers, C.R.: Software systems as complex networks: structure, function, and evolvability of software collaboration graphs. Phys. Rev. E **68**(4), 046,116 (2003). https://doi.org/10.1103/PhysRevE.68.046116
52. Newman, M.E.J.: Assortative mixing in networks. Phys. Rev. Lett. **89**, 208,701 (2002). https://doi.org/10.1103/PhysRevLett.89.208701
53. Oyetoyan, T.D., Cruzes, D.S., Conradi, R.: A study of cyclic dependencies on defect profile of software components. J. Syst. Softw. **86**(12), 3162–3182 (2013). https://doi.org/10.1016/j.jss.2013.07.039
54. Pan, W., Li, B., Ma, Y., Liu, J.: Multi-granularity evolution analysis of software using complex network theory. J. Syst. Sci. Complex. **24**(6), 1068–1082 (2011). https://doi.org/10.1007/s11424-011-0319-z
55. Parnas, D.L.: Designing software for ease of extension and contraction. IEEE Trans. Softw. Eng. **SE-5**(2), 128–138 (1979). https://doi.org/10.1109/TSE.1979.234169
56. Paymal, P., Patil, R., Bhomwick, S., Siy, H.: Empirical study of software evolution using community detection. Technical Report, Department of Computer Science, University of Nebraska, Omaha, USA (2011)
57. Potanin, A., Noble, J., Frean, M., Biddle, R.: Scale-free geometry in OO programs. Commun. ACM **48**, 99–103 (2005). https://doi.org/10.1145/1060710.1060716
58. Puppin, D., Silvestri, F.: The social network of Java classes. In: Proceedings of the 2006 ACM Symposium on Applied Computing, SAC '06, pp. 1409–1413. ACM, New York (2006). https://doi.org/10.1145/1141277.1141605
59. Qu, Y., Guan, X., Zheng, Q., Liu, T., Wang, L., Hou, Y., Yang, Z.: Exploring community structure of software call graph and its applications in class cohesion measurement. J. Syst. Softw. **108**, 193–210 (2015). https://doi.org/10.1016/j.jss.2015.06.015
60. Qu, Y., Guan, X., Zheng, Q., Liu, T., Zhou, J., Li, J.: Calling network: a new method for modeling software runtime behaviors. ACM SIGSOFT Softw. Eng. Notes **40**(1), 1–8 (2015). https://doi.org/10.1145/2693208.2693223
61. Radicchi, F., Castellano, C., Cecconi, F., Loreto, V., Parisi, D.: Defining and identifying communities in networks. Proc. Natl. Acad. Sci. **101**(9), 2658–2663 (2004). https://doi.org/10.1073/pnas.0400054101

62. Radjenović, D., Heričko, M., Torkar, R., Živkovič, A.: Software fault prediction metrics. Inf. Softw. Technol. **55**(8), 1397–1418 (2013). https://doi.org/10.1016/j.infsof.2013.02.009
63. Rakić, G.: Extendable and adaptable framework for input language independent static analysis. Ph.D. thesis, University of Novi Sad, Faculty of Sciences (2015)
64. Rakić, G., Budimac, Z.: Introducing enriched concrete syntax trees. In: Proceedings of the 14th International Multiconference on Information Society (IS), Collaboration, Software And Services In Information Society (CSS), pp. 211–214 (2011)
65. Redner, S.: Citation Statistics from 110 Years of Physical Review. Phys. Today **58**(6), 49–54 (2005). https://doi.org/10.1063/1.1996475
66. Savić, M., Ivanović, M.: Graph clustering evaluation metrics as software metrics. In: Proceedings of the 3rd Workshop on Software Quality Analysis, Monitoring, Improvement and Applications (SQAMIA 2014), Lovran, Croatia, September 19–22, 2014, vol. 1266, pp. 81–89 (2014). http://CEUR-WS.org
67. Savić, M., Ivanović, M.: Validation of static program analysis tools by self-application: a case study. In: Proceedings of the 4th Workshop on Software Quality Analysis, Monitoring, Improvement and Applications (SQAMIA 2015), Maribor, Slovenia, June 8–10, 2015, vol. 1375, pp. 61–68 (2015). http://CEUR-WS.org
68. Savić, M., Ivanović, M., Radovanović, M.: Characteristics of class collaboration networks in large Java software projects. Inf. Technol. Control **40**(1), 48–58 (2011). https://doi.org/10.5755/j01.itc.40.1.192
69. Savić, M., Ivanović, M., Radovanović, M.: Connectivity properties of the Apache Ant class collaboration network. In: 15th International Conference on System Theory, Control and Computing, pp. 544–549 (2011). http://ieeexplore.ieee.org/document/6085650/
70. Savić, M., Ivanović, M., Radovanović, M.: Analysis of high structural class coupling in object-oriented software systems. Computing **99**(11), 1055–1079 (2017). https://doi.org/10.1007/s00607-017-0549-6
71. Savić, M., Radovanović, M., Ivanović, M.: Community detection and analysis of community evolution in Apache Ant class collaboration networks. In: Proceedings of the Fifth Balkan Conference in Informatics, BCI '12, pp. 229–234. ACM, New York (2012). https://doi.org/10.1145/2371316.2371361
72. Savić, M., Rakić, G., Budimac, Z.: Translation of Tempura specifications to eCST. AIP Conf. Proc. **1738**(1), 240,009 (2016). https://doi.org/10.1063/1.4952028
73. Savić, M., Rakić, G., Budimac, Z., Ivanović, M.: Extractor of software networks from enriched concrete syntax trees. AIP Conf. Proc. **1479**(1), 486–489 (2012). https://doi.org/10.1063/1.4756172
74. Savić, M., Rakić, G., Budimac, Z., Ivanović, M.: A language-independent approach to the extraction of dependencies between source code entities. Inf. Softw. Technol. **56**(10), 1268–1288 (2014). https://doi.org/10.1016/j.infsof.2014.04.011
75. Sora, I.: A PageRank based recommender system for identifying key classes in software systems. In: 10th IEEE Jubilee International Symposium on Applied Computational Intelligence and Informatics, pp. 495–500 (2015). https://doi.org/10.1109/SACI.2015.7208254
76. Steidl, D., Hummel, B., Juergens, E.: Using network analysis for recommendation of central software classes. In: 19th Working Conference on Reverse Engineering, pp. 93–102 (2012). https://doi.org/10.1109/WCRE.2012.19
77. Stumpf, M.P.H., Porter, M.A.: Critical truths about power laws. Science **335**(6069), 665–666 (2012). https://doi.org/10.1126/science.1216142
78. Subramanyam, R., Krishnan, M.: Empirical analysis of ck metrics for object-oriented design complexity: implications for software defects. IEEE Trans. Softw. Eng. **29**(4), 297–310 (2003). https://doi.org/10.1109/TSE.2003.1191795
79. Sudeikat, J., Renz, W.: On complex networks in software: how agentorientation effects software structures. In: H.D. Burkhard, G. Lindemann, R. Verbrugge, L.Z. Varga (eds.) Multi-Agent Systems and Applications V. Lecture Notes in Computer Science, vol. 4696, pp. 215–224. Springer, Berlin (2007). https://doi.org/10.1007/978-3-540-75254-7_22

80. Sun, S., Xia, C., Chen, Z., Sun, J., Wang, L.: On structural properties of large-scale software systems: from the perspective of complex networks. In: 2009 Sixth International Conference on Fuzzy Systems and Knowledge Discovery, vol. 7, pp. 309–313 (2009). https://doi.org/10.1109/FSKD.2009.635

81. Tarjan, R.: Depth-first search and linear graph algorithms. SIAM J. Comput. **1**(2), 146–160 (1972). https://doi.org/10.1137/0201010

82. Taube-Schock, C., Walker, R.J., Witten, I.H.: Can we avoid high coupling? In: M. Mezini (ed.) ECOOP 2011 Object-Oriented Programming. Lecture Notes in Computer Science, vol. 6813, pp. 204–228. Springer, Berlin (2011). https://doi.org/10.1007/978-3-642-22655-7_10

83. Turnu, I., Marchesi, M., Tonelli, R.: Entropy of the degree distribution and object-oriented software quality. In: 3rd International Workshop on Emerging Trends in Software Metrics (WETSoM), pp. 77–82 (2012). https://doi.org/10.1109/WETSoM.2012.6226997

84. Valverde, S., Cancho, R.F., Solé, R.V.: Scale-free networks from optimal design. EPL (Europhys. Lett.) **60**(4), 512–517 (2002). https://doi.org/10.1209/epl/i2002-00248-2

85. Valverde, S., Solé, V.: Hierarchical small worlds in software architecure. Dyn. Contin. Discret. Impuls. Syst. Ser. B Appl. Algorithms **14(S6)**, 305–315 (2007)

86. Vasa, R., Schneider, J.G., Nierstrasz, O.: The inevitable stability of software change. In: 2007 IEEE International Conference on Software Maintenance, pp. 4–13 (2007). https://doi.org/10.1109/ICSM.2007.4362613

87. Vasa, R., Schneider, J.G., Woodward, C., Cain, A.: Detecting structural changes in object oriented software systems. In: 2005 International Symposium on Empirical Software Engineering, pp. 8 pp.– (2005). 10.1109/ISESE.2005.1541855

88. Šubelj, L., Bajec, M.: Community structure of complex software systems: analysis and applications. Phys. A Stat. Mech. Appl. **390**(16), 2968–2975 (2011). https://doi.org/10.1016/j.physa.2011.03.036

89. Wang, L., Wang, Z., Yang, C., Zhang, L., Ye, Q.: Linux kernels as complex networks: a novel method to study evolution. IEEE International Conference on Software Maintenance (ICSM 2009) pp. 41–50 (2009). https://doi.org/10.1109/ICSM.2009.5306348

90. Wang, L., Yu, P., Wang, Z., Yang, C., Ye, Q.: On the evolution of Linux kernels: a complex network perspective. J. Softw. Evol. Process **25**(5), 439–458 (2013). https://doi.org/10.1002/smr.1550

91. Watts, D.J., Strogatz, S.H.: Collective dynamics of "small-world" networks. Nature **393**, 440–442 (1998). https://doi.org/10.1038/30918

92. Wen, H., DSouza, R.M., Saul, Z.M., Filkov, V.: Evolution of apache open source software. In: N. Ganguly, A. Deutsch, A. Mukherjee (eds.) Dynamics On and Of Complex Networks, Modeling and Simulation in Science, Engineering and Technology, pp. 199–215. Birkhuser Boston (2009). https://doi.org/10.1007/978-0-8176-4751-3_12

93. Wen, L., Dromey, R.G., Kirk, D.: Software engineering and scale-free networks. IEEE Trans. Syst. Man Cybern. Part B Cybern. **39**, 845–854 (2009). https://doi.org/10.1109/TSMCB.2009.2020206

94. Wheeldon, R., Counsell, S.: Power law distributions in class relationships. In: Proceedings of the Third IEEE International Workshop on Source Code Analysis and Manipulation, pp. 45–54 (2003). https://doi.org/10.1109/SCAM.2003.1238030

95. Yourdon, E., Constantine, L.L.: Structured Design: Fundamentals of a Discipline of Computer Program and Systems Design, 1st edn. Prentice-Hall Inc, Upper Saddle River, NJ, USA (1979)

96. Yuan, P., Jin, H., Deng, K., Chen, Q.: Analyzing software component graphs of grid middleware: hint to performance improvement. In: Proceedings of the 8th Internationsl Conference on Algorithms and Architectures for Parallel Processing (ICA3PP), pp. 305–315 (2008). https://doi.org/10.1007/978-3-540-69501-1_32

97. Zaidman, A., Demeyer, S.: Automatic identification of key classes in a software system using webmining techniques. J. Softw. Maint. Evol. Res. Pract. **20**(6), 387–417 (2008). https://doi.org/10.1002/smr.370

98. Zanetti, M.S., Schweitzer, F.: A network perspective on software modularity. In: ARCS 2012 Workshops, pp. 175–186 (2012)

99. Zanetti, M.S., Tessone, C.J., Scholtes, I., Schweitzer, F.: Automated software remodularization based on move refactoring: a complex systems approach. In: Proceedings of the 13th International Conference on Modularity, MODULARITY '14, pp. 73–84. ACM, New York (2014). https://doi.org/10.1145/2577080.2577097
100. Zheng, X., Zeng, D., Li, H., Wang, F.: Analyzing open-source software systems as complex networks. Phys. A. Stat. Mech. Appl. **387**(24), 6190–6200 (2008). https://doi.org/10.1016/j.physa.2008.06.050

Chapter 4
Analysis of Ontology Networks

Abstract In computer and information sciences, an ontology is, in its essence, a named set of axioms encoding a network of relationships and dependencies between ontological entities present in a knowledge domain. With the rise of Semantic Web technologies, real-world ontologies have become considerably large leading to complex ontology networks. In this chapter we firstly present an overview of previous research works dealing with analysis of ontology networks. Nodes of ontology networks can be enriched with various metrics reflecting complexity and quality attributes of corresponding ontological entities. On a case study involving one large-scale modularized ontology, we demonstrate that analysis of enriched ontology networks can help ontology engineers not only to understand the structural complexity of ontologies, but also to evaluate their quality with respect to well-established modular design principles.

Ontology networks are directed graphs showing relationships and dependencies between ontological entities – ontologies and terms defined within ontologies such as classes (concepts) and properties (roles). The meaning of an ontological entity is dependent on the meanings of those ontological entities used in its definition. A change in the meaning of an ontological entity may affect the meanings of those entities that are directly or indirectly dependent on it [7]. Therefore, coupling between ontological entities is among the most fundamental issues related to ontology development, maintenance and evaluation. Analysis of ontology networks can help us to understand and quantify the complexity of coupling structures among ontological entities, thus providing a valuable instrument for ontology evaluation. For example, a large number of isolated nodes or the absence of a giant weakly connected component in an ontology network implies that the corresponding ontology defines poorly inter-linked ontological terms. Such fragmented ontology networks indicate either non-fully operational ontologies that are in an early phase of development or large ontologies aggregating ontological terms from unrelated knowledge domains. Consequently, analysis of weakly connected components in ontology networks may be very beneficial for ontology engineers when deciding whether to reuse an ontology, or when selecting an ontology from a set of candidate ontologies describing the same knowledge domain.

There are two principal design approaches for creating an ontology describing a complex knowledge domain. The first one is to gather all captured ontological entities and axioms into a monolithic ontology. The content of a monolithic ontology is given in a single document, e.g. one OWL file in the case of semantic web ontologies described using the OWL language. Maintenance, scalability and reusability problems associated with large-scale monolithic semantic web ontologies have lead to an increasing research interest in ontology modularization [12, 14, 46]. Modularized ontologies bring many benefits including easier ontology understanding and maintenance, more efficient reasoning and reuse, and enhanced collaborative ontology development. In the modular approach to ontology engineering, a knowledge domain D is described by a set of related ontologies called ontology modules, where each module corresponds to a sub-domain of D. Modularization of semantic web ontologies is enabled by the import construct of the OWL language [4, 11]: an ontology A by importing an ontology B gains access to all ontological entities and axioms present in B. The coupling of an ontology (resp. ontology module) is the degree of its interdependence with other ontologies (resp. ontology modules). Ontology cohesion refers to the degree of relatedness of ontological entities defined within an ontology or ontology module. An ontology module should be logically self-contained which means that it should be loosely coupled to other ontology modules and, at the same time, it should exhibit a high degree of cohesion. Therefore, the principle of "low coupling and high cohesion" emphasizes a good modularization of a knowledge domain into ontology modules [15, 35].

The principle of avoiding cyclic dependencies is another important design principle in software engineering. This principle can also be applied to modular ontology designs. Large and dense strongly connected components in ontology networks are indicators of poor ontology modularization since large cyclic dependencies negatively impact the following quality attributes:

1. understandability – to understand an ontology module belonging to a strongly connected component S an ontology engineer effectively has to examine all modules from S.
2. reusability – reuse of an ontology module from a strongly connected component S forces reuse of all modules from S.
3. maintainability – changes in an ontology module from a strongly connected component S may affect all modules from S. In the most drastic case, changes that cause inconsistencies in one module from S make all modules from S inconsistent.

In the last ten years, a growing number of ontology metrics have been suggested for ontology quality evaluation [18, 45, 50]. Nodes of ontology networks can be enriched with ontology metrics in order to enhance analysis of coupling and cohesion of ontological entities. Such ontology networks we call *enriched ontology networks*. After reviewing previous research works dealing with analysis of "pure" ontology networks (Sect. 4.2), we present the results of analysis of enriched ontology networks representing a large-scale modularized ontology. We followed basically the same methodology we proposed for studying the structure of enriched software networks (Chap. 3). Section 4.3 gives necessary methodological remarks and explains our case study. The obtained results are presented and discussed in Sect. 4.4. The last section summarizes our main findings.

4.1 Preliminaries and Definitions

The Web Ontology Language (OWL) is an ontology language for semantic web applications that is designed and recommended by W3C (World Wide Web Consortium). The latest version of the language is called OWL2 and dates from December 2012. OWL extends capabilities of the Resource Description Framework (RDF) which is a language for describing web accessible resources and relationships between them. OWL is a knowledge representation language based on description logic. This means that knowledge expressed in OWL is amenable to decision procedures that can be used to check its consistency and derive implicit knowledge.

OWL ontologies and ontological entities are identified using Internationalized Resource Identifiers (IRIs) which have a global scope. Ontological entities can be classified as:

- Classes (concepts) - sets of objects,
- Objects (individuals) - instances of classes,
- Data types - sets of data values (literals) such as strings and numbers,
- Object properties (roles) - relations between objects,
- Data properties (attributes) - relations between objects and literals,
- Annotation properties - relations between ontological entities and annotation values (IRIs, literals and anonymous objects – objects that do not have IRIs).

OWL ontologies besides import statements and annotations (human-readable descriptions/comments) contain axioms. OWL axioms can be divided into the following categories:

- Declaration axioms state the existence of ontological entities and associate them to particular types.
- Class axioms describe relations (equivalence, subsumption and disjointness) between classes and/or class expressions. A class expression (also known as a complex concept, class description, anonymous class or class constructor) describes a set of objects using set operations, enumerations and restrictions relying on existing classes, objects and properties.
- Object property axioms describe relations between object properties, relations between object properties and classes, or characterize object properties (specify whether an object property is reflexive, symmetric, transitive or functional). An object property is functional if it connects an object to at most one other object.

- Data property axioms describe relations between data properties, relations between data properties and classes, relations between data properties and data types or characterize data properties as functional.
- Assertion axioms, also called *facts*, describe relations between objects (e.g. same objects, different objects), state that two objects are associated by an object property, associate objects and literals by data properties, or state that an object is an instance of a particular class.
- Annotation axioms describe relations between annotation properties or associations between annotation properties and other ontological entities.
- Data type definition axioms map data types to data ranges.

OWL ontologies can be represented by directed graphs that show explicitly stated relations between entities appearing in OWL axioms. An ontology graph describing a modularized OWL ontology can be defined as follows.

Definition 4.1 (*Ontology graph*) An ontology graph G representing a modularized OWL ontology M is a directed, typed graph in which both nodes and links have types. The nodes of G represent different types of ontological entities: ontologies, classes, objects, object properties, annotation properties, data properties, data types and literals. The links present in G belong to the following categories:

- SUB links representing subsumption relations between ontological entities of the same type. SUB links may connect classes (as stated by SubClassOf axioms), object properties (as stated by SubObjectPropertyOf axioms), data properties (as stated by SubDataPropertyOf axioms) and annotation properties (as stated by SubAnnotationPropertyOf axioms).
- ASSERTION links representing associations between ontological entities induced by assertion axioms.
- EQUIVALENT links denote that two ontological entities are equivalent. Those links can be established between two classes (as stated by EquivalentClasses axioms), two objects (as stated by SameIndividuals axioms) and two object properties (as stated by EquivalentObjectProperties axioms). EQUIVALENT links are reciprocal, i.e. if a points to b by an EQUIVALENT link then b also points to a by another EQUIVALENT link.
- DISJOINT links denote that two ontological entities are not equivalent. Similarly to EQUIVALENT links, DISJOINT links connect two classes, two objects or two object properties.
- REFERENCES links connect anonymous classes with named classes. Let e be a complex class expression. Then e is represented by an anonymous class node A in G. A is connected via REFERENCES links to all named classes appearing in e.
- CONTAINS links associate ontologies or ontology modules to other ontological entities. A CONTAINS link $O \rightarrow e$ states that an ontological entity e is part of an ontology or ontology module O.
- IMPORTS links denote dependencies between ontology modules. Two ontologies or ontology modules A and B are connected by an IMPORT link $A \rightarrow B$ if A imports B.

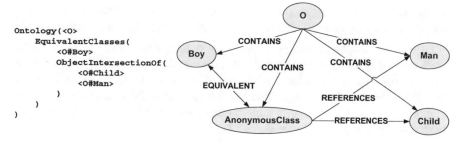

```
Ontology(<O>
    EquivalentClasses(
        <O#Boy>
        ObjectIntersectionOf(
            <O#Child>
            <O#Man>
        )
    )
)
```

Fig. 4.1 The ontology graph of a simple ontology containing an axiom with a class expression

- Links representing user-defined associations between classes determined by relevant pairs of ObjectPropertyDomain and ObjectPropertyRange axioms. This category of links also includes user-defined association stated by assertion axioms involving object, data and annotation properties.

As pointed out, class expressions are also represented by nodes in ontology graphs. An example is given in Fig. 4.1. The ontology O contains one axiom which states that two classes are equivalent. This axiom has the class expression as the second argument (ObjectIntersectionOf) that is represented by the "AnonymousClass" node in the ontology graph. "AnoymousClass" references all named classes that appear in the class expression (classes Child and Man). Therefore, there is an indirect coupling between the first argument of the axiom (class Boy) and the classes that are involved in the class expression. The ObjectIntersetionOf class expression is contained in the ontology O and consequently there is the CONTAINS link between O and the node representing the class expression.

A variety of ontological networks can be induced from the ontology graph representation.

Definition 4.2 (*Ontology module network, OMN*) The ontology module network of a modularized OWL ontology O is a subgraph of the ontology graph of O that contains all nodes representing ontologies (modules) and existing IMPORTS links.

Definition 4.3 (*Ontology class network, OCN*) The ontology class network of a (modularized) OWL ontology O is a subgraph of the ontology graph of O induced by nodes representing classes (including anonymous classes too) without restrictions on link types. In other words, the ontology class network of O shows dependencies between concepts present in the knowledge domain described by O.

Definition 4.4 (*Ontology subsumption network, OSN*) The ontology subsumption network of a (modularized) OWL ontology O is a subgraph of the ontology graph of O induced by classes (including anonymous classes too) and SUB links. This network represents the taxonomy of concepts present in O.

Definition 4.5 (*Ontology object network, OON*) The ontology object network of a (modularized) OWL ontology O is a subgraph of the ontology graph of O induced by objects without restrictions on link types. In other words, this network shows associations among concrete objects from the knowledge domain described by O.

There is one important difference between ontology object networks and other above defined types of ontology networks. An ontology object network describes facts – a particular state of affairs in a knowledge domain. Other ontology networks are oriented towards knowledge domains themselves: OCNs and OSNs depict the structure of conceptualizations within knowledge domains, while OMNs describe the modularization of knowledge domains.

4.2 Related Work

The first study dealing with analysis of ontology networks under the framework of complex network theory was conducted by Gil et al. [22]. The authors extracted a complex RDF network from 282 ontologies contained in the DAML ontology library and then analyzed its structure using the Pajek tool [33]. The obtained results show that the network exhibits the small-world property in the Watts-Strogatz sense:

1. the average path length of the network is equal to 4.37 and it is lower than the average path length of comparable random graphs, and
2. the clustering coefficient of the network is three orders of magnitude higher than the clustering coefficient of comparable random graphs.

Gil et al. also investigated whether the degree distribution of the network follows a power-law by applying linear regression to the log-log plot of the complementary cumulative degree distribution. The authors obtained a small regression error and concluded that the analyzed network exhibits the scale-free property.

Various benefits of applying centrality measures to ontology networks were demonstrated by Hoser et al. [28], Alani et al. [1–3], Zhang et al. [53] and Queiroz-Sousa et al. [37]. Hoser et al. [28] utilized the degree, betweenness and eigenvector centrality measures to identify core concepts and roles present in the SWRC (Semantic Web for Research Communities) and SUMO (Suggested Upper Merged Ontology) ontologies. The previously mentioned centrality metrics were applied to ontology networks formed using an ontology network extractor based on the KAON Ontology API [5]. Alani et al. [3] demonstrated that communities of practice of individuals present in an ontology can be identified by ontology network analysis based on the eigenvector centrality measure. Alani, Brewster and Shadbolt [1, 2] proposed AKTiveRank, an algorithm for ranking ontologies retrieved by an ontology search engine. AKTiveRank relies on network-based centrality measures to quantify the extent to which a class is representative of an ontology. Finally, Zhang et al. [53] and Queiroz-Sousa et al. [37] proposed ontology summarization methods based on centrality measures applied to ontology networks.

Theoharis et al. [49] analyzed degree distributions of ontology subsumption networks and ontology property networks (ontology class networks without subsumption relations) for 83 different ontologies each of them containing more than 100 classes. Similarly to Gil et al. [22], the authors applied linear regression to log-log transformed degree distributions to fit power-laws. The main finding of the study is that total degree distributions of examined property networks and in-degree distributions of class subsumption networks can be very well approximated by power-laws for a majority of investigated ontologies. The authors also showed that classes having a high degree centrality in ontology property networks tend to be highly abstract (located at higher levels in class subsumption networks). In their subsequent study [48], the authors exploited identified structural patterns to generate synthetic ontologies that can be used for benchmarking of ontology repositories and query languages implementations.

Cheng and Qu [7] extracted and analyzed a large-scale ontology network reflecting relationships between classes and properties present in 3090 unrelated semantic web ontologies collected by the Falcons semantic web search engine [8]. The analyzed ontology network contains more than one million of nodes and more than seven million of links. The network was formed using an ontology network extractor based on the Jena library [31]. The authors analyzed in-degree, out-degree and total degree distributions of the network, as well as reachability and connectivity of nodes. The main findings of the study are:

- The in-degree distribution of the network follows a power-law.
- About half of nodes have a small eccentricity (less than 6), but there is a relatively large fraction of nodes (about 10% of the total number) that have relatively high eccentricity (higher than 25). Nodes exhibiting high eccentricity tend to be highly specialized (positioned at bottom levels in the hierarchy of classes).
- The network contains relatively large strongly connected components implying the presence of large cyclic dependencies between ontological entities.

The authors also examined the structure of an ontology module network that shows dependencies between collected ontologies. This network becomes extremely fragmented after removing only 4 language-level ontologies indicating that the semantic web at the time of the study was far from being a cohesive network of inter-linked ontologies.

Total degree distributions of ontology class networks associated to seven biological and biomedical ontologies were examined by Zhang [51]. Similarly to Gil et al. [22] and Theoharis et al. [49], Zhang applied linear regression to log-log plots to check whether examined distributions follow power-laws. The obtained values of the coefficient of determination, ranging from 0.91 to 0.99, indicate that analyzed ontology networks belong to the class of scale-free networks. In his subsequent study, Zhang et al. [52] proposed an ontology metrics suite for measuring design complexity of semantic web ontologies based on ontology class networks. The authors performed an empirical analysis of proposed metrics also investigating properties of in-degree and out-degree distributions of ontology class networks associated to three ontologies (the full GALEN ontology of medical terms, the Gene ontology and NCI Thesaurus) using linear regression on log-log plots. The same methodology was also employed to investigate distributions of metric values for other two proposed class-level metrics: NOC – the number of children of a class in the ontology subsumption network, and DIT – the distance of a class from the root concept in the ontology subsumption network. The authors concluded that all examined distributions can be approximated by power-laws. A power-law scaling in degree distributions of two ontology class networks was also reported by Ma and Chen [30].

Ge et al. [20] analyzed a large-scale ontology object network describing relations between objects present in ontologies crawled by the Falcons ontology search engine. The authors found that the network has a giant connected component and exhibits the scale-free and small-world properties. The authors investigated also the evolution of the network by comparing the structure of its two evolutionary snapshots. The first evolutionary snapshot was constructed taking into account ontologies collected until 2008 (inclusive), while the second one was formed considering ontologies collected until 2009 (inclusive). The average node degree increased between two evolutionary snapshots implying that connections between class instances had densified during the evolution of the network. The authors also noticed that the diameter of the network decreased which means that the growth of the network caused a stronger small-world effect.

Ding et al. [13] examined the structure of a SameAs network extracted from the BTC (Billion Triple Challenge) 2010 dataset. SameAs networks are undirected subgraphs of object ontology networks induced by equivalence relations between objects. In other words, connected components in SameAs networks contain equivalent objects. The authors observed that a vast majority of connected components in the examined network have a small size concentrated around 2.4 objects. However, the network also contains a small fraction of unusually large components each of them encompassing more than one hundred objects. Moreover, the analyzed network contains two components with more than thousand nodes. The authors reported that the degree distribution of the network follows a power-law implying that the network exhibits the scale-free property.

Guéret et al. [24] examined the structure of a RDF network derived from approximately 10 million RDF triples crawled from the Web of Data in 2009. The authors analyzed the total degree distribution of the network concluding that it follows a power-law. A power-law scaling behavior in RDF networks was also reported by Luczak-Rösch and Tolksdorf [29], Färber and Rettinger [16] and Fernândez et al. [17] in more recent studies. Guéret et al. in [24] also investigated the network of RDF datasets (the Linked Data cloud network). The application of different centrality measures showed that the Web of Data in 2009 was strongly dependent on a small number of topic-oriented hubs. In their subsequent study [25], the authors compared two evolutionary snapshots of the Linked Data cloud network corresponding to years 2009 and 2010. The obtained results show that connectivity, cohesiveness and compactness of the Web of Data significantly increased in just one year. The authors also proposed an evolutionary algorithm for identifying links that can substantially increase the robustness of the Web of Data [23]. The studies by Rodriguez [42] and Caraballo et al. [6] showed that the Linked Data cloud network exhibits a strong community structure and disassortative mixing patterns. The presence of strong communities in RDF graphs and ontology class networks was also observed by Coskun et al. [10] who investigated the applicability of community detection techniques for concept grouping.

4.3 Analysis of Enriched Ontology Networks: A Case Study

Nodes in enriched ontology networks are augmented with various types of ontology metrics including also domain-independent metrics used in complex network analysis (e.g. different node centrality measures). In Chap. 3 we proposed the methodology to analyze enriched software networks. The main purpose of this chapter is to show that the same methodology can be applied to enriched ontology networks in order to (1) evaluate coupling and cohesion of ontological entities, and (2) assess the design quality of modularized ontologies.

The central component of our methodology for studying enriched software networks is the metric-based comparison test. This test relies on the Mann-Whitney U test and probabilities of superiority to compare two independent groups of nodes in an enriched network and determine whether one of the groups tend to exhibit systematically higher values of metrics associated to nodes (see Chap. 3, Sect. 3.4.1). In the same way as for enriched software networks, the metric-based comparison can be utilized to determine distinctive characteristics of highly coupled ontological entities and ontological entities involved in cyclic dependencies. To identify highly coupled ontological entities we use the same node classifier as for highly coupled software entities (see Chap. 3, Sect. 3.4.3, Definition 3.5).

Previous empirical studies of ontology networks [7, 13, 16, 17, 22, 24, 29, 30, 49, 51, 52] suggest that their degree distributions follow power-laws. However, the main weakness of previous studies regarding degree distribution analysis of ontology networks is that only power-laws were taken into account to model empirically observed degree distributions. Secondly, power-laws were fitted by biased methods (linear regression to log-log plots of degree distributions). On the other hand, our methodology for studying enriched software/ontology networks relies on the statistically robust power-law test introduced by Clauset et al. [9] to fit power-laws, check their plausibility and determine whether alternative theoretical models provide better fits.

In Chap. 3 we proposed graph clustering evaluation (GCE) metrics as software cohesion metrics. GCE metrics are domain-independent metrics and, consequently, they can also be applied to partitioned ontology networks. Let us suppose that we have a modular ontology M consisting of k ontology modules denoted by $O_1, O_2, \ldots,$ and O_k. Let G_M denote a directed graph showing dependencies between ontological entities defined in these k modules. We can distinguish between two types of links in G_M:

1. intra-module links (links between ontological entities defined in the same ontology module), and
2. inter-module links (links between ontological entities defined in different modules).

If O_i is a highly cohesive ontology module then entities defined in O_i must inevitably form a cluster in G_M, i.e. entities defined in O_i must be densely connected among themselves (via intra-module links) and at the same time sparsely connected to entities defined in other modules (via inter-module links). Consequently, we can instrument GCE metrics to G_M partitioned according to ontology modules in order to assess the cohesiveness of individual modules. Existing ontology cohesion metrics proposed in the literature estimate cohesiveness of ontology modules in isolation [15, 35] which means that they do not take into account inter-module links. However, inter-module links also have a high importance when measuring cohesiveness of ontology modules:

1. An ontology module whose entities have a higher number of inter-module than intra-module links hardly can be considered highly cohesive regardless of the density of intra-module links.
2. Let A and B be two ontology modules that have the same density of intra-module links. If entities in A have a smaller number of inter-module links compared to entities in B then A can be considered more cohesive, and vice versa.

The main advantage of GCE metrics compared to existing ontology module cohesion metrics proposed in the literature is that they do not ignore inter-module links. In the same way as for software modules, two GCE metrics, conductance and Flake-ODF, can be utilized to classify ontology modules according to the Radicchi et al. definitions of graph clusters [38] into three groups:

1. ontology modules having a very strong cohesion (Radicchi strong clusters in G_M),
2. ontology modules exhibiting a good degree of cohesion (Radicchi weak clusters in G_M), and
3. poorly cohesive ontology modules (modules which are not Radicchi strong nor Radicchi weak clusters in G_M).

To extract enriched ontology networks representing Semantic Web ontologies we used the ONGRAM tool [39, 43]. ONGRAM relies on the enriched Concrete Syntax Tree (eCST) representation [39, 40, 44] of ontological descriptions to extract ontology networks and compute various ontology metrics at the level of ontology modules and classes. ONGRAM forms three types of enriched ontology networks: enriched ontology modules networks (OMNs, Definition 4.4), enriched ontology class networks (OCNs, Definition 4.3) and enriched ontology subsumption networks (OSNs, Definition 4.4). The complete list of ontology metrics computed by ONGRAM is given in Table 4.1. Domain-independent coupling and centrality metrics are computed for both ontology modules and classes. At the class level ONGRAM also computes inheritance metrics from the Chidamber–Kemerer object-oriented metric suite adapted for ontologies [52]. Nodes in enriched ontology module networks extracted by ONGRAM are annotated with the following categories of metrics:

Table 4.1 The list of metrics attached to nodes of enriched ontology networks extracted by the ONGRAM tool

Metric	Abbr.	Level	Type
Lines of code	LOC	Module	Internal complexity
Total expression complexity	TEXPR	Module	Internal complexity
Average expression complexity	AEXPR	Module	Internal complexity
The number of axioms	AXM	Module	Internal complexity
Halstead volume	HVOL	Module	Internal complexity
Halstead difficulty	HDIF	Module	Internal complexity
The number of classes	NCLASS	Module	Internal complexity
The number of instances	NINST	Module	Internal complexity
In-degree	IN	Module, class	Coupling
Out-degree	OUT	Module, class	Coupling
The number of children	NOC	Class	Inheritance
The depth in inheritance tree	DIT	Class	Inheritance
Betweenness centrality	BET	Module, class	Importance
Page rank	PR	Module, class	Importance
HITS authority score	HITSA	Module, class	Importance
HITS hub score	HITSH	Module, class	Importance
Henry-Kafura complexity	HK	Module	Hybrid complexity
The average population of classes	AP	Module	Diversity
Class richness	CR	Module	Diversity
Relationship richness	RR	Module	Diversity
The number of external classes	NEC	Module	Coupling
References to external classes	REC	Module	Coupling
Conductance	CON	Module	Cohesion
Expansion	EXP	Module	Cohesion
Cut-ratio	CUTR	Module	Cohesion
Average out-degree fraction	AVGODF	Module	Cohesion
Maximum out-degree fraction	MAXODF	Module	Cohesion
Flake out-degree fraction	FODF	Module	Cohesion
Internal density	DEN	Module	Cohesion
Internal connectedness	COMP	Module	Cohesion

1. metrics quantifying complexity of ontological descriptions including software metrics adapted for ontologies (the Henry-Kafura complexity [27] and the Halstead's metrics [26]) and metrics quantifying the complexity of class expressions [43],
2. coupling metrics quantifying the strength of efferent coupling (NEC and REC) proposed by Orme et al. [36],

3. population and richness metrics (AP, CR and RR) proposed by Tartir et al. [47], and
4. GCE metrics.

The LOC (resp. AXM) metric quantifies the internal complexity of an ontology module by measuring the number of lines (resp. axioms) in the description of the ontology module. The eCST representation of OWL axioms enabled us to translate the cyclomatic complexity software metric [32] into the realm of ontologies. The cyclomatic complexity reflects the complexity of nested (syntactically recursive) statements such as functions, procedures or methods. Classes in OWL can be defined using class expressions which are syntactically recursive constructs of the OWL language. We define the expression complexity of a class expression R as the number of class expressions nested in R increased by one. The expression complexity of an axiom is equal to the sum of expression complexities of all class expressions contained in the axiom. The total expression complexity (TEXPR) of an ontology module is the sum of expression complexities of all axioms contained in the module, while the average expression complexity (AEXPR) is defined as $AEXPR(O) = TEXPR(O)/AXM(O)$. The Henry-Kafura complexity of an ontology module O is the product of its internal complexity measured by LOC and the square of the product of its afferent and efferent coupling measured by the in-degree and out-degree in the corresponding ontology module network, i.e.

$$HK(O) = LOC(O)(IN(O) \cdot OUT(O))^2 \qquad (4.1)$$

ONGRAM also computes the Halstead complexity metrics adapted for ontologies. To derive Halstead metrics, each token in an ontological description has to be marked either as an operator or as an operand. ONGRAM relies on the following translation scheme: operator tokens are those tokens introduced by OWL language designers (keywords and separators), while operands are tokens introduced by knowledge engineers (names of ontological entities). Then, the Halstead metrics can be trivially computed by counting the total number of operator and operand tokens, the cardinality of the set containing all operator tokens and the cardinality of the set containing all operand tokens in an ontology module.

Besides GCE metrics, ONGRAM also computes two additional metrics of ontology module cohesion, internal density (DEN) and internal connectedness (COMP), that are based solely on internal dependencies (intra-module links). Let $G_I(O)$ denote the graph representing internal dependencies between ontological entities defined within an ontology module O. Then, the internal density of O is the density of $G_I(O)$, while the internal connectedness of O is the number of weakly connected components in $G_I(O)$.

In our case study, we analyze the structure of enriched ontology networks representing the SWEET ontology [41] in version 2.2. The SWEET ontology is a modular ontology that defines more than 6000 concepts related to earth and environmental sciences in 204 ontology modules. SWEET contains an umbrella ontology module named "sweetAll.owl" that imports all SWEET ontology modules. By loading the

Table 4.2 Basic characteristics of enriched ontology networks extracted from the SWEET ontology

Network	Abbr.	The number of nodes	The number of links
Ontology module network	OMN	203	1138
Ontology class network	OCN	6374	8483
Ontology subsumption network	OSN	6003	6202

umbrella ontology module into the Protégé ontology editor [21, 34] we collected the rest of SWEET ontology modules and converted them into the OWL2 functional-style syntax that is supported by the ONGRAM tool. The module "sweetAll.owl" was omitted during the extraction of SWEET enriched ontology networks since this module does not contain any knowledge – its only purpose is to enable the complete retrieval of SWEET ontology modules. Basic structural characteristics (the number of nodes and links) of enriched SWEET ontology networks are shown in Table 4.2. Ontology subsumption networks extracted by ONGRAM do not contain isolated nodes, i.e. classes not involved in subsumption relationships are not included in ontology subsumption networks.

4.4   Results and Discussion

Enriched ontology networks are directed graphs. Thus, they contain both weakly and strongly connected components. We firstly identified and investigated characteristics of weakly connected components (WCCs) in the SWEET enriched ontology networks. The SWEET ontology module network (OMN) and ontology class network (OCN) are weakly connected directed graphs, i.e. both networks contain exactly one weakly connected component encompassing all present nodes. On the other hand, the ontology subsumption network (OSN) has 36 non-trivial WCCs (we recall that SWEET classes not involved in class subsumption relations are not included in the SWEET OSN). One of WCCs present in the SWEET OSN is a giant weakly connected component occupying more than 90% of nodes and 90% of links in the network. The second largest WCC in the SWEET OSN contains 86 classes which is approximately 1.4% of the total number of nodes in the network. The number of weakly connected components in the SWEET OMN indicates that SWEET exhibits a good modular cohesion since there are no isolated ontology modules. SWEET also shows a high degree of class cohesion: there are no isolated classes and a vast majority of SWEET classes form a compact taxonomy of concepts. The existence of a single/giant weakly connected component in analyzed networks is quite expected since the SWEET ontology describes a domain-specific terminology. This property is also not surprising from a theoretical standpoint: the average node degree in all three networks is greater than one which is the critical threshold for the emergence of a giant connected component in the Erdős-Renyi model of random graphs.

Table 4.3 Weakly connected components in the SWEET ontology networks. #WCC – the number of weakly connected components, LWCCN – the fraction of nodes in the largest WCC, LWCCL – the fraction of links in the largest WCC, SW – the small-world coefficient, SW-rnd – the small-world coefficient of a comparable random graph, CC – the clustering coefficient, CC-rnd – the clustering coefficient of a comparable random graph, A – the Newman assortativity index

Network	#WCC	LWCCN [%]	LWCCL [%]	SW	SW-rnd	CC	CC-rnd	A
OMN	1	100	100	2.55	2.22	0.15	0.028	0.023
OCN	1	100	100	9.51	9.74	0.007	0.00021	−0.158
OSN	36	93.35	94.11	11.8	11.74	0.001	0.00017	−0.171

Table 4.3 also shows the small-world coefficient, clustering coefficient and assortativity index of examined ontology networks. It can be observed that all three networks exhibit the Watts-Strogatz small-world property: the small-world coefficients are close to the predictions made by the Erdős-Renyi model (SW $\approx$ SW-rnd) and at the same time the clustering coefficients are drastically higher than the clustering coefficients of comparable random graphs (CC $\gg$ CC-rnd). SWEET exhibits different types of assortativity mixing patterns at different levels of abstraction. The ontology module network shows an extremely weak assortative mixing. On the other hand, SWEET class-level networks exhibit a significant disassortative mixing which means that highly coupled SWEET classes do not tend to be directly connected among themselves. Therefore, we can conclude that SWEET class-level networks do not have a hub-like core encompassing highly coupled ontological concepts.

4.4.1 Strongly Connected Components and Cyclic Dependencies

We determined strongly connected components (SCCs) in the SWEET enriched ontology networks using the Tarjan algorithm. Table 4.4 shows the basic structural characteristics of identified SCCs: their total number, the percentage of nodes and links in the largest SCC, the percentage of nodes involved in cyclic dependencies, the link reciprocity (the probability that an ontology module A directly depends on an ontology module B when B is directly dependent on A), the normalized link reciprocity by Garlaschelli and Loffredo [19], the path reciprocity (the probability that an ontology module A directly or indirectly depends on an ontology modules B when B is directly or indirectly dependent on A) and the percentage of SCCs that are pure cycles (SCCs in which the number of nodes is equal to the number of links). It can be observed that the SWEET ontology module network contains three SCCs. The largest SCC can be considered giant since it encompasses more than 60% of SWEET ontology modules and more than 60% of ontology module dependencies.

Table 4.4 Strongly connected components in the SWEET ontology networks. #SCC – the total number of strongly connected components, LSCCN – the percentage of nodes in the largest SCC, LSCCL – the percentage of links in the largest SCC, S – the percentage of nodes contained in all SCCs, R – link reciprocity, R_n – normalized link reciprocity, R_p – path reciprocity, C – the percentage of SCCs that are pure cycles

Network	#SCC	LSCCN [%]	LSCCL [%]	S [%]	R	R_n	R_p	C [%]
OMN	3	61.57	60.63	64.53	0.0545	0.0275	0.608	33.33
OCN	410	0.17	0.20	15.05	0.1214	0.1212	0.0136	80.24
OSN	1	0.03	0.03	0.03	0.0004	0.0003	0.0001	100

The presence of a giant SCC implies that there are large cyclic dependencies among SWEET modules. The other two SCCs are small size components:

- the second largest SCC contains only 4 mutually dependent SWEET ontology modules connected by 6 links, and
- the smallest SCC contains 2 SWEET ontology modules connected by a pair of reciprocal links.

The reciprocity of links in the SWEET OMN is relatively small implying that a small percentage of SWEET ontology modules are mutually directly dependent. However, the reciprocity of links in the network is still higher than it could be expected by a random chance ($R_n > 0$). In contrast to the link reciprocity, the path reciprocity in the SWEET OMN is extremely high: more than 60% of all dependencies among ontological modules (both direct and indirect) are cyclic dependencies. Large cyclic dependencies between SWEET ontology modules mean that the modular structure of SWEET strongly deviates from a layered organization, i.e. SWEET ontology modules cannot be topological sorted and grouped into hierarchically organized layers. Therefore, the presence of large cyclic dependencies between SWEET modules can be considered as an indicator of a weak modularization with respect to the effort needed to comprehend the SWEET ontology design and modules involved in cyclic dependencies. The presence of large cyclic dependencies also prevents efficient external reuse of SWEET ontology modules: an external reuse of an ontology module that belongs to the giant SCC implies that all modules from the SCC have to be externally reused.

The SWEET OCN has a large number of SCCs, but all of them are relatively small. Less than 1% of SWEET classes are involved in cyclic dependencies. The largest SCC in the SWEET OCN encompasses 11 mutually dependent SWEET classes connected by 17 links. The reciprocity of links in the SWEET OCN is significantly higher than the reciprocity of paths implying that a majority of existing cyclic relationships between SWEET classes are direct reciprocal links stated by EquivalentClasses and DisjointClasses axioms. Consequently, we can conclude that SCCs in the SWEET OCN can be easily comprehend by ontology engineers since large cyclic class dependencies are absent. Moreover, a vast majority of SCCs in the SWEET OCN (more than 80% of the total number) have trivial complexity which means that the number of nodes in a SCC is equal to the number of links (i.e. SCCs are pure circles of nodes).

The SWEET OSN has only one SCC encompassing two SWEET classes. Classes "RadiativeForcing" and "RadiantFlux" defined in ontology module "quanEnergyFlux.owl" are mutually dependent in the taxonomy of SWEET classes: "RadiativeForcing" subsumes "RadiantFlux" and vice versa. This means that those two classes actually represent equivalent concepts but an axiom stating the equivalence relation is not present in the SWEET ontology. Therefore, the taxonomy of SWEET classes minimally deviates from a hierarchical structure.

We applied the metric-based comparison test to determine distinctive characteristics of SWEET ontology modules contained in the giant strongly connected component (GSCC) of the SWEET ontology module network. The obtained results are summarized in Table 4.5. It can be observed that the null hypothesis of the MWU test (no statistically significant differences between ontology modules located in the GSCC and the rest of ontology modules) is accepted for two metrics of internal complexity: TEXPR (the total expression complexity of axioms in an ontology module) and AEXPR (the average expression complexity of axioms in an ontology module). On the other hand, ontology modules located in the GSCC slightly tend to exhibit higher values of other metrics of internal complexity (LOC – the number of lines in the corresponding OWL file, AXM – the number of OWL axioms, HVOL – the Halstead volume, and HDIF – the Halstead difficulty). Consequently, we can conclude that SWEET ontology modules located in the GSCC tend to be slightly more voluminous compared to the rest of modules, but they do not tend to contain class expressions that are more complex than class expressions present in modules not located in the GSCC.

The null hypothesis of the MWU test is also accepted for ontology metrics from the OntoQA ontology metric suite proposed by Tartir et al. [47] (AP – average population, CR – class richness, RR – relationship richness), as well as for the NINST metric (the number of class instances declared in an ontology module). The same situation is also with graph clustering evaluation (GCE) metrics. This means that SWEET ontology modules from the GSCC do not tend to (1) contain more class instances, (2) exhibit a higher diversity of ontology relationships, and (3) be less cohesive than SWEET ontology modules not involved in cyclic dependencies. There are also no statistically significant differences between SWEET ontology modules from the GSCC and the rest of SWEET modules regarding their efferent coupling: the null hypothesis of the MWU test is accepted for the OUT metric (the number of imported ontology modules) and two efferent coupling metrics from the ontology metric suite introduced by Orme et al. [36] (NEC – the number of referenced external classes and REC – the number of references to external classes).

A high value of PS_1 in Table 4.5 indicates that SWEET ontology modules located in GSCC strongly tend to be superior with respect to the corresponding metric compared to the rest of SWEET ontology modules. We can see that SWEET ontology modules located in GSCC strongly tend to have higher afferent coupling compared to the rest of modules:

Table 4.5 The results of the metric-based comparison test for the giant strongly connected component (GSCC) in the SWEET ontology module network. Avg(GSCC) is the average value of the corresponding ontology metric for nodes in the GSCC, while Avg(Rest) represents the average value for the rest of ontology modules. U – the Mann-Whitney U statistic, p – the significance probability of the MWU test statistic. The column "NullHyp" denotes whether the null hypothesis of the MWU test is accepted or not. PS_1 is the probability of superiority of ontology modules in the GSCC over the rest of ontology modules with respect to the corresponding ontology metric, while PS_2 denotes the inverse probability of superiority. Bold PS values indicate a strong superiority with respect to the corresponding ontology metric

Metric	Avg(GSCC)	Avg(Rest)	U	p	NullHyp	PS_1	PS_2
LOC	106.8	101.1	6029	0.0045	rejected	0.61	0.38
TEXPR	5.26	4.11	5412	0.1872	**accepted**	0.5	0.39
AEXPR	0.068	0.071	5058	0.6522	**accepted**	0.49	0.45
AXM	92.8	87.3	6033	0.0044	rejected	0.61	0.38
HVOL	2905	2855.5	6048	0.0039	rejected	0.62	0.38
HDIF	20.5	17.7	6376	0.0002	rejected	0.65	0.35
NCLASS	34.06	27.04	5773	0.0274	rejected	0.58	0.4
NINST	9.24	13.97	5007	0.7448	**accepted**	0.24	0.27
IN	8.34	1.22	9110	$<10^{-4}$	rejected	**0.91**	0.04
OUT	5.77	5.35	5002	0.7541	**accepted**	0.47	0.46
TOT	14.1	6.55	8057	$<10^{-4}$	rejected	**0.81**	0.15
BET	870.8	20.6	8781	$<10^{-4}$	rejected	**0.89**	0.09
PR	0.0066	0.0022	8971	$<10^{-4}$	rejected	**0.92**	0.08
HITSH	0.0642	0.0467	6048	0.0039	rejected	0.62	0.38
HITSA	0.0549	0.0064	9414	$<10^{-4}$	rejected	**0.97**	0.03
HK	717545.28	7888.64	8959	$<10^{-4}$	rejected	**0.92**	0.08
AP	1.74	1.28	4892	0.9666	**accepted**	0.24	0.23
CR	0.11	0.09	5104	0.5729	**accepted**	0.3	0.25
RR	0.23	0.23	5025	0.7125	**accepted**	0.5	0.47
NEC	5.12	4.68	4962	0.8298	**accepted**	0.46	0.44
REC	9.49	8.76	4981	0.7946	**accepted**	0.47	0.49
CON	0.21	0.22	5470	0.1438	**accepted**	0.44	0.56
EXP	0.29	0.31	5498	0.1259	**accepted**	0.43	0.56
CUTR	0.000027	0.000029	5497	0.1266	**accepted**	0.44	0.56
AVGODF	0.24	0.27	5471	0.1432	**accepted**	0.44	0.56
MAXODF	0.94	0.95	4946	0.8615	**accepted**	0.08	0.1
FODF	0.76	0.73	5395	0.2015	**accepted**	0.55	0.44

1. the average in-degree of SWEET modules locates in the GSCC is four times higher than the average in-degree of the rest of SWEET modules, and
2. the probability that a randomly selected SWEET module located in the GSCC has a higher number of in-coming links compared to a randomly selected SWEET module outside GSCC is equal to 0.91, while the inverse probability of superiority is close to zero ($PS_2 = 0.04$).

Consequently, we can conclude that the most internally reused SWEET ontology modules are located in the GSCC. The presence of large differences in the total coupling and Henry-Kafura complexity are caused by large differences in afferent coupling since those two metrics incorporate in-degree as their constituent factor. The metric-based comparison test also revealed that there are statistically significant differences considering node centrality metrics for compared groups of modules. SWEET ontology modules located in GSCC strongly tend to have higher values of betweeness centrality, page rank and HITS authority score ($PS_1 \geq 0.89$). This means that the most important SWEET modules are involved in cyclic dependencies and that the SWEET ontology module network exhibits a core-periphery structure with a strongly connected core containing the most reused ontology modules. This can be considered as an indicator of a poor knowledge modularization: one has to reorganize the most reused and the most important SWEET ontology modules in order to eliminate or drastically reduce existing cyclic dependencies.

4.4.2   Correlation Based Analysis of Ontology Modules

The nodes in the enriched ontology module network of SWEET are augmented with a large number of metrics including both metrics of complexity of ontological descriptions and ontology design metrics. Figure 4.2 shows the Spearman correlation matrix for metrics of internal complexity of ontology modules. It can be observed that there arc strong Sperman correlations ($\rho \geq 0.65$) between the size of an ontology module (LOC), the number of axioms in an ontology module (AXM), Halstead's difficulty (HDIF) and Halstead's volume (HVOL). Strong correlations between the previously mentioned metrics on one side and two metrics of class expression complexity on the other side are absent. This means that highly complex class expressions can be found in both small-size and large-size SWEET ontology modules.

Strong Sperman correlations ($|\rho| \geq 0.65$) between ontology design metrics are shown in Table 4.6. It can be observed that in-degree (IN) and total degree (TOT) strongly correlate with page rank (PR), HITS authority score (HITSA) and betweenness centrality (BET) implying that the most coupled and the most internally reused ontological modules tend to be the most central nodes in the enriched SWEET ontology module network. It is also important to notice that there is a strong correlation between in-degree and total degree, while a strong correlation between out-degree and total degree is absent. This suggests that there are no significant differences between highly and loosely coupled SWEET ontology modules regarding their effer-

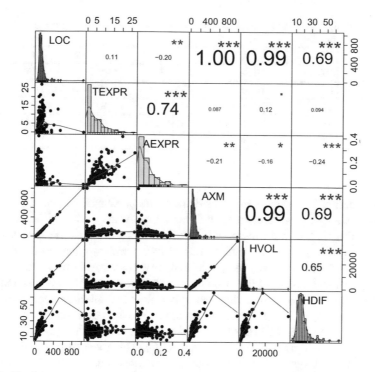

Fig. 4.2 The Spearman correlation matrix for metrics of internal complexity of ontology modules

Table 4.6 Strong Spearman correlations between ontology design metrics

Metric 1	Metric 2	ρ	Metric 1	Metric 2	ρ
IN	HITSA	0.96	IN	TOT	0.75
IN	PR	0.95	IN	BET	0.75
NEC	REC	0.95	OUT	NEC	0.72
PR	HITSA	0.89	PR	BET	0.72
OUT	HITSH	0.84	TOT	BET	0.71
TOT	HITSA	0.78	OUT	REC	0.68
BET	HITSA	0.76	TOT	PR	0.66

ent coupling. A Strong correlations is also present between two metrics of the strength of efferent ontology module coupling from the Orme et al. metrics suite [36] (NEC and REC). Additionally, out-degree strongly correlates with both NEC and REC which means that the strength of efferent coupling of ontology modules increases with the number of imported modules.

Finally, we investigated correlations between metrics of internal complexity of ontology modules and ontology design metrics. Here we also included the Henry-Kafura complexity (HK) which is a hybrid complexity metric. Table 4.7 shows

Table 4.7 Strong Spearman correlations between ontology metrics of internal complexity and ontology design metrics

Metric 1	Metric 2	ρ	Metric 1	Metric 2	ρ
HK	HITSA	0.88	TEXPR	REC	0.77
HK	IN	0.87	TEXPR	NEC	0.74
HK	BET	0.85	TEXPR	OUT	0.69
HK	TOT	0.85	TEXPR	HITSH	0.66
HK	PR	0.78			

ontology metrics for which strong correlations are present. It can be seen that HK strongly correlates with in-degree, total-degree and node centrality metrics (BET, PR) indicating that the most important SWEET ontology modules tend to have high values of the Henry-Kafura complexity. A strong correlation between HK and in-degree is not surprising since in-degree is one of constituent components of HK. On the other hand, HK does not strongly correlate with other two constituent metrics, LOC and out-degree, implying that only afferent coupling impacts the Henry-Kafura complexity of SWEET ontology modules. The total expression complexity of ontology modules (TEXPR) strongly correlates with metrics that quantify ontology module efferent coupling (OUT, NEC and REC). This means that ontology modules containing a large amount of complex class expressions tend to establish strong connections to other SWEET ontology modules. TEXPR also exhibits a strong Spearman correlation with HITS hub score (HITSH) which means that important SWEET ontology modules referencing other important SWEET ontology modules tend to contain complex class expressions.

4.4.3   Degree Distribution Analysis

The scale-free property of ontology networks was reported in several previous studies reviewed in Sect. 4.2. However, the main characteristic of previous studies dealing with degree distribution analysis of ontology networks is that only power-laws were considered to model empirically observed degree distributions. We employed the power-law test introduced by Clauset et al. [9] to investigate whether SWEET ontology networks belong to the class of scale-free networks. The best power-law fits are compared to the best fits of exponential and log-normal distributions. The results of the power-law test for in-degree, out-degree and total degree distributions of the SWEET OMN, OCN and OSN are summarized in Table 4.8. The lower bound of the best power-law fit is denoted by x_m, M is the maximal degree value, α denotes the power-law scaling exponent, and p-value is the statistical significance of the power-law fit. p-value smaller than 0.1 implies that the power-law model is not plausible for an empirically observed degree distribution [9]. The power-law hypothesis is

Table 4.8 The results of the power-law test for degree distributions of the SWEET ontology networks. OMN, OCN and OSN denote ontology module network, ontology class network and ontology subsumption network, respectively. Bold p-values indicate that the power-law hypothesis is not rejected

Network	Distribution	x_m	M	α	p-value	R_{ln}	$p(R_{ln})$	R_e	$p(R_e)$
OMN	Total degree	11	56	3.24	**0.31**	−4.09	$<10^{-4}$	9.18	$<10^{-4}$
	In degree	8	51	2.78	**0.44**	−8.19	$<10^{-4}$	15.95	$<10^{-4}$
	Out degree	8	17	4.69	0.01				
OCN	Total degree	2	54	2.54	$<10^{-4}$				
	In degree	1	53	2.3	$<10^{-4}$				
	Out degree	5	17	6.26	**0.99**	−0.22	0.82	1.01	0.31
OSN	Total degree	3	54	2.72	$<10^{-4}$				
	In degree	1	53	2.39	$<10^{-4}$				
	Out degree	3	17	5.67	$<10^{-4}$				

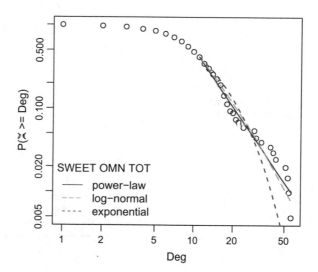

Fig. 4.3 The complementary cumulative total degree distribution of the SWEET ontology module network with the best fits of power-law, log-normal and exponential distributions

rejected for a majority of empirically observed distributions. Power-laws are plausible statistical models in tails ($x_m > 1$) of three degree distributions:

1. the total degree distribution of the SWEET OMN (Fig. 4.3),
2. the in-degree distribution of the SWEET OMN (Fig. 4.4), and
3. the out-degree distribution of the SWEET OCN (Fig. 4.5).

Table 4.8 also shows the results of the likelihood ratio test in the case that a power-law is plausible. The likelihood ratio compares the best power-law fit with the best fits of alternative theoretical models. The value of the log likelihood ratio is denoted by

Fig. 4.4 The
complementary cumulative
in-degree distribution of the
SWEET ontology module
network with the best fits of
power-law, log-normal and
exponential distributions

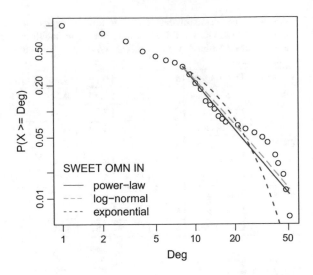

Fig. 4.5 The
complementary cumulative
out-degree distribution of the
SWEET ontology class
network with the best fits of
power-law, log-normal and
exponential distributions

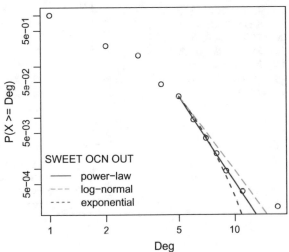

R_d where d is an alternative statistical model ("ln" – log-normal, "e" – exponential).
A positive and statistically significant value of R_d ($R_d > 0$, $p(R_d) < 0.1$) indicates
that the power-law fit is better than the best fit of d, while a negative and statistically
significant value of R_d ($R_d < 0$, $p(R_d) < 0.1$) indicates that d provides a better fit.
We can see that the log-normal model provides better fits to total degree and in-degree
distributions of the SWEET OMN compared to the power-law model ($R_{ln} < 0$,
$p(R_{ln}) < 0.1$). On the other hand, all three theoretical models are plausible in the
tail of the out-degree distribution of the SWEET OCN. However, the size of the tail
exhibiting the power-law scaling is extremely small suggesting that basically any
theoretical distribution can be very well fitted to a such small range of degree values.

Table 4.9 The results of the log likelihood test for total degree distributions of examined ontology networks

Network	$R\left(\frac{ps}{ln}\right)$	p	$R\left(\frac{ps}{e}\right)$	p	$R\left(\frac{ln}{e}\right)$	p	Best fit
OMN	-3.74	0.0001	-3.39	0.0007	2.59	0.009	Log-normal
OCN	-6.65	$<10^{-4}$	-6.11	$<10^{-4}$	10.07	$<10^{-4}$	Log-normal
OSN	-4.84	$<10^{-4}$	-1.57	0.11	15.11	$<10^{-4}$	Log-normal

Summarizing all obtained results, we can conclude that examined ontology networks do not possess the scale-free property:

1. the power-law hypothesis is rejected for 6 out of 9 degree distributions, and
2. the log-normal model provides better fits in tails of the three degree distributions for which the power-law model is plausible.

Do examined ontology networks contain hubs (highly coupled ontology modules and classes)? To answer this question we compared the best fits of log-normal, exponential and Poisson distribution to empirically observed total degree distributions of examined ontology networks considering the whole range of degree values. If the Poisson distribution provides the best fit to an empirically observed total degree distribution then the corresponding ontology network possesses node connectivity characteristic to Erdős-Renyi random graphs and, consequently, does not contain hubs. The results of the log likelihood test are shown in Table 4.9. The log likelihood ratio is denoted by $R\left(\frac{d_1}{d_2}\right)$, where d_1 and d_2 are two theoretical models ("ps" – Poisson, "ln" – log-normal, "e" – exponential). A positive and statistically significant values of R ($R\left(\frac{d_1}{d_2}\right) > 0$, $p < 0.1$) indicates that d_1 provides a significantly better fit than d_2, while a negative and statistically significant value of R means exactly the opposite. In the case that R is not statistically significant ($p \geq 0.1$) then d_1 and d_2 are equally plausible models for empirical data. The obtained results show that (1) the Poisson distribution is never preferred over other two theoretical models, and (2) the log-normal model always provides the best fit to degree distributions of examined ontology networks. Additionally, the coefficient of variation and skewness of empirically observed total degree distributions are at least two times higher than the same quantities of Poisson fits (Table 4.10). Therefore, it can be concluded that examined networks contain hubs due to highly skewed total degree distributions.

4.4.4 Highly Coupled Ontological Entities

The degree distribution analysis presented in the previous section revealed that examined enriched ontology networks contain hubs (highly coupled ontology modules and classes). To separate hubs from non-hubs in enriched ontology networks we use the same principle as in the analysis of enriched software networks: hubs form the mini-

Table 4.10 The coefficient of variation (c_v), skewness (G_1) and the average value (μ) of total degree distributions for examined ontology networks. The coefficient of variation and skewness of the Poisson fit are equal to $\mu^{-0.5}$

Network	c_v	G_1	μ	$\mu^{-0.5}$
OMN	0.81	2.68	11.2	0.3
OCN	1.21	5.32	2.66	0.61
OSN	1.38	5.92	2.08	0.69

Table 4.11 The percentage of hubs and the minimal total degree of hub nodes (H_d) in SWEET ontology networks

Network	The percentage of hubs	H_d
Ontology module network	25.62	14
Ontology class network	29.26	3
Ontology subsumption network	22.36	3

mal subset of nodes H such that the sum of total degrees of nodes in H is higher than the sum of total degrees of the rest of nodes. The size of hub sets and the minimal total degree of hub nodes for SWEET ontology networks are shown in Table 4.11. It can be observed that the minimal total degree of hub classes is relatively low – more than 70% of classes contained in the SWEET class-level networks are classes whose total degree is less than 3 implying that a vast majority of SWEET classes are extremely loosely coupled.

The top five highest coupled ontological entities in SWEET ontology networks are given in Table 4.12. It can be observed that in-degree values are significantly higher than out-degree values indicating that highly coupled SWEET ontology modules and classes are, similarly to highly coupled classes in object-oriented software systems examined in Chap. 3, dominantly caused by internal reuse.

In the same way as in the analysis of enriched software networks, we investigated the degree of disbalance between in-degree and out-degree for hubs by measuring two quantities:

1. C_k – the average ratio of in-degree to total-degree for nodes whose total-degree is higher than or equal to k, and
2. P_k which is the probability that a randomly selected node whose total degree is higher than or equal to k has two times higher in-degree than out-degree.

Figure 4.6 shows the values of C_k and P_k for SWEET ontology networks. It can be observed that both C_k and P_k increase with k. This means that the disbalance between in-degree and out-degree becomes more drastic as the total coupling of ontological entities increases. In other words, highly coupled SWEET modules and classes are dominantly caused by their internal reuse, while internal aggregation of other ontological entities does not significantly contribute to high coupling. There-fore, metrics reflecting coupling of ontological entities (such as adapted CBO from

Table 4.12 The top five highest coupled nodes in SWEET ontology networks. TOT – total degree, IN – in-degree, OUT – out-degree

Network	Entity	TOT	IN	OUT
OMN	realm.owl	56	48	8
	quan.owl	54	46	8
	matr.owl	53	51	2
	phen.owl	51	42	9
	reprMath.owl	43	40	3
OCN	phenAtmoWindMesoscale.owl#MesoscaleWind	54	53	1
	human.owl#HumanActivity	46	46	0
	quanTemperature.owl#Temperature	41	36	6
	realm.owl#Ocean	36	31	5
	quanPressure.owl#Pressure	36	28	8
OSN	phenAtmoWindMesoscale.owl#MesoscaleWind	54	53	1
	human.owl#HumanActivity	44	44	0
	quanFraction.owl#FractionalProperty	33	30	3
	phenAtmoCloud.owl#Cloud	32	27	5
	phenAtmo.owl#MeteorologicalPhenomena	31	30	1

the Chidamber–Kemerer metric suite) can indicate only negative aspects of extensive internal reuse, not negative aspects of extensive internal aggregation.

We applied the metric based comparison test to determine characteristics of highly coupled SWEET modules. The results of the test are shown in Table 4.13. It can be seen that highly coupled ontology modules in SWEET strongly tend to have higher values of LOC (lines of code), AXM (the number of axioms), HVOL (Halstead volume), NCLASS (the number of classes), IN (in-degree), BET (betweenness centrality), PR (page rank), HITSA (HITS authority score) and HK (Henry-Kafura complexity) compared to non-hub modules. This means that highly coupled SWEET modules strongly tend to be larger and more functionally important than non-hub SWEET modules. The null hypothesis of the MWU test is accepted for all GCE metrics implying that hub modules exhibit a similar degree of cohesiveness as non-hub modules. Statistically significant differences between hub and non-hub modules are absent with respect to the AEXPR metric implying that individual axioms contained in hub modules do not tend to be more or less complex than individual axioms in loosely coupled modules.

4.4.5 Cohesiveness of Ontology Modules

To analyze cohesiveness of ontology modules we proposed the usage of graph clustering evaluation (GCE) metrics. The ONGRAM tools enriches nodes of ontology module networks with 6 different GCE metrics: CON (conductance), EXP (expan-

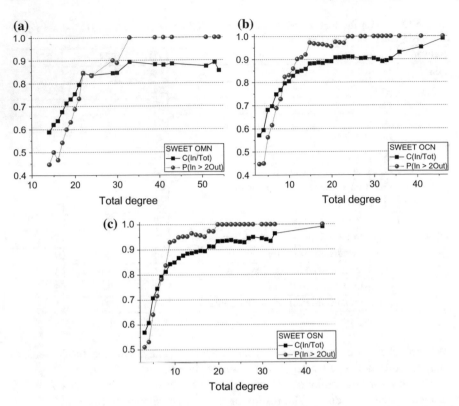

Fig. 4.6 The disbalance between afferent and efferent coupling for highly coupled ontological entities in **a** the SWEET OMN, **b** the SWEET OCN and **c** the SWEET OSN

sion), CUTR (cut-ratio), AODF (the average out-degree fraction), MODF (the maximal out-degree fraction) and FODF (the Flake out-degree fraction). It is important to recall that the FODF metric reflects ontology module cohesion, while the rest of GCE metrics reflect the lack of cohesion. Figure 4.7 shows the Spearman correlation matrix for the previously mentioned GCE metrics computed considering SWEET ontology modules. It can be observed that there are strong Sperman correlations between different GCE metrics with one exception: the MODF metric weakly correlates with the rest of GCE metrics. The examination of MODF metric values for individual SWEET ontology modules revealed that 184 SWEET modules (90.64% of the total number) have MODF equal to 1 which is the maximal possible value of the MODF metric. This means a vast majority of SWEET modules contain at least one ontological entity which does not reference any other entity from the same module, but solely entities from other modules.

One of main characteristics of GCE metrics is that they do not ignore external ontological dependencies (dependencies between ontological entities belonging to different ontology modules). The cohesiveness of an ontology module can also be estimated in isolation from the rest of ontology modules present in the system rely-

Table 4.13 The results of the metric-based comparison test for hubs in the SWEET ontology module network

Metric	Avg(Hubs)	Avg(Rest)	U	p	NullHyp	PS_1	PS_2
LOC	138.4	93	6185	$<10^{-4}$	rejected	**0.79**	0.21
TEXPR	7.6	3.9	5449	$<10^{-4}$	rejected	0.66	0.27
AEXPR	0.076	0.068	4579	0.07	**accepted**	0.57	0.4
AXM	122.5	79.7	6097	$<10^{-4}$	rejected	**0.77**	0.22
HVOL	3931.3	2526.1	6237	$<10^{-4}$	rejected	**0.79**	0.21
HDIF	23.1	18.2	5797	$<10^{-4}$	rejected	0.74	0.26
NCLASS	46.6	26.1	5947	$<10^{-4}$	rejected	**0.75**	0.24
NINST	8.8	11.8	4316	0.28	**accepted**	0.19	0.29
IN	14.7	2.5	7214	$<10^{-4}$	rejected	**0.9**	0.06
OUT	7.4	5	5175	0.0006	rejected	0.62	0.31
BET	1438.7	236.01	6815	$<10^{-4}$	rejected	**0.87**	0.13
PR	0.0128	0.0022	6737	$<10^{-4}$	rejected	**0.86**	0.14
HITSA	0.09	0.01	7326	$<10^{-4}$	rejected	**0.93**	0.07
HITSH	0.08	0.04	5435	$<10^{-4}$	rejected	0.69	0.31
HK	1681429.9	19033.9	7688	$<10^{-4}$	rejected	**0.98**	0.02
AP	0.83	1.82	4198	0.45	**accepted**	0.19	0.26
CR	0.07	0.11	3926	0.99	**accepted**	0.29	0.28
RR	0.28	0.22	4845	0.01	rejected	0.61	0.38
NEC	7	4.2	5257	0.0003	rejected	0.63	0.29
REC	13.2	7.8	5019	0.003	rejected	0.62	0.34
CON	0.19	0.22	4450	0.15	**accepted**	0.44	0.57
EXP	0.26	0.31	4409	0.18	**accepted**	0.44	0.56
CUTR	0.000024	0.000029	4404	0.19	**accepted**	0.44	0.56
AVGODF	0.22	0.26	4359	0.24	**accepted**	0.44	0.55
MAXODF	0.96	0.94	4042	0.75	**accepted**	0.09	0.07
FODF	0.78	0.74	4245	0.38	**accepted**	0.53	0.45

ing only on internal dependencies (dependencies between ontological entities from the same module). The ONGRAM tool enriches nodes of ontology module networks with such two metrics: DEN (the density of a graph showing dependencies between ontological entities present in an ontology module) and COMP (the number of weakly connected components in the previously mentioned graph). Table 4.14 shows the values of the Spearman correlation coefficient between GCE metrics and metrics of internal ontology module density (DEN) and connectedness (COMP). It can be observed that strong correlations between those two types of cohesion measures are absent indicating that ontology cohesion measures based only on internal dependencies are unable to identify ontology modules whose constituent elements form strong clusters in ontology networks.

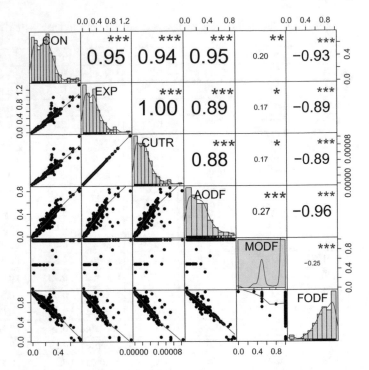

Fig. 4.7 The Spearman correlation matrix for GCE metrics

Table 4.14 The values of the Spearman correlation coefficient for GCE metrics and metrics of internal ontology module density (DEN) and connectedness (COMP)

	EXP	CON	CUTR	AODF	FODF
DEN	−0.035	−0.056	−0.038	−0.109	0.076
COMP	0.174	0.265	0.175	0.253	−0.235

We classified SWEET ontology modules into three categories according to obtained values of GCE metrics: Radicchi strong modules (Flake-ODF equal to 1), Radicchi weak modules (conductance smaller than 0.5) and poorly cohesive modules (conductance greater than or equal to 0.5). Eighteen SWEET ontology modules (8.87% of the total number) are Radicchi strong which means that they exhibit a very strong cohesion. A vast majority of SWEET ontology modules (196 of them which is 96.08% of the total number) are Raddichi weak implying that they possess a satisfactory degree of cohesion. The SWEET ontology modules that are not Raddichi weak are listed in Table 4.15. The small percentage of Radicchi non-weak SWEET modules suggests that the knowledge contained in SWEET is well-modularized with respect to the principle of high cohesion.

Table 4.15 Poorly cohesive SWEET modules. *A* denotes the average of value of the corresponding metric considering all SWEET modules

Module	LOC	TEXPR	IN	OUT	PR	BET	CON
stateSpaceConfiguration.owl	106	0	1	2	0.0016	12	0.75
stateTimeFrequency.owl	72	0	2	7	0.0021	260	0.75
quanTimeAverage.owl	89	1	3	8	0.0012	451	0.74
stateSpace.owl	70	0	0	5	0.0010	0	0.65
realmAtmoWeather.owl	61	4	0	7	0.0010	0	0.65
reprSpaceDirection.owl	97	0	9	2	0.0045	16	0.61
phenOcean.owl	15	1	2	2	0.0014	13	0.6
stateTime.owl	83	5	3	5	0.0015	7	0.5
A	104.6	4.82	5.6	5.6	0.0049	544	0.22

SWEET ontology modules listed in Table 4.15 are candidates for ontology refactoring, i.e. they should be examined in order to determine whether their cohesion can be improved by moving some axioms to other ontology modules. It can be seen that poorly cohesive SWEET modules have at least two times higher conductance compared to the average conductance. Poorly cohesive SWEET modules also tend to be smaller in size and less internally complex compared to an average SWEET ontology module. Additionally, poorly cohesive SWEET modules are mostly located on the periphery of the ontology module network (betweenness centrality equal to zero or significantly less than the average betweenness centrality) and have a smaller functional importance compared to the rest of SWEET modules (the page rank of poorly cohesive modules is smaller than the average page rank). Therefore, we can conclude that SWEET exhibit a good degree of ontology cohesion: a very small number of ontology modules are poorly cohesive and poorly cohesive SWEET modules are not central SWEET modules.

4.5 Conclusions

In this chapter we demonstrated that our methodology for analysis of enriched software networks presented in one of the previous chapters can also be applied to enriched ontology networks. Using the ONGRAM tool we extracted and investigated the structure of enriched ontology module, class and subsumption networks representing the SWEET modularized ontology. The studied ontology describes terminology related to earth and environmental sciences in approximately 200 ontology modules. We showed that all three examined networks have either single or a giant weakly connected component exhibiting the small-world property. However, SWEET ontology networks exhibit different patterns of assortativity and strong connectivity at the different levels of abstraction. Using the metric-based comparison test we showed that the SWEET ontology module network contains a strongly connected core encompassing the most functionally important and the most internally reused ontology modules.

Contrary to previous findings, the results of degree distribution analysis revealed that examined ontology networks are not scale-free. However, analyzed networks contain hubs (highly coupled ontology modules and classes). Similarly to highly coupled software entities, highly coupled ontological entities in SWEET are caused dominantly by internal reuse of ontological entities. The application of the metric-based comparison test revealed that highly coupled SWEET modules tend to be more voluminous and more functionally important compared to loosely coupled SWEET modules.

We proposed graph clustering evaluation (GCE) metrics to evaluate cohesiveness of ontology modules. The empirical evaluation of GCE metrics revealed that they do not strongly correlate with metrics quantifying cohesion of ontology modules in isolation. This result further implies that ontology cohesion metrics based only on internal dependencies within a module are unable to identify ontology modules whose constituent elements form strong clusters in ontology networks.

A good ontology modularization should conform to the principle of low coupling and high cohesion. Another important aspect of ontology modularization is to enable efficient reuse of ontological modules and better understandability of the whole ontology. Our case study demonstrates that analysis of enriched ontology module networks can help us not only to understand how knowledge from a certain domain is actually modularized, but also to evaluate the quality of ontology modularization. The analysis of the SWEET enriched ontology module network revealed that nearly all SWEET modules tend to be Radicchi weak clusters which means that they possess a good degree of cohesion. Regarding coupling, we showed that SWEET contains highly coupled modules. However, highly coupled SWEET modules tend to be dominantly caused by internal reuse. Internal reuse of ontology modules can be considered as a good ontology engineering practice and cannot cause problems if an ontology module being reused was previously verified to be logically consistent. Taking into account everything previously said it could be concluded that SWEET possesses a good degree of ontology modularization. However, we also showed that the SWEET enriched ontology module network contains a strongly connected core encompassing more than half of SWEET modules including also the most internally reused and functionally the most important modules. Ontology modules involved in large cyclic dependencies are hard to comprehend since they cannot be organized into hierarchical layers that can be examined independently. If someone wants to understand one ontology module from the SWEET strongly connected core then he/she has to be fully aware of all ontology modules from the strongly connected core. Secondly, an external reuse of a module from the SWEET strongly connected core implies that the whole core has to be externally reused. Therefore, it can be concluded that the SWEET modularization does not enable an easy comprehension and efficient external reuse of the most important SWEET modules.

References

1. Alani, H., Brewster, C.: Ontology ranking based on the analysis of concept structures. In: Proceedings of the 3rd International Conference on Knowledge Capture, K-CAP '05, pp. 51–58. ACM, New York, NY, USA (2005). https://doi.org/10.1145/1088622.1088633
2. Alani, H., Brewster, C., Shadbolt, N.: Ranking ontologies with AKTiveRank. In: Proceedings of the 5th International Conference on The Semantic Web, ISWC'06, pp. 1–15. Springer, Berlin (2006). https://doi.org/10.1007/11926078_1
3. Alani, H., Dasmahapatra, S., O'Hara, K., Shadbolt, N.: Identifying communities of practice through ontology network analysis. IEEE Intell. Syst. **18**(2), 18–25 (2003). https://doi.org/10.1109/MIS.2003.1193653
4. Bao, J., Caragea, D., Honavar, V.: On the semantics of linking and importing in modular ontologies. In: Cruz, I., Decker, S., Allemang, D., Preist, C., Schwabe, D., Mika, P., Uschold, M., Aroyo, L. (eds.), The Semantic Web - ISWC 2006. Lecture Notes in Computer Science, vol. 4273, pp. 72–86. Springer, Berlin (2006). https://doi.org/10.1007/11926078_6
5. Bozsak, E., Ehrig, M., Handschuh, S., Hotho, A., Maedche, A., Motik, B., Oberle, D., Schmitz, C., Staab, S., Stojanovic, L., Stojanovic, N., Studer, R., Stumme, G., Sure, Y., Tane, J., Volz, R., Zacharias, V.: Kaon - towards a large scale semantic web. In: Proceedings of the Third International Conference on E-Commerce and Web Technologies, EC-WEB '02, pp. 304–313. Springer, London (2002)
6. Caraballo, A.A.M., Nunes, B.P., Lopes, G.R., Leme, L.A.P.P., Casanova, M.A.: Automatic Creation and Analysis of a Linked Data Cloud Diagram, pp. 417–432. Springer International Publishing, Cham (2016). https://doi.org/10.1007/978-3-319-48740-3_31
7. Cheng, G., Qu, Y.: Term dependence on the semantic web. In: Proceedings of the 7th International Conference on The Semantic Web, ISWC '08, pp. 665–680. Springer, Berlin (2008). https://doi.org/10.1007/978-3-540-88564-1_42
8. Cheng, G., Ge, W., Qu, Y.: Falcons: Searching and browsing entities on the semantic web. In: Proceedings of the 17th International Conference on World Wide Web, WWW '08, pp. 1101–1102. ACM, New York, NY, USA (2008). https://doi.org/10.1145/1367497.1367676
9. Clauset, A., Shalizi, C., Newman, M.: Power-law distributions in empirical data. SIAM Rev. **51**(4), 661–703 (2009). https://doi.org/10.1137/070710111
10. Coskun, G., Rothe, M., Teymourian, K., Paschke, A.: Applying community detection algorithms on ontologies for identifying concept groups. In: Kutz, O., Schneider, T. (eds.) Workshop on Modular Ontologies, vol. 230, pp. 12–24. IOS Press (2011). https://doi.org/10.3233/978-1-60750-799-4-12
11. Cuenca Grau, B., Parsia, B., Sirin, E.: Ontology integration using e-connections. In: Stuckenschmidt, H., Parent, C., Spaccapietra, S. (eds.) Modular Ontologies. Lecture Notes in Computer Science, vol. 5445, pp. 293–320. Springer, Berlin (2009). https://doi.org/10.1007/978-3-642-01907-4_14
12. d'Aquin, M.: Modularizing Ontologies, pp. 213–233. Springer, Berlin (2012). https://doi.org/10.1007/978-3-642-24794-1_10
13. Ding, L., Shinavier, J., Shangguan, Z., McGuinness, D.: SameAs Networks and Beyond: analyzing deployment status and implications of owl:sameAs in linked data. In: Patel-Schneider, P.F., Pan, Y., Hitzler, P., Mika, P., Zhang, L., Pan, J.Z., Horrocks, I., Glimm, B. (eds.) The Semantic Web ISWC 2010. Lecture Notes in Computer Science, vol. 6496, pp. 145–160. Springer, Berlin (2010). https://doi.org/10.1007/978-3-642-17746-0_10
14. Ensan, F., Du, W.: A knowledge encapsulation approach to ontology modularization. Knowl. Inf. Syst. **26**(2), 249–283 (2011). https://doi.org/10.1007/s10115-009-0279-y
15. Ensan, F., Du, W.: A semantic metrics suite for evaluating modular ontologies. Inf. Syst. **38**(5), 745–770 (2013). https://doi.org/10.1016/j.is.2012.11.012
16. Färber, M., Rettinger, A.: A statistical comparison of current knowledge bases. In: Joint Proceedings of the Posters and Demos Track of 11th International Conference on Semantic Systems - SEMANTiCS 2015 and 1st Workshop on Data Science: Methods, Technology and Applications (DSci15), pp. 18–21 (2015). http://ceur-ws.org/Vol-1481/paper6.pdf

17. Fernândez, J.D., Martînez-Prieto, M.A., de la Fuente Redondo, P., Gutiêrrez, C.: Characterising rdf data sets. J. Inf. Sci. (2017). https://doi.org/10.1177/0165551516677945

18. García, J., García-Peñalvo, F.J., Therón, R.: A Survey on Ontology Metrics, pp. 22–27. Springer, Berlin (2010). https://doi.org/10.1007/978-3-642-16318-0_4

19. Garlaschelli, D., Loffredo, M.: Patterns of link reciprocity in directed networks. Phys. Rev. Lett. **93**, 268,701 (2004). https://doi.org/10.1103/PhysRevLett.93.268701

20. Ge, W., Chen, J., Hu, W., Qu, Y.: Object link structure in the semantic web. In: Aroyo, L., Antoniou, G., Hyvnen, E., ten Teije,A., Stuckenschmidt, H., Cabral, L., Tudorache, T. (eds.) The Semantic Web: Research and Applications. Lecture Notes in Computer Science, vol. 6089, pp. 257–271. Springer, Berlin (2010). https://doi.org/10.1007/978-3-642-13489-0_18

21. Gennari, J.H., Musen, M.A., Fergerson, R.W., Grosso, W.E., Crubzy, M., Eriksson, H., Noy, N.F., Tu, S.W.: The evolution of Protégé: an environment for knowledge-based systems development. Int. J. Hum.-Comput. Stud. **58**(1), 89 – 123 (2003). https://doi.org/10.1016/S1071-5819(02)00127-1

22. Gil, R., Garca, R., Delgado, J.: Measuring the semantic web. AIS SIGSEMIS Bull. **1**(2), 69–72 (2004)

23. Guéret, C., Groth, P., van Harmelen, F., Schlobach, S.: Finding the achilles heel of the web of data: Using network analysis for link-recommendation. In: Proceedings of the 9th International Semantic Web Conference on the Semantic Web - Volume Part I, ISWC'10, pp. 289–304. Springer, Berlin (2010)

24. Guéret, C., Wang, S., Schlobach, S.: The web of data is a complex system - first insight into its multi-scale network properties. In: The European Conference on Complex Systems, ECCS 2010, pp. 1–12 (2010)

25. Guéret, C., Wang, S., Groth, P., Schlobach, S.: Multi-scale analysis of the web of data: a challenge to the complex system's community. Adv. Complex Syst. **14**(04), 587–609 (2011). https://doi.org/10.1142/S0219525911003153

26. Halstead, M.H.: Elements of Software Science (Operating and Programming Systems Series). Elsevier Science Inc., New York (1977)

27. Henry, S., Kafura, D.: Software structure metrics based on information flow. IEEE Trans. Softw. Eng. **SE-7**(5), 510–518 (1981). https://doi.org/10.1109/TSE.1981.231113

28. Hoser, B., Hotho, A., Jäschke, R., Schmitz, C., Stumme, G.: Semantic network analysis of ontologies. In: Proceedings of the 3rd European Conference on The Semantic Web: Research and Applications, ESWC'06, pp. 514–529. Springer, Berlin (2006). https://doi.org/10.1007/11762256_38

29. Luczak-Rösch, M., Tolksdorf, R.: On the topology of the web of data. In: Proceedings of the 24th ACM Conference on Hypertext and Social Media, HT '13, pp. 253–257. ACM, New York, NY, USA (2013). https://doi.org/10.1145/2481492.2481526

30. Ma, J., Chen, H.: Complex network analysis on TCMLS sub-ontologies. In: Third International Conference on Semantics, Knowledge and Grid, pp. 551–553 (2007). https://doi.org/10.1109/SKG.2007.25

31. McBride, B.: Jena: a semantic web toolkit. IEEE Internet Comput. **6**(6), 55–59 (2002). https://doi.org/10.1109/MIC.2002.1067737

32. McCabe, T.J.: A complexity measure. IEEE Trans. Softw. Eng. **2**(4), 308–320 (1976). https://doi.org/10.1109/TSE.1976.233837

33. Mrvar, A., Batagelj, V.: Analysis and visualization of large networks with program package Pajek. Complex Adapt. Syst. Model. **4**(1), 6 (2016). https://doi.org/10.1186/s40294-016-0017-8

34. Noy, N.F., Sintek, M., Decker, S., Crubezy, M., Fergerson, R.W., Musen, M.A.: Creating semantic web contents with Protégé-2000. IEEE Intell. Syst. **16**(2), 60–71 (2001). https://doi.org/10.1109/5254.920601

35. Oh, S., Yeom, H.Y., Ahn, J.: Cohesion and coupling metrics for ontology modules. Inf. Technol. Manag. **12**(2), 81–96 (2011). https://doi.org/10.1007/s10799-011-0094-5

36. Orme, A., Tao, H., Etzkorn, L.: Coupling metrics for ontology-based system. IEEE Softw. **23**(2), 102–108 (2006). https://doi.org/10.1109/MS.2006.46

37. Queiroz-Sousa, P.O., Salgado, A.C., Pires, C.E.S.: A method for building personalized ontology summaries. J. Inf. Data Manag. **4**(3), 236–250 (2013)
38. Radicchi, F., Castellano, C., Cecconi, F., Loreto, V., Parisi, D.: Defining and identifying communities in networks. Proc. Natl. Acad. Sci. **101**(9), 2658–2663 (2004). https://doi.org/10.1073/pnas.0400054101
39. Rakić, G.: Extendable and adaptable framework for input language independent static analysis. Ph.D. thesis, University of Novi Sad, Faculty of Sciences (2015)
40. Rakić, G., Budimac, Z.: Introducing enriched concrete syntax trees. In: Proceedings of the 14th International Multiconference on Information Society (IS), Collaboration, Software And Services In Information Society (CSS), pp. 211–214 (2011)
41. Raskin, R.G., Pan, M.J.: Knowledge representation in the semantic web for Earth and environmental terminology (SWEET). Comput. Geosci. **31**(9), 1119–1125 (2005). https://doi.org/10.1016/j.cageo.2004.12.004
42. Rodriguez, M.A.: A graph analysis of the linked data cloud. CoRR (2009). arXiv:abs/0903.0194
43. Savić, M., Budimac, Z., Rakić, G., Ivanović, M., Heričko, M.: SSQSA ontology metrics front-end. In: Proceedings of the 2nd Workshop on Software Quality Analysis, Monitoring, Improvement, and Applications, Novi Sad, Serbia, September 15–17, 2013, pp. 95–101 (2013). http://ceur-ws.org/Vol-1053/sqamia2013paper12.pdf
44. Savić, M., Rakić, G., Budimac, Z.: Translation of Tempura specifications to eCST. AIP Conf. Proc. **1738**(1), 240,009 (2016). https://doi.org/10.1063/1.4952028
45. Sicilia, M., Rodrguez, D., Garca-Barriocanal, E., Sinchez-Alonso, S.: Empirical findings on ontology metrics. Expert Syst. Appl. **39**(8), 6706 – 6711 (2012). https://doi.org/10.1016/j.eswa.2011.11.094
46. Stuckenschmidt, H., Parent, C., Spaccapietra, S. (eds.): Modular Ontologies: Concepts, Theories and Techniques for Knowledge Modularization. Lecture Notes in Computer Science, vol. 5445. Springer, Berlin (2009). https://doi.org/10.1007/978-3-642-01907-4
47. Tartir, S., Arpinar, I.B., Moore, M., Sheth, A.P., Aleman-Meza, B.: OntoQA: Metric-based ontology quality analysis. In: IEEE Workshop on Knowledge Acquisition from Distributed, Autonomous, Semantically Heterogeneous Data and Knowledge Sources (2005)
48. Theoharis, Y., Georgakopoulos, G., Christophides, V.: On the synthetic generation of semantic web schemas. In: Christophides, V., Collard, M., Gutierrez, C. (eds.) Semantic Web, Ontologies and Databases. Lecture Notes in Computer Science, vol. 5005, pp. 98–116. Springer, Berlin (2008). https://doi.org/10.1007/978-3-540-70960-2_6
49. Theoharis, Y., Tzitzikas, Y., Kotzinos, D., Christophides, V.: On graph features of semantic web schemas. IEEE Trans. Knowl. Data Eng. **20**(5), 692–702 (2008). https://doi.org/10.1109/TKDE.2007.190735
50. Vrandečić, D.: Ontology Evaluation, pp. 293–313. Springer, Berlin (2009). https://doi.org/10.1007/978-3-540-92673-3_13
51. Zhang, H.: The scale-free nature of semantic web ontology. In: Proceedings of the 17th International Conference on World Wide Web, WWW '08, pp. 1047–1048. ACM, New York, NY, USA (2008). https://doi.org/10.1145/1367497.1367649
52. Zhang, H., Li, Y.F., Tan, H.B.K.: Measuring design complexity of semantic web ontologies. J. Syst. Softw. **83**(5), 803–814 (2010). https://doi.org/10.1016/j.jss.2009.11.735
53. Zhang, X., Cheng, G., Qu, Y.: Ontology summarization based on RDF sentence graph. In: Proceedings of the 16th International Conference on World Wide Web, WWW '07, pp. 707–716. ACM, New York, NY, USA (2007). https://doi.org/10.1145/1242572.1242668

Part III
Co-authorship Networks: Social Networks of Research Collaboration

Chapter 5
Co-authorship Networks:
An Introduction

Abstract In this chapter we introduce and formally define co-authorship networks. Formal definitions of co-authorship networks as undirected graphs, directed graphs and hypergraphs are given. Different schemes to assign weights to co-authorship links are also discussed. Then, we give a classification of co-authorship networks according to the type of research collaboration they represent. Finally, the main applications of co-authorship networks are outlined.

Research collaboration is one of the key social features of modern science [5, 10, 24]. From a social perspective, modern science can be viewed as a complex self-organizing social system since it is mostly done in collaboration due to:

- the increasing growth of the scientific community,
- the professionalization and institutionalization of scientific research practice,
- the rationalization of scientific manpower,
- information and knowledge overload,
- the increasing complexity of contemporary scientific problems,
- the increasing specialization of researchers,
- the increasing mobility of researchers,
- the demands of complex instrumentation,
- the need to make progress more rapidly,
- the need to produce accurate research, reduce errors and mistakes and find flaws more efficiently,
- the need to gain experience and learn new skills, techniques or tacit knowledge from others,
- the desire to increase cross-fertilization of ideas,
- research funding, and many other socio-psychological and socio-economic factors.

Katz emphasizes that "scientific collaboration is a social process and probably there are as many reasons for researchers to collaborate as there are reasons for people to communicate" [15]. Additionally, society recognized the importance of research collaboration and there are numerous initiatives, governmental policies and international exchange programs aimed at stimulating mobility of researchers and collaborative research [8, 14, 29, 35].

Empirical studies investigating scientific collaboration date back to the 1960s. Derek J. de Solla Price in his book "Little Science Big Science" published in 1963 noted that the proportion of multi-authored publications in chemistry had rapidly increased since the beginning of the 20th century in a way that if the same trend continued there would be no single-authored publications by 1980 [28]. A more recent resurgence of interest in the field was sparked by the observation of the small-world and scale-free phenomena in various types of real-world networks including also complex networks of research collaboration extracted from massive bibliographic databases [3, 25, 26]. Research collaboration can be investigated at various levels: intra-institutional (within an institution), inter-institutional (between institutions), national (within a country), international (between countries), disciplinary (within a scientific discipline), inter-disciplinary (between scientific disciplines), and so on. At all of these levels, major research questions are how research collaboration is structured, how it evolves, and how it is related to research productivity and the impact of multi-authored publications.

Research collaboration is difficult to rigorously define since it may manifest in various formal and informal forms [16, 24]. However, research collaboration can be reliably captured by so-called *co-authorship networks* since co-authorship is one of the most visible and well-documented manifestation of scientific collaboration [10]. Although co-authorship is no more than a partial indicator of collaboration, there are several advantages of relying on co-authorship as a proxy to research collaboration including its verifiability, stability over time, data availability and ease of measurement [16, 40]. Co-authorship is a form of association in which two or more researchers jointly report their research results on some topic. Therefore, co-authorship networks can be viewed as social networks encompassing researchers that reflect collaboration among them. Researchers are represented by nodes in co-authorship networks. Two nodes are connected if corresponding researchers co-authored at least one publication together with or without other co-authors. Additionally, link weights can be introduced in order to express the strength of research collaboration.

It should be emphasized that co-authorship networks are quite different from so-called *citation networks* [31, 39] which are another important type of complex networks related to scientometrics. Nodes in citation networks represent publications that are connected by directed links reflecting citations between publications. Therefore, citation networks are information networks showing the structure of scientific knowledge, while co-authorship networks are social networks depicting the structure of academic societies. By aggregating those two types of complex networks we are able to analyze and model interactions and mutual influences between collaboration and citation practices [6, 23, 43].

The rest of the chapter is structured as follows. Formal definitions of co-authorship networks as undirected graphs, directed graphs and hypergraphs are discussed in Sects. 5.1, 5.2 and 5.3, respectively. In Sect. 5.4 we give a classification of co-authorship networks according to the type of research collaboration they represent. The main applications of co-authorship networks are outlined in the last section.

5.1   Co-authorship Networks as Undirected Graphs

Let U be a set of bibliographic units (journal and conference publications, books, and so on), and let A be the set of authors appearing in U. The co-authorship network corresponding to U is most commonly defined as an undirected and weighted graph $G = (V, E)$ with the following properties:

- The set of nodes V corresponds to the set of authors A, i.e. each author from A is represented by one node in G.
- Two authors x and y are connected by an undirected link e ($e \in E$) if there is at least one bibliographic unit in U jointly co-authored by x and y (with or without other co-authors). In other words, co-authorship networks are author-centered one-mode projections of bipartite networks linking researchers to bibliographic units they (co-)authored.
- Link weights express the strength of collaboration between connected authors.

Three weighting schemes are commonly used to assign weights to co-authorship links: the straight, Newman and Salton scheme. In the straight weighting scheme, two authors are connected by an undirected link of weight w if they co-authored exactly w different bibliographic units from U [4]. The Salton scheme assigns a weight w to a link connecting authors x and y by the following formula:

$$w = \frac{h_{x,y}}{\sqrt{h_x \cdot h_y}} \tag{5.1}$$

where $h_{x,y}$ is the number of bibliographic units co-authored by x and y, h_x is the number of bibliographic units authored by x, and h_y is the number of bibliographic units authored by y [21]. It can be seen that the Salton scheme is actually a normalized variant of the straight weighting scheme. Namely, w satisfies $0 < w \leq 1$ and the maximal value of w ($w = 1$) is achieved when the set of joint bibliographic units of x and y is equal to the set of bibliographic units authored by x and to the set of bibliographic units authored by y.

The Newman weighting scheme takes into account the total number of authors in multi-authored bibliographic units [27]. Let J denote the set of publications jointly authored by x and y. Then x and y are connected by a link of weight w that is determined by the following formula:

$$w = \sum_{k \in J} \frac{1}{n_k - 1} \tag{5.2}$$

where n_k is the number of authors of a bibliographic unit k. In other words, each joint bibliographic unit of x and y adds some weight to the overall strength of research collaboration between x and y. However, the more authors a joint bibliographic unit has the less weight is added.

One of the ways to extend the most commonly used definition of co-authorship networks is to introduce node attributes. Those attributes can express different characteristics of authors such as productivity (in terms of one or more productivity measures), impact (e.g. h-index, the total number of received citations), career longevity (the time passed from the publication of the first to the publication of the last bibliographic unit of an author), and so on.

5.2 Co-authorship Networks as Directed Graphs

It is also possible to define co-authorship networks as directed graphs. Yoshikane et al. proposed the following definition: two researchers x and y are connected by a directed link $x \rightarrow y$ of weight w if x and y co-authored exactly w different bibliographic units in which y is the first author [47]. This definition emphasizes the leading role of the first author of an article. Additionally, it enables the application of directed centrality measures to rank authors with respect to leading and following behavior in research collaboration. The authors proposed a modification of HITS hub and authority scores that takes into account link weights. The proposed measures enable the identification of research authorities (researchers having a high weighted authority score), as well as important followers of influential authorities (those are researchers having a high weighted hub score). The weighted hub score, denoted by $H(x)$, and the weighted authority score, denoted by $A(x)$, are mutually recursive measures of centrality that can be expressed as follows:

$$H(x) = \sum_{y \in O(x)} \big(A(y) \times weight(x \rightarrow y) \big) \tag{5.3}$$

$$A(x) = \sum_{z \in I(x)} \big(H(z) \times weight(z \rightarrow x) \big) \tag{5.4}$$

where $I(x)$ is the set of nodes pointing to x and $O(x)$ is the set of nodes to which x points. In the same way as for the original HITS hub and authority, weighted hub and authority scores can be computed by successive approximations starting from an initial assignment in which all nodes have equal values of those two measures.

Liu et al. took slightly different approach inspired by a notion of exclusivity of research collaboration [20]. In their definition of co-authorship networks, two co-authors are connected by a pair of reciprocal directed links. The weight of directed co-authorship links is determined according to the exclusivity of research collaboration based on the two following principles:

1. links connecting frequent collaborators should have higher weights than links connecting less frequent collaborators, and

2. the total number of authors of an article determines the exclusivity of co-authorship of two researchers on this particular article in the sense that if the article has many co-authors each individual co-authorship relation should be weighted less.

Formally, the weight of a link $x \rightarrow y$ is computed by the following formula

$$w(x \rightarrow y) = \frac{c(x, y)}{\sum_{k \in V \setminus \{x\}} c(x, k)} \qquad (5.5)$$

where V is the set of nodes in the network and $c(x, y)$ quantifies the co-authorship frequency of x with y on all articles authored by x. The co-authorship frequency takes into account the exclusivity of research collaboration, i.e.

$$c(x, y) = \sum_{p \in P_x} e(x, y, p) \qquad (5.6)$$

where P_x denotes the set of articles authored by x and $e(x, y, p)$ is the exclusivity of research collaboration between x and y on article p. The exclusivity of research collaboration between two authors on an article is defined according to the number of co-authors of the article, i.e.

$$e(x, y, p) = 1/(n_p - 1) \qquad (5.7)$$

where n_p is the number of authors of article p. It can be noticed that link weights are in the interval $(0, 1]$ and that the weights of links emanating from a node sum to one.

Liu et al. emphasized that their definition of co-authorship link weights corresponds to the probability distribution of a random walk on the co-authorship network: the weight of $x \rightarrow y$ can be interpreted as the probability that the random walker moves from x to y in one random walk step. Consequently, they proposed a modification of the page rank measure called *author rank* that takes into account link weights. More formally, the author rank of x can be expressed as

$$AR(x) = (1 - d) + d \sum_{z \in I_x} \left(AR(z) \times weight(z \rightarrow x) \right) \qquad (5.8)$$

where I_x is the set of nodes pointing to x in the co-authorship network and d is the damping factor. Similar to the page rank measure, the author rank measure can be computed by successive approximations starting from an initial assignment in which all nodes have an equal author rank.

5.3 Co-authorship Networks as Hypergraphs

Co-authorship networks can be also modeled by hypergraphs. A hypergraph is a generalization of a graph such that a link can connect any number of nodes. Let P be a set of bibliographic units and let A denote the set of authors of publications in P. Then, the co-authorship network corresponding to P can be represented as a hypergraph $G = (V, E)$ where

- The set of nodes V is equal to A, i.e each researcher is represented by one node in G.
- A hyperlink $e = \{a_1, a_2, ..., a_k\}$, $a_1, a_2, ..., a_k \in V$, represents all bibliographic units from P authored exactly by $a_1, a_2, ..., a_k$. The weight of e is the number of such bibliographic units.

Han et al. [11] exploited the hypergraph representation of scientific collaboration to define a co-authorship supportiveness measure and a supportiveness-based author ranking scheme. The co-authorship supportiveness measure reflects how important is the collaboration between two researchers to each of them. The authors firstly introduced a measure called contribution. This measure, denoted by $\mathrm{cont}(a \leftarrow b)$, quantifies how much the collaboration between two researchers a and b contributes to the total work of a. It is computed by the following formula

$$\mathrm{cont}(a \leftarrow b) = \frac{\sum_{e \in E(a,b)} \mathrm{weight}(e)}{L(a)} \tag{5.9}$$

where

- $E(a, b)$ is the subset of hyperlinks containing both a and b, and
- $L(a)$ is the total number of papers co-authored by a.

Relying on the contribution measure, the authors defined the closeness between a and b as the harmonic mean of $\mathrm{cont}(a \leftarrow b)$ and $\mathrm{cont}(b \leftarrow a)$. Then, the distance between a and b can be defined as the inverse of their supportiveness-based closeness. Finally, the supportiveness of author a is defined as

$$\mathrm{sup}(a) = \sum_{b \in \mathrm{RNN}(a)} |\mathrm{NN}(b)|^{-1} \tag{5.10}$$

where

- $\mathrm{NN}(x)$ is the set of nearest neighbors of x according to the previously defined distance measure, and
- $\mathrm{RNN}(x)$ is the set of reverse nearest neighbors of x, i.e. $RNN(x) = \{y \in V : x \in \mathrm{NN}(y)\}$

The supportiveness measure mimics a simple voting process. Each researcher has exactly one vote. If x has exactly one nearest neighbor y then x votes for y. Otherwise,

x splits its vote evenly to each of its nearest neighbors. Then, the supportiveness of a researcher is the number of received votes. The authors also extended the previously given definition of supportiveness taking into account k-nearest neighbors. They also developed a fast algorithm for identifying top-n most supportive authors in a network.

5.4 Types of Co-authorship Networks

The scope of a co-authorship network is determined by the set of nodes contained in the network. Considering this aspect several types of co-authorship networks can be distinguished.

Field co-authorship networks. Field co-authorship networks represent collaboration among researchers working in a specific scientific field (such as physics, mathematics, computer science, and so on) or some more narrower research discipline (e.g., genetic programming in computer science or graph theory in mathematics). Usually, field co-authorship networks are extracted using data provided by massive bibliography databases (e.g., Web of Science, DBLP) or (preprint) digital libraries (e.g., arXiv).

National co-authorship networks. A national or domestic co-authorship network represents research collaboration among researchers affiliated to institutions from one country. Here we can separate between co-authorship networks that include all (registered) researchers in a country and co-authorship networks that encompass only a subset of scholars tied to one specific or more related scientific fields (e.g. co-authorship network of Hungarian mathematicians).

Publication venue co-authorship networks. These co-authorship networks depict research collaboration in a community of researchers that publish their results in specific journals or conferences.

Institutional co-authorship networks represent collaboration among researchers employed in the same institution or some organizational unit within an institution. Those network enable analysis of intra-institutional research collaboration.

Networks representing inter-institutional and international research collaboration can be derived from co-authorship networks when information about institutional affiliation of researchers is available. Let us a suppose that we have a field co-authorship network G such that we know the institutional affiliation of each node in G. Let I denote the set of institutions (resp. countries) appearing in G. The network of inter-institutional (resp. international) collaboration derived from G can be represented as an undirected, weighted graph $G^I = (I, E^I)$ such that

1. Two institutions (resp. countries) A and B ($A, B \in I$) are connected in G^I ($\{A, B\} \in E^I$) if there are at least two researchers a and b, a from A and b from B, that are connected in G.
2. The weight of the link connecting A and B can be defined as the sum of weights of all links $\{a, b\}$ in G where a is from A and b is from B.

5.5   Applications of Co-authorship Networks

The main application of co-authorship networks is to study the structure and evolution of scientific collaboration [3, 25, 26]. The degree of a node, as the most basic centrality metric of nodes in complex networks, has a very important application in the context of co-authorship networks. The degree of a node R in a co-authorship network is equal to the number of collaborators of R. Therefore, this measure quantifies the capacity of individual researchers to collaborate with other researchers. Additionally, co-authorship networks are weighted graphs thus we can also measure the strength of nodes, i.e. the total weight of links incident with a node. This measure reflects the strength of research collaboration of a researcher with his/her co-authors. By investigating statistical properties of node degree and strength distributions we are able to determine whether there are inequalities and gaps in research collaboration within a scientific field, country or institution (depending on the type of an analyzed co-authorship network). Moreover, by comparing statistical properties of node degree/strength distributions for co-authorship networks of the same type we can spot differences in patterns of scientific collaboration characteristic to different scientific fields, countries or institutions.

If a researcher is an isolated node in a co-authorship network then he/she has not established any research collaboration with other researchers present in the network. Two researchers belong to the same connected component of the co-authorship network if they are either directly or indirectly connected. The identification of isolated researchers and connected components in co-authorship networks has two important applications: (1) characteristics of isolated nodes and connected components indicate the overall cohesiveness of a community of researchers represented by the network, and (2) we can spot weak points where research collaboration can be considerably improved by stimulating collaboration between researchers belonging to different connected components.

In one of our previous works, we proposed a classification scheme of connected components in co-authorship networks into trivial and non-trivial connected components [36]. Trivial connected components are components that do not evolve, i.e. those are isolated completely-connected subgraphs (cliques) of co-authorship networks that reflect research collaboration on only one publication unit. The identification of trivial connected components depends on the scheme that is used to assign weights to co-authorship links. For example, if the straight weighting scheme is used then trivial connected components correspond to isolated cliques in which the weight of all links is equal to 1. In the same paper we showed that the proposed classification of connected components is particularly useful when studying the evolution of fragmented co-authorship networks (co-authorship networks in which giant connected components do not emerge).

Various centrality measures can be applied to a co-authorship network in order to rank and identify the most important researchers present in the network. We already emphasized that directed centrality measures, such as page rank and HITS hub and authority scores, can be employed when scientific collaboration is represented by a directed graph. In the case of undirected co-authorship networks, we can employ three most fundamental node centrality measures to rank researchers: betweenness, closeness and eigenvector centrality. Those measures can be also utilized when co-authorship networks are given as directed graphs. Another way to quantify the importance of researchers is to rely on weighted variants of previously mentioned node centrality metrics. In this way the strength of research collaboration is also taken into account when ranking researchers. The weight of a co-authorship link quantifies the strength of research collaboration between two researchers. Consequently, higher weights imply smaller distances between connected researchers. Therefore, weighted centrality measures based on shortest paths (e.g. betweenness and closeness centrality) should be computed on weighted co-authorship networks in which link weights are inverted (e.g. the weight of each link is transformed into $1/w$ where w is the original weight).

The evolution of co-authorship networks reveals temporal patterns and evolutionary trends in scientific collaboration [3, 36]. The evolution of a co-authorship network can be investigated at the microscopic, mesoscopic and macroscopic level [13]. At the microscopic level, we can examine how characteristics of individual nodes (researchers) and links (collaborations between researchers) evolve in time. Evolutionary analysis at the mesoscopic level deals with the evolution of network subgraphs representing research groups. At this level we can also investigate how collaboration between research groups change in time. Also, we are often interested to spot important evolutionary events such as merging and splitting of research groups. Analysis of co-authorship network evolution at the macroscopic level is related to evolutionary dynamics of global network properties. The evolution of a co-authorship network is usually investigated by constructing time-ordered yearly snapshots of the network and analyzing time series of some quantifiable property of nodes, links, subgraphs or the network as a whole. There are two principal approaches to construct the time-ordered snapshots of the network [7, 41]:

1. *Cumulative approach*: the snapshot of the network at year y is constructed taking into account all publication units published before or during y.
2. *Sliding window approach*: the snapshot of the network at year y is constructed considering publication units published during y or in the last k years prior to y, where k is the length of the sliding window.

Co-authorship networks can also be exploited to cluster researchers and determine research groups. The study of community structures in co-authorship networks is particularly important since research groups from various (sub-)disciplines might often display local properties that differ significantly from the properties of the network as a whole [22]. The problem of co-authorship network clustering is usually approached using community detection techniques since they are computationally efficient and do not require to define the number of clusters in advance. It is quite common that researchers when proposing new community detection techniques test them on co-authorship networks. For example, the Girvan–Newman algorithm was experimentally evaluated by the authors on four different complex networks including also one co-authorship network. Other community detection techniques whose implementations can be found in widely used complex network analysis libraries, frameworks and tools were also experimentally investigated by their authors on co-authorship networks. The classification of communities proposed by Radicchi et al. [30] can be utilized to assess the quality of clusters obtained by community detection techniques applied to weighted co-authorship networks. Namely, Radicchi strong clusters correspond to research groups in which each researcher established a stronger research collaboration with members of his/her group than with researchers belonging to other groups. On the other hand, a research group is a Radicchi weak cluster if the total strength of research collaboration within the group is strictly higher than the total strength of research collaboration between members of the group and researchers outside the group. Additionally, graph clustering evaluation metrics [18] can be employed to assess cohesiveness of research groups [34]. In one of our previous works, we compared five different community detection algorithms suitable for weighted graphs (the Girvan–Newman algorithm, Walktrap, Infomap, Label propagation and Louvain) on a co-authorship network of researchers publishing in Serbian mathematical journals [37]. The obtained results revealed that the Louvain method shows the best clustering performance – this algorithm achieves the highest value of the modularity measure and identifies the largest fraction of Radicchi strong clusters in the analyzed network. In a subsequent work, we proposed a new community detection technique for weighted co-authorship networks that is based on frequent collaborator cores [38]. The main idea of the algorithm is to determine so-called w-cores – maximal subgraphs of the network such that the weight of each link in these subgraphs is higher than some prespecified threshold. Community labels are assigned to nodes in w-cores such that two nodes belonging to the same w-core have the same label. After that, community labels are propagated to nodes not belonging to w-cores according to the following rules:

1. The safe label propagation rule: if all neighbors of an unlabeled node n have the same label l then n will be also labeled by l.
2. If the safe label propagation rule cannot be applied and there are unlabeled nodes then a tie resolution strategy is employed for unlabeled neighbors of labeled nodes. A tie node adopts the label of a community to which it has the strongest connection according to the weights of incident links. Additionally, tie nodes are resolved in the decreasing order of weighted degree because tie nodes may be connected among themselves.
3. The safe label propagation step continues after tie nodes are resolved.

The evaluation of the algorithm on a co-authorship network representing collaboration among researchers employed at the same research department showed that the algorithm identifies Radicchi strong clusters of researchers corresponding to research groups dealing with specific research topics. The comparison with 7 other community detection techniques provided by the iGraph library revealed that our method performs better or equally with respect to the cohesiveness of obtained clusters.

Bibliometric indicators such as the impact factor and h-index are commonly used to evaluate and compare scientific impact of journals. On the other hand, structural properties of journal co-authorship networks enable us to observe differences between journals that are related to the collaborative behavior of their authors rather than the scientific impact of their work. For example, in one of our previous work we compared the structure of co-authorship networks corresponding to 10 mathematical journals published in Serbia [36]. The obtained results enabled us to identify journals publishing mostly single-authored papers from journals which attract collaborative groups of authors.

Another important application of co-authorship networks is related to the prediction and recommendation of research collaboration. The research on this topic intensively started to develop after the seminal paper by Liben-Nowell and Kleinberg [19]. The authors proposed several link prediction metrics based on different node proximity notions and evaluated them on five co-authorship network extracted from bibliographic records contained in the arXiv preprint database. Link prediction metrics can be also exploited for the identification of missing links in co-authorship networks that are caused by missing publications in bibliographic databases [17]. The problem of predicting and recommending links in co-authorship networks can be also approached using supervised machine learning techniques based on features that include both domain-independent link prediction metrics (metrics derived solely from network topology) and domain-dependent node similarity measures [2, 12, 44].

Co-authorship networks can be also used to recommend reviewers for a manuscript [32]. The algorithm proposed by Rodriguez and Bollen [32] identifies appropriate referees for a manuscript by applying a particle-swarm propagation algorithm to a co-authorship network. Particles propagated through the network are initially placed into nodes representing authors cited in the bibliography of an article for which referees are requested. The authors empirically validated their algorithm using referee bid data from the 2005 Joint Conference on Digital Libraries and the co-authorship network extracted from bibliographic records contained in the DBLP database.

Finally, co-authorship networks can be exploited to predict scientific success of researchers. Several empirical studies indicated that there are strong correlations between node centrality metrics in co-authorship networks and the impact of corresponding researchers estimated by citation-based measures [1, 9, 42, 45, 46]. Moreover, Sarigöl et al. [33] showed that a random forest classifier based on features involving only centrality metrics computed on a co-authorship network is able to predict with a high precision whether an article will be highly cited in the future.

References

1. Abbasi, A., Chung, K.S.K., Hossain, L.: Egocentric analysis of co-authorship network structure, position and performance. Inf. Process. Manag. **48**(4), 671–679 (2012). https://doi.org/10.1016/j.ipm.2011.09.001
2. Backstrom, L., Leskovec, J.: Supervised random walks: Predicting and recommending links in social networks. In: Proceedings of the Fourth ACM International Conference on Web Search and Data Mining, WSDM '11, pp. 635–644. ACM, USA (2011). https://doi.org/10.1145/1935826.1935914

3. Barabasi, A.L., Jeong, H., Neda, Z., Ravasz, E., Schubert, A., Vicsek, T.: Evolution of the social network of scientific collaborations. Phys. A **311**, 590–614 (2002)
4. Batagelj, V., Cerinšek, M.: On bibliographic networks. Scientometrics **96**(3), 845–864 (2013). https://doi.org/10.1007/s11192-012-0940-1
5. Beaver, D.d., Rosen, R.: Studies in scientific collaboration. Scientometrics **1**(1), 65–84 (1978). https://doi.org/10.1007/BF02016840
6. Borner, K., Maru, J.T., Goldstone, R.L.: The simultaneous evolution of author and paper networks. Proc. Natl. Acad Sci. U. S. A. **101**(Suppl 1), 5266–5273 (2004). https://doi.org/10.1073/pnas.0307625100
7. Brunson, J.C., Fassino, S., McInnes, A., Narayan, M., Richardson, B., Franck, C., Ion, P., Laubenbacher, R.: Evolutionary events in a mathematical sciences research collaboration network. Scientometrics **99**(3), 973–998 (2014). https://doi.org/10.1007/s11192-013-1209-z
8. Defazio, D., Lockett, A., Wright, M.: Funding incentives, collaborative dynamics and scientific productivity: Evidence from the EU framework program. Res. Policy **38**(2), 293–305 (2009). https://doi.org/10.1016/j.respol.2008.11.008
9. Fischbach, K., Putzke, J., Schoder, D.: Co-authorship networks in electronic markets research. Electron. Mark. **21**(1), 19–40 (2011). https://doi.org/10.1007/s12525-011-0051-5
10. Glänzel, W., Schubert, A.: Analysing Scientific Networks Through Co-Authorship, pp. 257–276. Springer, Netherlands, Dordrecht (2005). https://doi.org/10.1007/1-4020-2755-9_12
11. Han, Y., Zhou, B., Pei, J., Jia, Y.: Understanding importance of collaborations in co-authorship networks: A supportiveness analysis approach. In: Proceedings of the 2009 SIAM International Conference on Data Mining, pp. 1112–1123 (2009). https://doi.org/10.1137/1.9781611972795.95
12. Hasan, M.A., Chaoji, V., Salem, S., Zaki, M.: Link prediction using supervised learning. In: In Proceedings of SDM 06 workshop on Link Analysis, Counterterrorism and Security (2006)
13. Huang, J., Zhuang, Z., Li, J., Giles, C.L.: Collaboration over time: Characterizing and modeling network evolution. In: Proceedings of the 2008 International Conference on Web Search and Data Mining, WSDM '08, pp. 107–116. ACM, New York, USA (2008). https://doi.org/10.1145/1341531.1341548
14. Jacob, M., Meek, V.L.: Scientific mobility and international research networks: trends and policy tools for promoting research excellence and capacity building. Stud. High. Educ. **38**(3), 331–344 (2013). https://doi.org/10.1080/03075079.2013.773789
15. Katz, J.: Geographical proximity and scientific collaboration. Scientometrics **31**(1), 31–43 (1994). https://doi.org/10.1007/BF02018100
16. Katz, J., Martin, B.R.: What is research collaboration?. Res. Policy **26**(1), 1–18 (1997). https://doi.org/10.1016/S0048-7333(96)00917-1
17. Lu, L., Zhou, T.: Link prediction in complex networks: A survey. Phys. A Stat. Mech. Appl. **390**(6), 1150–1170 (2011). https://doi.org/10.1016/j.physa.2010.11.027
18. Leskovec, J., Lang, K.J., Mahoney, M.: Empirical comparison of algorithms for network community detection. In: Proceedings of the 19th International Conference on World Wide Web, WWW '10, pp. 631–640. ACM, New York, NY, USA (2010)
19. Liben-Nowell, D., Kleinberg, J.: The link prediction problem for social networks. In: Proceedings of the Twelfth International Conference on Information and Knowledge Management, CIKM '03, pp. 556–559. ACM, New York, NY, USA (2003). https://doi.org/10.1145/956863.956972
20. Liu, X., Bollen, J., Nelson, M.L., Van de Sompel, H.: Co-authorship networks in the digital library research community. Information Processsesing and Management **41**(6), 1462–1480 (2005). https://doi.org/10.1016/j.ipm.2005.03.012
21. Lu, H., Feng, Y.: A measure of authors centrality in co-authorship networks based on the distribution of collaborative relationships. Scientometrics **81**(2), 499–511 (2009). https://doi.org/10.1007/s11192-008-2173-x
22. Mali, F., Kronegger, L., Doreian, P., Ferligoj, A.: Dynamic Scientific Co-Authorship Networks, pp. 195–232. Springer Berlin Heidelberg, Berlin, Heidelberg (2012). https://doi.org/10.1007/978-3-642-23068-4_6

23. Martin, T., Ball, B., Karrer, B., Newman, M.E.J.: Coauthorship and citation patterns in the Physical Review. Phys. Rev. E **88**, 012,814 (2013). https://doi.org/10.1103/PhysRevE.88.012814
24. Milojević, S.: Modes of collaboration in modern science: Beyond power laws and preferential attachment. J. Am. Soc. Inf. Sci. Tech. **61**(7), 1410–1423 (2010). https://doi.org/10.1002/asi.21331
25. Newman, M.E.J.: Scientific collaboration networks I: network construction and fundamental results. Phys. Rev. E **64**, 016131 (2001). https://doi.org/10.1103/PhysRevE.64.016131
26. Newman, M.E.J.: Scientific collaboration networks II: shortest paths, weighted networks, and centrality. Phys. Rev. E **64**, 016132 (2001). https://doi.org/10.1103/PhysRevE.64.016132
27. Newman, M.E.J.: Who is the best connected scientist? A study of scientific coauthorship networks. In: Ben-Naim E., Frauenfelder H., Toroczkai Z. (eds.) Complex Networks. Lecture Notes in Physics, vol. 650, pp. 337–370. Springer, Berlin, Heidelberg (2004). https://doi.org/10.1007/978-3-540-44485-5_16
28. Price, D.J.d.S.: Little Science, Big Science. Columbia Univeristy Press, New York (1963)
29. Protogerou, A., Caloghirou, Y., Siokas, E.: Policy-driven collaborative research networks in Europe. Econ. Innov. New Tech. **19**(4), 349–372 (2010). https://doi.org/10.1080/10438590902833665
30. Radicchi, F., Castellano, C., Cecconi, F., Loreto, V., Parisi, D.: Defining and identifying communities in networks. Proc. Natl. Acad. Sci. **101**(9), 2658–2663 (2004). https://doi.org/10.1073/pnas.0400054101
31. Radicchi, F., Fortunato, S., Vespignani, A.: Citation Networks, pp. 233–257. Springer, Berlin, Heidelberg (2012). https://doi.org/10.1007/978-3-642-23068-4_7
32. Rodriguez, M.A., Bollen, J.: An algorithm to determine peer-reviewers. In: Proceedings of the 17th ACM Conference on Information and Knowledge Management, CIKM '08, pp. 319–328. ACM, New York, USA (2008). https://doi.org/10.1145/1458082.1458127
33. Sarigöl, E., Pfitzner, R., Scholtes, I., Garas, A., Schweitzer, F.: Predicting scientific success based on coauthorship networks. EPJ Data Sci. **3**(1), 9 (2014). https://doi.org/10.1140/epjds/s13688-014-0009-x
34. Savić, M., Ivanović, M., Dimic Surla, B.: Analysis of intra-institutional research collaboration: a case of a Serbian faculty of sciences. Scientometrics **110**(1), 195–216 (2017). https://doi.org/10.1007/s11192-016-2167-z
35. Savić, M., Ivanović, M., Putnik, Z., Tütüncü, K., Budimac, Z., Smrikarova, S., Smrikarov, A.: Analysis of ERASMUS staff and student mobility network within a big European project. In: 40th International Convention on Information and Communication Technology, Electronics and Microelectronics, MIPRO 2017, Opatija, Croatia, 22–26 May 2017, pp. 613–618 (2017). https://doi.org/10.23919/MIPRO.2017.7973498
36. Savić, M., Ivanović, M., Radovanović, M., Ognjanović, Z., Pejović, A., Jakšić Kruger, T.: The structure and evolution of scientific collaboration in Serbian mathematical journals. Scientometrics **101**(3), 1805–1830 (2014). https://doi.org/10.1007/s11192-014-1295-6
37. Savić, M., Ivanović, M., Radovanović, M., Ognjanović, Z., Pejović, A., Jakšić Kruger, T.: Exploratory analysis of communities in co-authorship networks: A case study. In: Bogdanova A.M., Gjorgjevikj D. (eds.) ICT Innovations 2014. Advances in Intelligent Systems and Computing, vol. 311, pp. 55–64. Springer International Publishing, New York (2015). https://doi.org/10.1007/978-3-319-09879-1_6
38. Savić, M., Ivanović, M., Surla, B.D.: A community detection technique for research collaboration networks based on frequent collaborators cores. In: Proceedings of the 31st Annual ACM Symposium on Applied Computing, SAC '16, pp. 1090–1095. ACM, New York, USA (2016). https://doi.org/10.1145/2851613.2851809
39. de Solla Price, D.J.: Networks of scientific papers. Science **149**(3683), 510–515 (1965). https://doi.org/10.1126/science.149.3683.510
40. Subramanyam, K.: Bibliometric studies of research collaboration: a review. Inf. Sci. **6**(1), 33–38 (1983). https://doi.org/10.1177/016555158300600105
41. Tomasini, M., Luthi, L.: Empirical analysis of the evolution of a scientific collaboration network. Phys. A Stat. Mech. Appl. **385**(2), 750–764 (2007). https://doi.org/10.1016/j.physa.2007.07.028

42. Uddin, S., Hossain, L., Rasmussen, K.: Network effects on scientific collaborations. PLoS ONE **8**(2), e57546 (2013). https://doi.org/10.1371/journal.pone.0057546
43. Wallace, M.L., Larivire, V., Gingras, Y.: A Small World of Citations? The Influence of Collaboration Networks on Citation Practices. PLoS ONE **7**(3), e33339 (2012). https://doi.org/10.1371/journal.pone.0033339
44. Wang, C., Satuluri, V., Parthasarathy, S.: Local probabilistic models for link prediction. In: Seventh IEEE International Conference on Data Mining (ICDM 2007), pp. 322–331 (2007). https://doi.org/10.1109/ICDM.2007.108
45. Yan, E., Ding, Y.: Applying centrality measures to impact analysis: A coauthorship network analysis. Journal of the American Society for Information Science and Technology **60**(10), 2107–2118 (2009). https://doi.org/10.1002/asi.21128
46. Yan, E., Ding, Y., Zhu, Q.: Mapping library and information science in China: A coauthorship network analysis. Scientometrics **83**(1), 115–131 (2010). https://doi.org/10.1007/s11192-009-0027-9
47. Yoshikane, F., Nozawa, T., Tsuji, K.: Comparative analysis of co-authorship networks considering authors' roles in collaboration: Differences between the theoretical and application areas. Scientometrics **68**(3), 643–655 (2006). https://doi.org/10.1007/s11192-006-0113-1

Chapter 6
Extraction of Co-authorship Networks

Abstract The extraction of a co-authorship network from a set of bibliographic records in which articles and authors are uniquely identified is an easily solvable problem. However, in a vast majority of bibliographic databases authors are identified by their names. This causes the problem of correct identification of nodes in co-authorship networks due to ambiguous author names. In this chapter we present an overview of initial-based, heuristic and machine learning approaches to the name disambiguation problem. Then, we study the performance of various string similarity measures for detecting name synonyms in bibliographic records. After that, we propose a novel method for disambiguating author names that is based on reference similarity networks and community detection techniques. Finally, we present a case study investigating the impact of name disambiguation on the structure of co-authorship networks.

The most convenient way to construct a co-authorship network is to extract it from a set of bibliographic records provided by a bibliographic database or digital library. Other methods to construct co-authorship networks, such as interviews or circulating questionnaires, require much human effort and time, and usually result in networks that contain no more than a few tens or hundred of nodes [51] making the analysis of scientific collaboration less statistically accurate.

The development of the World Wide Web and accompanying services enabled the creation of massive online bibliographic databases. They are extremely important in scientific communities since they give scholars ability to search and discover publications relevant for their work. If bibliographic databases additionally provide the full-text accessibility of indexed content then we call them digital libraries. Bibliographic records stored in bibliographic databases are usually created through the aggregation of publication meta-data that is either directly or indirectly provided by publishers. Namely, either publishers directly submit information about published articles in a format required by a bibliographic database (e.g. in order to gain better visibility to potential audience), or provide mentioned information online that can be later retrieved and processed by data aggregation services of bibliographic information systems.

M. Savić et al., *Complex Networks in Software, Knowledge,
and Social Systems*, Intelligent Systems Reference Library 148,
https://doi.org/10.1007/978-3-319-91196-0_6

6.1 Bibliographic Databases

Bibliographic databases store bibliographic records. One bibliographic record represents and provides crucial information (such as title, authors, publication venue, the year of publication, publisher, and so on) about one bibliographic unit. Generally speaking, three types of bibliographic databases can be distinguished:

- *Article-centered*. In article-centered databases each article (bibliographic unit) has an unique identifier. The authors of an article are identified by their names. Articles are usually grouped by publication venues which are also uniquely identified.
- *People-centered*. In people-centered databases each author has an unique identifier to which a list of his/her articles is associated. Articles and publication venues do not have unique identifiers. Usually people-centered databases provide a registration form to authors who can submit, correct and update their bibliographies.
- *People-article-centered*. In those databases each publication and each author (individual) have unique identifiers. Author identifiers are used in bibliographic records instead of their names.

The extraction of co-authorship networks from people-article-centered bibliographic databases is a relatively easy task since authors are uniquely identified in bibliographic records. On the other hand, people-article-centered bibliography databases are rarely used in practice since they are hard to maintain. For example, if an author of a newly retrieved bibliographic unit cannot be automatically matched to exactly one author registered in the database then it is necessary to manually verify whether a new author should be created or it already exists among obtained candidates. On the other hand, article-centered and people-centered databases are much easier to maintain, but the extraction of co-authorship networks from those types of bibliographic databases poses several difficulties. In the case of article-centered databases, in which authors are identified by their names, the author name disambiguation problem appears manifesting in two different forms:

- *Name homonymy*: many different individuals may have the same name.
- *Name synonymy*: a single individual may appear under different names in bibliographic records due to spelling errors, orthographic variants, name transliterations and abbreviations, authors may use pen names, names can change in time (e.g. due to marriage), and so on.

The name disambiguation problem can be formally defined as follows.

Definition 6.1 (*Name Disambiguation Problem*) Let $R = \{r_1, r_2, ..., r_m\}$ be a set of bibliographic references. Each bibliographic reference contains at least author names, publication title and the title of publication venue. The objective is to produce a disambiguation function that partitions R into n overlapping reference sets $\{A_1, A_2, ..., A_n\}$, where n is the number of distinct persons appearing in R, such that each A_i encompasses all and ideally only all references of the same person.

To address scalability issues and avoid all-pairs comparisons of references contained in a large bibliography database, name disambiguation approaches usually rely on blocking functions [55]. A blocking function splits the set of references into overlapping blocks – groups of references of authors with the same or highly similar names. Then, a name disambiguation procedure can be applied to each block separately. Typically used blocking functions are [55]:

- initial-based blocking functions, e.g. references of authors with the same last name and the same initials of the first name are put in the same block,
- the shared-token block function – references of authors whose names have at least one token in common are put in the same block,
- the shared-n-gram block function – references of authors whose names have at least one n-gram (n consecutive characters) in common are put in the same block.

In other words, a blocking function forms an inverted index of bibliographic references and name disambiguation is performed over references having the same key in the index.

For people-centered bibliography databases there are two additional issues relevant to the extraction of co-authorship networks. The first one is related to the boundary of co-authorship networks. Namely, a researcher registered in a people-centered bibliographic database can have publications that are joint works with researchers not registered in the database. Therefore, the question is whether to include unregistered researchers in the network or not. If non-registered researchers are included then the author name disambiguation problem appears since registered researchers can have non-registered co-authors in common. Additionally, bibliographies of unregistered researchers may be incomplete (since they are recovered from profiles of registered co-authors) and consequently some co-authorship links between non-registered researchers will be missing.

A co-authorship link between two registered researchers in a people-centered bibliography database is established if there is at least one publication associated to both of them. However, one publication may appear in different forms due to different citation conventions, spelling errors, different or missing information, and so on. In other words, for people-centered databases there is also the citation disambiguation problem. As showed in [44, 59], the citation disambiguation problem can be accurately handled using string similarity measures (the Tanimoto string similarity metric over bi-grams was used in [59], while the Levenstein string distance was used in [44]).

We performed a literature review in order to observe how co-authorship networks are extracted in studies dealing with their analysis. We employed the following manual snowball sampling procedure to collect relevant papers:

- The paper "The structure of scientific collaboration networks" [50] by Mark E. Newman was added as the first to the pool of relevant papers since this paper is the first and the most cited paper in the field.
- Using the Google Scholar web service we examined titles and excerpts from abstracts of papers that cite the Newman's article. In a majority cases just this information provided enough support to discard an article as irrelevant. Namely, many studies in the field of complex networks cite the Newman's paper although

they are not closely related work, i.e. they do not directly deal with co-authorship network analysis, but with other types of complex networks or complex networks in general (such as theoretical studies of complex network metrics and models). When a title or an abstract suggested that the corresponding study is relevant then the full text was examined.

- When a paper from the pool was examined in details then the references given in the paper were checked in order to see whether they are relevant studies not previously included in the pool of relevant papers.

In this way we obtained exactly 76 studies which is a large enough body of research works to observe general trends. To our surprise in 31 papers the name disambiguation problem is not discussed at all, although in a vast majority of those papers studied networks were extracted from bibliography databases that are not people-article centered. In 11 studies the name disambiguation problem is mentioned as important, in some cases examples of ambiguous names are given, but those studies do not clearly explain how the problem was systematically tackled. In the rest of studies the name disambiguation problem was approached in the following ways:

- through a manual inspection of bibliographic records,
- using the strict matching of author name labels which means that the problem was simply ignored or it was not relevant for a particular study (i.e. studies in which co-authorship networks were extracted from people-article-centered databases),
- using simple initial-based methods, and
- using author similarity heuristics.

Only in one study the problem was approached using more advanced, machine learning techniques.

6.2 Extraction of Co-authorship Networks from People-Article-Centered Bibliography Databases

In people-article-centered bibliography databases each author and each publication are uniquely identified. From one such database it is easy to extract the complete list of publications L whose elements are in the form:

$$P = (p, k, a_1, a_2, ..., a_k) \tag{6.1}$$

where p is the identifier of a publication P, k is the number of authors of P, and a_i is the identifier of the i-th author of P. In this section we explain how co-authorship network whose nodes are enriched with standard researcher productivity metrics can be constructed from L.

The productivity of researchers is commonly quantified using one of the following three counting schemes [38]:

- The normal counting scheme: each co-author of a publication receives the full credit (one point) for the publication.
- The straight counting scheme: only the first author of a publication receives the full credit for the publication.
- The fractional counting scheme: the credit for a publication is equally divided among the co-authors.

Only one iteration through the list L is necessary to form the co-authorship network and enrich nodes with previously described researcher productivity metrics. For each author identifier a_i in P ($1 \leq i \leq P.k$), it is checked whether the node corresponding to a_i is already present in the network. If not, a new node representing a_i is added to the network. After that, for each pair of author identifiers in P it is checked whether a link connecting them is present in the network. If such link does not exist then a new link is created. The whole procedure is described in Algorithm 6.1.

The previously described algorithm can be used to extract co-authorship networks from data stored in modern current research information systems (CRIS). For example, we used it in our previous studies to extract co-authorship networks from a people-article-centered bibliography database developed according to the euro-CRIS[1] standards and recommendations [63, 65]. To apply the algorithm on bibliographic records contained in article-centered bibliographic databases it is necessarily to uniquely identify authors in bibliographic records using some name disambiguation method. In the next sections of this chapter we will explain different approaches to the name disambiguation problem.

6.3 Initial-Based Name Disambiguation Approaches

In initial-based approaches to the name disambiguation problem each author is identified by his/her surname and initial(s) of the first name. There are two basic initial-based methods, the first initial method and the all initials method, proposed by Mark Newman in his seminal studies of large-scale field co-authorship networks [48–50].

In the first initial method, each author is identified by his/her surname and the first initial of the first name. Newman emphasized that this approach is "clearly prone to confusing two people for one, but will rarely fail to identify two names which genuinely refer to the same person". This method efficiently handles name synonyms when spelling errors and other inconsistencies occur in first names, but it is sensitive to spelling errors in surnames. Also, this method does not take into the account the problem of name homonyms.

In the all initials approach, each author is identified by his/her surname and all initials of the first name. This method will identify a researcher who is inconsistently reporting his middle initials as two or more different individuals. The all initials

---

[1] http://www.eurocris.org/.

Algorithm 6.1: The extraction of an enriched co-authorship network from a list
of publications contained in a people-article-centered bibliography database

---

Input : L - a list of publications
Output: G - the enriched co-authorship network extracted from L

G = an empty graph

foreach $P \in L$ **do**
    for $i = 1$ **to** $P.k$ **do**
        Node $n = G$.findNode($P.a_i$)
        if n does not exist **then**
            $n = G$.createNode($P.a_i$)
            n.normalCount = 1
            n.fractionalCount = $1/ P.k$
            if $i = 1$ **then**
                n.straightCount = 1
            else
                n.straightCount = 0
        else
            n.normalCount = n.normalCount + 1
            n.fractionalCount = n.fractionalCount + $1/ P.k$
            if $i = 1$ **then**
                n.straightCount = n.straightCount + 1

    for $j = 2$ **to** $P.k$ **do**
        for $i = 1$ **to** $j - 1$ **do**
            $w = 1 / (P.k - 1)$
            Link $l = G$.findLinkConnecting($P.a_i$, $P.a_j$)
            if l does not exist **then**
                $l = G$.createLinkConnecting($P.a_i$, $P.a_j$)
                l.weight = w
            else
                l.weight = l.weight + w

---

method also does not take into account the problem of name homonyms – two
author having the same last name and all initials of the first name may be two
different persons.

In the Mark Newman's seminal studies of large-scale field co-authorship networks both initial-based methods were used to identify researchers. For each analyzed bibliographic dataset, Newman constructed two co-authorship networks – one
obtained by the first initial author disambiguation method and another by the all initials method. Newman computed and compared various structural metrics (the average degree, characteristic path length, clustering coefficient, the size of the largest
connected component, and so on) of paired co-authorship networks concluding that
co-authorship networks constructed by different initial-based methods possess similar structural characteristics. However, Newman noticed that initial based methods

are in particularly sensitive to authors of Japanese and Chinese descent. For example, the most productive authors in the biomedicine dataset analyzed by Newman were Japanese and Chinese scholars with frequent last names (e.g. Suzuki, Nakamura, Tanaka, Takahashi, Wang). Secondly, the most productive author published 1679 papers in a five year period (1995–1999) which is an extremely large output for one person (a paper per day including weekends and holidays) indicating that the name label of the most productive author actually represents two or more different researchers.

The accuracy of the Newman's initial-based methods was investigated by Milojević [46] using artificially generated datasets in which true identities of authors are known. Milojević selected five unrelated scientific disciplines and built five real bibliographic datasets using the Thomson Reuters' Web of Science service. For each real dataset, Milojević estimated the number of authors in the discipline assuming that different name labels correspond to different persons, the frequency distribution of last names, the frequency distributions of the first and middle initials, the intrinsic rate of middle initials, the reporting rate of middle initials, and the distribution of researcher productivity measured by the straight counting scheme. Then, she built artificial datasets which mimic statistical properties of real datasets using the following procedure:

- Artificial authors are created according to the estimated number of authors in the real dataset. The first initial, last name and number of publications of an artificial author are assigned according to empirically observed frequency distributions. It is randomly chosen whether an author has middle initials or not according to the empirically estimated intrinsic rate of middle initials. Finally, middle initials are assigned according to the frequency distribution of middle initials.
- A certain number of artificial papers is created for each artificial author according to the empirically observed researcher productivity distribution. If an artificial author has middle initials then it is randomly decided whether middle initials are reported or not according to the empirically observed reporting rate of middle initials.

Milojević then performed author name disambiguation on artificially generated datasets using initial-based methods. The obtained results indicate that the first initial method correctly identifies 97% of authors on average and that the all initials method is typically less accurate than the first initial method except for mathematics and economy which are scientific disciplines characterized by a high reporting rate of middle initials and a relatively low productivity (a lower productivity decreases chances of inconsistent reporting of middle initials). Following obtained results, Milojević proposed a hybrid initial-based method that takes into account the frequency distribution of last names and the size of the dataset. Milojević analyzed the accuracy of her hybrid method showing that it slightly outperforms the Newman's initial-based methods.

Contrary to the findings by Milojević, the studies by Fegley and Torvik [16] and Kim and Diesner [32, 33] indicate that the initial-based methods significantly distort the structure of co-authorship networks. Fegley and Torvik [16] analyzed the impact of node splitting (one person represented by two or more nodes in a co-authorship network due to name variants) and lumping (more than one person represented by one node in a co-authorship network due to common names) on statistical properties of two large-scale co-authorship networks extracted from the MEDLINE bibliography database and the USPTO patent database. The authors formed and compared three different variants of co-authorship networks: one where author names were disambiguated using the Authority disambiguation method [75, 76], one where the first initial method was used to identify authors, and one where the all initials method was employed. The main findings of the study are:

1. The initial-based author name disambiguation methods drastically underestimate the number of authors compared to the Authority name disambiguation method. For example, the number of authors in the MEDLINE co-authorship network disambiguated by the Authority method, the first initial approach and the all initials approach are $3.17 \cdot 10^6$, $1.56 \cdot 10^6$ and $2.8 \cdot 10^6$, respectively.
2. Name homonyms caused by initial-based methods drastically change the structure of examined co-authorship networks. Lumping effects caused by the initial-based identification of authors are reflected by a significantly higher average node degree and significantly smaller values of the clustering, small-world and assortativity coefficient.

The similar findings were also reported by Kim and Diesner [33] who used bibliography datasets from the ISI Web of Science in their experiments. Additionally, Kim and Diesner investigated the impact of the initial-based methods on evolutionary properties of the DBLP co-authorship network. The obtained results suggest that the degree of distortion of global network characteristics (the number of nodes and links, the average node degree, characteristic path length, the clustering coefficient, the size of the largest connected component) increases as the network evolves [32]. Due to the differences in results reported by Milojević [46] on the one side and Fegley and Torvik [16] and Kim and Diesner [32, 33] on the other side, it can be concluded that the accuracy of initial-based methods is still an open research question that demands a more comprehensive analysis.

6.4   Heuristic Name Disambiguation Approaches

The main characteristic of heuristic approaches to the name disambiguation problem is that they are based on simple and easy implementable name or reference approximate matching functions or rules.

Moody [47] proposed a heuristic name disambiguation method based on the frequency of first and last names. A first or last name is considered common if it appears in 15 or more references contained in a bibliography database. If a name has less than 15 appearances then it is treated as an uncommon name. The method assumes that identical author name labels represent the same individual (consequently, the Moody's method does not take into account the problem of name homonyms). Two different name labels are considered to represent the same individual if and only if the following two criteria hold:

1. they differ only in their middle initials, and
2. either the first or the last name is uncommon.

After a co-authorship network is constructed according to the above stated name disambiguation rules, it is checked whether the network contains structurally equivalent nodes. Two nodes A and B are structurally equivalent if they are connected to the same nodes, i.e. $n(A) = n(B)$, where $n(X)$ is the set of nodes to which X is connected. Two structurally equivalent nodes with name labels that differ only in the middle initials are also treated as the same individual. Clearly, the Moody's method is able to identify authors who inconsistently report their middle initials, but the method, besides name homonyms, does not take into account possible spelling errors in first and last names.

An approach similar to Moody's was used by Chen et al. [7] to construct the co-authorship network of authors publishing in the *Scientometrics* journal. Namely, two authors whose name labels differ only in the middle initial are considered to be the same individual if they have identical affiliations. The authors manually examined present affiliations in order to unify institutions whose labels appear in different forms due to abbreviations and misspellings.

Bird et al. [3] investigated the structure and dynamic of research collaboration in several computer science disciplines relying on co-authorship networks extracted from the DBLP bibliography database. The authors emphasized that the DBLP data is fairly accurate (due to massive human efforts invested in the maintenance of the database), but that the database still suffers to some degree from the name disambiguation problem. The authors used several heuristics such as string similarity of author names, the number of co-authors in common, the number of publication venues in common and publication dates to identify name label pairs that are likely to represent same authors. The candidates obtained by previously mentioned heuristics are manually examined in order to identify name synonyms. String similarity measures were also used to identify name synonyms in the extraction of the co-authorship network of the ED-MEDIA conference [52]. It should be emphasized that string similarity measures are commonly used in name matching tasks [9] which is the same problem as the identification of name synonyms in a set of bibliographic records.

In one of our previous publications, we proposed a semi-automatic approach to the extraction of co-authorship networks from medium-size article-centered bibliography databases [64]. The method is based on the analysis of author name labels using string similarity metrics which is performed twice during the extraction of a co-authorship network. More specifically, our method consists of the following six steps:

1. A preliminary co-authorship network is constructed from a set of bibliographic references using the strict name label matching which means that each distinct name label corresponds to exactly one node in the network. Then, nodes that are articulation points in their ego-networks are located. Such nodes are manually

examined in order to identify potential name homonyms. This step is motivated
by the following reasoning: if one node in the network represents two or more
distinct authors then the removal of the node from its own ego-network will
separate research groups to which different individuals belong.

2. Similar author name labels are identified by different string similarity metrics.
   Extremely similar name labels are manually examined in order to detect name
   synonyms and form a lookup table that is used to correct author names.
3. The input set of bibliographic references is transformed according to identified
   name synonyms and homonyms. Then, an inverted index mapping authors to their
   bibliographic references is constructed.
4. The co-authorship network is formed from the previously constructed inverted
   index I by the following two rules: (1) the set of nodes in the network corresponds to the set of keys in I, and (2) two authors A and B are connected in
   the network if $I(A) \cap I(B) \neq \emptyset$, where $I(A)$ denotes the set of bibliographic
   references authored by A.
5. Then, the second analysis of name labels is performed. Firstly, connected components in the network are identified. If the network does not contain a giant
   connected component then string similarity functions are applied on name labels
   of nodes belonging to the same connected component. In the case that the network
   has a giant connected component, string similarity functions are computed at the
   level of extended ego networks (the extended ego network of a node is the subnet-
   work induced by the node, its neighbors and the neighbors of its neighbors). This
   step is motivated by the following reasoning: if two nodes having similar name
   labels represent the same individual then there is a high probability that such two
   nodes will be extremely close to each other in the network.
6. Steps 3 and 4 are repeated in order to obtain the final co-authorship network.

Martin et al. [42] proposed an author name disambiguation approach that relies
on the similarity of author names, collaboration patterns and institutional affiliations.
In contrast to the previously described name disambiguation approaches, the starting
assumption of the proposed method is that each author of each paper is a different
individual. This means that two authors with identical name labels are not initially
treated as the same person. In the first step of the method, a string similarity measure is
computed for each pair of affiliation strings. Affiliation strings that are similar enough
are assumed to represent the same institution. All authors that have the same name
and at least one shared affiliation are treated as one individual. Then, author pairs with
similar names are identified. Two author names are considered similar if they have
the same last name and compatible first and middle names (identical if fully given,
or the same initials). A hybrid similarity measure based on the number of shared
affiliations, the number of shared co-authors and the number of joint publication
venues is computed for authors with similar names. Finally, two authors with a high
hybrid similarity are treated as the same person.

6.5 Comparison of String Similarity Metrics for Name Disambiguation Tasks

String similarity metrics are functions that map a pair of strings to a real number r in the interval $[0..1]$ such that

- $r = 1$ indicates two identical strings, and
- a higher value of r indicates a higher similarity between compared strings.

In this section we analyze the performance of standard string similarity metrics for identifying name synonyms in bibliographic records stored in article-centered bibliography databases. The analysis is based on bibliographic records contained in eLib – the electronic library of the Mathematical Institute of the Serbian Academy of Sciences and Arts [45, 64]. ELib is an article-centered bibliography database in which authors are not uniquely identified.

6.5.1 Analyzed String Similarity Metrics

The analysis presented here covers three commonly used string similarity functions: Jaccard, Jaro-Winkler and TF-IDF. The Jaccard and TF-IDF similarity metrics are computed at the level of tokens and n-grams ($n = 2$). The complete list of analyzed string similarity metrics is given in Table 6.1.

The Jaccard string similarity metric belongs to the class of set based string similarity functions. The main idea of set based string similarity functions is that the similarity between two strings can be quantified by the degree of the overlap between sets of tokens or n-grams derived from compared strings. Let p and q denote two arbitrary strings. Let T_s be the set of tokens contained in a string s, i.e. the set of substrings of s that are separated by delimiters (one or more white space characters). Let $N_{s,n}$ denote the set of n-grams contained in s. A n-gram of s is a contiguous sequence of n characters in s. Then, the Jaccard string similarity metric at the token level is defined as

$$\text{Jaccard}(p, q) = \frac{|T_p \cap T_q|}{|T_p \cup T_q|} \tag{6.2}$$

Table 6.1 The list of analyzed string similarity metrics

String similarity metric	Abbreviation
The Jaccard metric at the token level	JT
The Jaccard metric at the n-gram level ($n = 2$)	JN
The Jaro-Winkler metric	JW
The TF-IDF metric at the token level	TIT
The TF-IDF metric at the n-gram level ($n = 2$)	TIN

At the n-gram level we have the following formula:

$$\mathrm{Jaccard}(p, q, n) = \frac{|N_{p,n} \cap N_{q,n}|}{|N_{p,n} \cup N_{q,n}|} \tag{6.3}$$

where n is some fixed value (2 for bigrams, 3 for trigrams, and so on).

The Jaro–Winkler string metric was originally introduced by Matthew Jaro in 1989 for the purpose of matching individuals in the 1985 Census of Tampa (Florida) to individuals in a later independent post-enumeration survey [27]. The Jaro metric was extended by William Winkler in [81] to include the length of the common prefix of compared strings. This modification was inspired by the fact that typing errors are much more likely to occur toward the end of a string. The original Jaro metric is based on the notion of *common* or *matching* characters and the notion of the transposition of matching characters. Let $p = p_1 p_2 ... p_P$ be a string of length P where p_i denotes the i-th character in p. For p_i we say that is common with another string q of length Q if and only if

$$(\exists j)\ p_i = q_j \wedge i - D \le j \le i + D \tag{6.4}$$

where

$$D = \frac{\min(P,\ Q)}{2} \tag{6.5}$$

In other words, two identical characters from compared strings are treated as matching characters if the distance of their positions within strings is less than the half of the length of the shorter string. Let M be the number of matching characters for two strings p and q. From both p and q we delete all non-matching characters. Let p' and q' denote strings obtained from p and q after deleting non-matching characters. Two identical characters p'_i and q'_j are *transposed* if $i \ne j$. Let T denote the half of the number of transpositions. Then, the Jaro string similarity metric is defined as

$$\mathrm{Jaro}(p, q) = \frac{1}{3}\left(\frac{M}{P} + \frac{M}{Q} + \frac{M - T}{M}\right) \tag{6.6}$$

$\mathrm{Jaro}(p, q)$ is defined to be zero when $M = 0$, $P = 0$ or $Q = 0$. From the definition of the Jaro metric it can be observed that this metric represents the average value of

- the portion of the first string that is matched to the second string,
- the portion of the second string that is matched to the first string,
- the portion of untransposed common characters.

The Winkler's modification of the Jaro metric incorporates the length of the common prefix of compared strings if the value of the Jaro metric is equal to or higher than some previously given threshold T (Winkler proposed T to be 0.7 and this value is usually used in implementations of the metric in string distance libraries). Let J denote the value of the Jaro metric for strings p and q. Then, the Jaro–Winkler string similarity is defined as

$$\text{Jaro-Winkler}(p, q) = \begin{cases} J, J < T \\ J + pL(1 - J), J \geq T \end{cases} \tag{6.7}$$

where L is the length of the common prefix up to the maximum of 4 characters and p is a constant scaling factor lower than 0.25 (usually p is set to 0.1).

The TF-IDF metric is a supervised string similarity function since it has to be trained on a corpus of textual documents (usually the training set corresponds to the set of textual documents in which we look for similar strings). TF-IDF can be computed at the level of tokens and n-grams. The main idea of the TF-IDF metric is that two strings can be considered similar if they contain common distinguishing terms (terms are tokens or n-grams). The TF-IDF metric is based on the notions of *term frequency* and *inverse document frequency*. Term frequency, denoted by $\text{TF}(t, s)$, measures how frequently a term t appears in a string s. Raw frequencies can be used, but they are typically either normalized (the probability that t appears in s) or logarithmically scaled. Inverse document frequency, denoted by $\text{IDF}(t, C)$, measures how much important or informative t is considering a training set C. The importance of t depends on the inverse frequency of textual documents from C that contain t. Mathematically speaking,

$$\text{IDF}(t, C) = \log \frac{N}{|\{d \in C : t \in d\}|} \tag{6.8}$$

where N is the number of documents in the training set (in our case one document is one author name label), while $|\{d \in C : t \in d\}|$ is the number of documents from the training set that contain t. Then, the TF-IDF score of t with respect to s and C is defined as the product of $\text{TF}(t, s)$ and $\text{IDF}(t, C)$: $\text{TF-IDF}(t, s, C) = \text{TF}(t, s)\,\text{IDF}(t, C)$. Therefore, the TF-IDF score of a term contained in a string increases with

1. the number of appearances of the term in the string, and
2. the rarity of the term in the training set.

Each string can be viewed as a bag of words – a D-dimensional vector whose components correspond to terms from the training set. D is the number of terms in the training set. If t appears in s then the value of the corresponding component is equal to $\text{TF-IDF}(t, s, C)$, otherwise it is set to zero. Now, the TF-IDF similarity of two strings p and q is defined as the cosine similarity between their TF-IDF vectors P and Q. More formally,

$$\text{TF-IDF}(p, q) = \frac{P \cdot Q}{\|P\| \|Q\|} = \frac{\sum\limits_{i=1}^{D} P_i Q_i}{\sqrt{\sum\limits_{i=1}^{D} P_i^2} \sqrt{\sum\limits_{i=1}^{D} Q_i^2}} \qquad (6.9)$$

$$= \frac{\sum\limits_{t \in T(p) \cap T(q)} \text{TF-IDF}(t, p, C) \, \text{TF-IDF}(t, q, C)}{\sqrt{\sum\limits_{t \in T(p)} \text{TF-IDF}(t, p, C)^2} \sqrt{\sum\limits_{t \in T(q)} \text{TF-IDF}(t, q, C)^2}} \qquad (6.10)$$

where $T(x)$ is the set of terms contained in a string x.

6.5.2   Dataset

The dataset used in the analysis covers 6480 bibliographic references corresponding to articles published in Serbian mathematical journals. Author name labels contained in the dataset can be divided into two categories: full names (both the full first name and the full last name of an author are given) and short names (the first name of an author is reduced to the first letters). The dataset contains 8842 name labels in total, where 5192 name labels (58.72%) are full names, while 3650 name labels (41.28%) are short names. We computed five previously mentioned string similarity metrics for all pairs of author name labels in the dataset and manually examined those pairs whose similarity by at least one of the metrics was higher than 0.6. In this way we identified 690 name label pairs which refer to same individuals where

- 206 (29.85%) name label pairs are pairs of full names,
- 369 (53.47%) name label pairs are pairs of full and short names, and
- 115 (16.66%) name label pairs are pairs of short names.

Some typical examples of different name label pairs representing same individuals are shown in Table 6.2. It can be observed that common errors and inconsistencies in author names in the dataset are due to:

- the inversion of an author's first and last name (the first row in Table 6.2).
- spelling errors (the second row),
- the anglicization of personal names (the third row),
- the presence of middle names (the fourth row),
- the addition of marital surname for female authors (the fifth row),
- name abbreviations in which the first name of an author is shortened to the first letter (the sixth row),
- the presence of separators (the seventh row),
- the presence of titles, usually the PhD title ("dr") is added to the name of an author (the eight row).

Table 6.2 Examples of different name label pairs representing same individuals and the corresponding values of string similarity metrics

Name label 1	Name label 2	JT	JN	JW	TIT	TIN
Nikola Hajdin	Hajdin Nikola	1	0.71	0.46	1	0.8
Todorqević Stevo	Todorčević Stevo	0.33	0.75	0.98	0.44	0.77
Petronievics Branislav	Petronijević Branislav	0.33	0.67	0.93	0.38	0.74
Nisheva-Pavlova Maria	Nisheva-Pavlova Maria M.	0.67	0.95	0.98	0.95	0.98
Rajter-Ćirić Danijela	Rajter Danijela	0.5	0.62	0.92	0.8	0.71
Milogradov-Turin J.	Milogradov-Turin Jelena	0.5	0.74	0.95	0.78	0.87
Lin C.-S.	Lin C. S.	0.8	0.67	0.98	0.91	0.69
Kočinac dr Ljubiša	Kočinac Ljubiša	0.67	0.82	0.97	1	0.89
Van Gulck S.	Van Gulck Stefan	0.5	0.67	0.92	0.73	0.83
Gulck Stefan Van	Van Gulck Stefan	1	0.87	0.72	1	0.93
Gulck S. Van	Van Gulck Stefan	0.5	0.47	0.69	0.73	0.69

Also it can be observed that one person can be represented by multiple name labels (the last three rows in Table 6.2).

6.5.3 Evaluation Methodology

Let $P = \langle N_1, N_2 \rangle$ be a pair of two different author name labels and let S denote an arbitrary string similarity function. If N_1 and N_2 are identical or similar strings then they are likely to refer to the same person. Consequently, string similarity functions can be employed to identify potential name synonyms: N_1 and N_2 are potential name synonyms if $S(N_1, N_2) \geq t$, where t denotes an acceptance threshold. We varied the acceptance threshold from 0.6 to 0.95 with a 0.01 step size and computed the precision and recall of analyzed string similarity functions. Precision and recall are measures commonly used in information retrieval and data mining to evaluate the performance of search engines and classifiers, respectively. In our case, string similarity functions can be viewed as binary classifiers since they decide whether a name label pair refers to the same individual or to two different individuals.

For a given string similarity function S and acceptance threshold t, we say that $P = \langle N_1, N_2 \rangle$ is

- *relevant* if N_1 and N_2 refer to the same author (otherwise P is irrelevant),
- *retrieved* if $S(N_1, N_2) \geq t$ (otherwise P is unretrieved).

Let L denote the number of relevant name pairs that were retrieved. Then, the precision of S is equal to L divided by the total number of retrieved name pairs. On the other hand, the recall of S is equal to L divided by the total number of relevant name pairs. In other words, precision indicates how many name pairs having string similarity score higher than or equal to t represent same individuals, while recall indicates how many name pairs representing same individuals can be obtained at acceptance threshold t.

Precision and recall are always analyzed together since they reflect two complementary aspects of classifier performance. The F_β measure combines precision and recall into a single score of classifier performance. It is defined as

$$F_\beta = \frac{(\beta^2 + 1) \cdot \text{Precision} \cdot \text{Recall}}{\beta^2 \cdot \text{Precision} + \text{Recall}} \tag{6.11}$$

where β is the tuning parameter of the measure ($\beta \geq 0$) which balance between the relative importance of precision and recall. Considering β three cases are possible:

1. $\beta < 1$: the more importance is given to precision. In the extreme case when $\beta = 0$ we have that $F_0 = \text{Precision}$.
2. $\beta = 1$: precision and recall are equally important and F_1 is the harmonic mean of precision and recall.
3. $\beta > 1$: the more importance is given to recall. In the extreme case when $\beta = \infty$, by the application of the L'Hôpital's rule, we obtain that $F_\infty = \text{Recall}$.

6.5.4 Results and Discussion

Analyzed string similarity metrics are computed independently for:

1. author name label pairs from the set of full name labels appearing in the dataset (full-full name pairs),
2. author name label pairs from the set of short name labels (short-short name pairs),
3. author name label pairs in which the first element of a pair is a full name label, the second element is a short name label (full-short name pairs) and the first element is the only name label in the dataset that can be reduced (shortened) to the second element.

Figure 6.1 shows precision and recall curves for full-full name pairs. It can be observed that token based string similarity metrics have a very high precision, but at the same time they have an extremely small recall. The Jaro–Winkler metric exhibits the best recall for full-full name pairs. It can be seen that an increase of the acceptance threshold results in a higher precision and lower recall for Jaro–Winkler and token based metrics. On the other hand, for n-gram measures we have critical acceptance thresholds above which both precision and recall start to decrease. This means that n-gram measures above critical acceptance thresholds are extremely inefficient for detecting author name synonyms.

In order to determine optimal values of the acceptance threshold we computed the maximal F_1 score for each analyzed string similarity metric. The obtained results are summarized in Table 6.3. It can be seen the Jaccard metric at the token level is

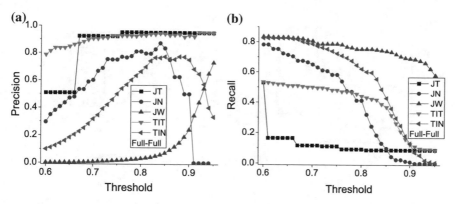

Fig. 6.1 Precision **a** and recall **b** of string similarity measures for full-full name pairs

Table 6.3 The maximal F_1 scores for full-full, full-short and short-short name pairs. One star indicates the worst performing metric, while two stars indicate the best performing metric for identifying name synonyms

Category	Metric	Maximal F_1	Threshold	Precision	Recall
Full-full	JT*	0.249	0.6	0.507	0.165
	JN**	0.672	0.73	0.753	0.607
	JW	0.645	0.95	0.73	0.578
	TIT	0.645	0.62	0.837	0.524
	TIN	0.661	0.82	0.741	0.597
Full-short	JT*	0.123	0.66	0.231	0.084
	JN	0.295	0.6	0.473	0.214
	JW	0.449	0.9	0.429	0.471
	TIT	0.229	0.6	0.329	0.176
	TIN**	0.482	0.65	0.418	0.569
Short-short	JT	0.44	0.83	0.698	0.322
	JN**	0.446	0.7	0.357	0.591
	JW*	0.376	0.95	0.253	0.73
	TIT	0.466	0.88	0.527	0.417
	TIN**	0.446	0.8	0.365	0.574

the worst performing metric, while the Jaccard metric at the n-gram level is the most effective metric for identifying name synonyms in full-full name label pairs. The Jaccard n-gram metric at the optimal threshold ($t = 0.73$) achieves the coverage of 60% (60% of different name labels representing same individuals can be identified at the optimal threshold) with an error rate of 25% (25% of the total number of name pairs that have Jaccard n-gram similarity above the optimal threshold represent different persons).

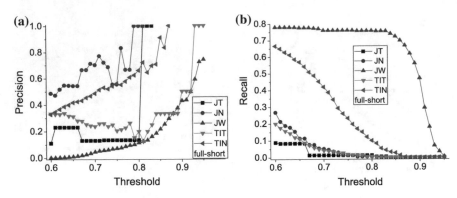

Fig. 6.2 Precision **a** and recall **b** of string similarity measures for full-short name pairs

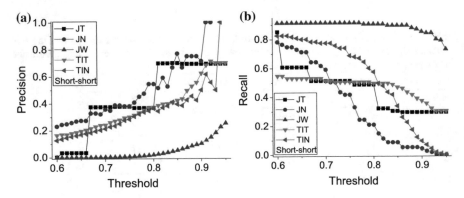

Fig. 6.3 Precision **a** and recall **b** of string similarity measures for short-short name pairs

Regarding full-short and short-short name pairs, the Jaro–Winkler string similarity function shows the highest recall but an extremely low precision (see Figs. 6.2 and 6.3). On the opposite side, the Jaccard metric at the n-gram level possesses the highest precision, but at the same time it has the smallest recall. Also, it could be observed that for full-short name pairs all metrics except Jaro–Winkler at some point reach the maximum precision. However, at the points of maximum precision those metrics exhibit an extremely small recall which means that only a small fraction of author name synonyms can be identified automatically without errors.

The Jaro–Winkler metric exhibits the worst performance for short-short name pairs (see Table 6.3). The TF-IDF metric at n-gram level is the best performing string similarity function for both full-short and short-short name pairs. Also, it is interesting to observe that both n-gram metrics achieve the same maximal F_1 score but at different acceptance thresholds and with close values of precision and recall. Since n-gram metrics exhibit the best performance in all categories of name label pairs, it can be concluded that they are superior to the rest of analyzed metrics for identifying name synonyms. However, a relatively small values of precision and recall (lower than 0.7) at the optimal acceptance thresholds indicate that those metrics can not be used to perform an automatic identification of name synonyms in bibliographic data without significant errors.

6.6 Machine Learning Name Disambiguation Approaches

Machine learning name disambiguation approaches are rarely used to extract co-authorship networks in studies dealing with analysis of research collaboration. In our literature review, which covered more than 70 papers on the subject, we observed just one article where machine learning techniques were employed to disambiguate author names in bibliographic records. Namely, Huang et al. [26] studied the evolution of a co-authorship network of computer scientists that was extracted from bibliographic records contained in the CiteSeer digital library. To determine the nodes of the network, the authors used a name disambiguation method they previously proposed in [25]. The main characteristics of their method are:

- All name labels appearing in bibliographic records are grouped into name classes such that one name class encompasses similar name labels. Two name labels are considered similar if they have identical last names and compatible first and middle names.
- A blocking function is used to form blocks of references. Each block corresponds to one name class and contains papers authored by all name labels from the name class. Name disambiguation is performed on each block separately, which drastically increases the time efficiency of the method.
- The DBSCAN clustering algorithm [14] is used to detect clusters in a block of references. Distinct authors are identified by obtained clusters. This means that the identification of authors is performed by the following rule: all references from a cluster are authored by the same person, while references from different clusters are authored by different persons.
- A similarity function used by the clustering algorithm is learned by LASVM [5] – an online learning SVM (support vector machine) algorithm. This means that the method is supervised (it requires a training set) and adaptable (the similarity function can be learned incrementally).

The idea that clusters of references correspond to different persons implies that the reference similarity function used by the clustering technique has to quantify the extent to which two references are authored by the same person, i.e. a high value of the reference similarity function indicates that two references are authored by the same person, whereas a low value indicates exactly the opposite. Huang et al. used different string similarity measures for different reference attributes: the edit distance for emails and URLs, the token-based Jaccard similarity for affiliations and addresses and the Soft-TFIDF string similarity measure for author names. The authors manually disambiguated 3335 references authored by 10 ambiguous name labels representing 490 different persons. Then, they formed a training set for LASVM that contains labeled reference similarity vectors where labels indicate whether a similarity vector refers to two publications authored by the same person or not.

The authors also investigated the accuracy of their method. Reference pairs associated to 3 ambiguous name labels were used to learn the similarity function while reference pairs associated to the rest of ambiguous name labels were used to evaluate the method. The obtained results show that:

- 63.8% of identified reference clusters are completely correct, i.e. they contain all references authored by the same individual without any reference authored by a person having similar name.
- The average pairwise precision of the method is 0.873 which implies that 1 in 10 reference pairs is incorrectly marked as being authored by the same person.
- The average pairwise recall of the method is 0.944 which means that about 5% of references pairs authored by the same person are not identified as being authored by the same person.
- The ratio between the number of obtained reference clusters and the number of ground truth clusters is 0.944. This means that the method slightly underestimates the real number of distinct authors.

A survey of the state of the art machine learning methods for author name disambiguation can be found in the article by Ferreira et al. [17]. The authors divided existing machine learning approaches to the name disambiguation problem into two categories:

1. author grouping methods and,
2. author assignment methods.

6.6.1 Author Grouping Methods

Author grouping methods rely on clustering techniques and they can be either supervised or unsupervised. Their main idea is that different individuals having ambiguous names can be identified by clustering their references where obtained clusters directly correspond to different individuals. Various clustering techniques are used in existing author grouping methods including agglomerative clustering [2, 6, 10, 29, 34, 36, 39, 66, 71, 73, 75, 82], K-means clustering [23, 79], spectral clustering [24, 68], graph partitioning [20, 28, 54, 56], affinity propagation [15] and already mentioned density-based clustering [25, 30, 31]. Additionally, some of approaches rely on two-stage clustering techniques in which a rule-based clustering is firstly performed according to simple name or reference matching rules in order to obtain very pure, but fragmented clusters of references [10, 36]. Then, an agglomerative clustering technique is applied to fragmented clusters in order to obtain the final partitioning of the set of references. There are also author grouping methods based on

- two-stage clustering in which a corrective clustering is performed in the second stage according to an implicit evidence gathered from the Web [84].

- multi-stage agglomerative clustering in which a set of references is firstly clustered according to co-author names, then obtained clusters are merged according to the similarity of publication titles, and then merged according to the similarity of publication venues [40].

To apply a clustering technique to a set of references it is necessary to define a function which quantifies the similarity between two references. Reference similarity functions can be divided into two categories:

1. predefined reference similarity functions, and
2. learned reference similarity functions.

Unsupervised author grouping methods use predefined reference similarity functions, while in supervised author grouping methods reference similarity functions are learned from training datasets.

Predefined similarity functions rely on existing string similarity measures that are either computed on whole references or aggregated from string similarities of reference attributes (e.g. one string similarity function is used for titles and some other for author names). Additionally, graph-based similarity functions applied to co-authorship networks formed by the strict name label matching can be used to quantify the similarity of author names. Let G denote a co-authorship graph formed from an input set of references assuming that different author name labels represent different individuals. This means that G is formed without disambiguating author name labels. Then, the similarity between two nodes a and b representing two different name labels can be expressed as:

- The similarity of their ego-networks which can be quantified by structural node similarity metrics (e.g. the number of common neighbors, the Jaccard coefficient, the Adamic-Adar metric) [2, 29]. Besides previously mentioned structural node similarity metrics, which are frequently used to identify missing links and predict future links in complex networks, On et al. [53] proposed a structural node similarity metric for name disambiguation tasks based on the notion of γ-quasi-cliques. A γ-quasi-clique is a subgraph S of G such that the degree of each node in S is at least $\gamma(n-1)$, where n is the number of nodes in S. The similarity of two ego-networks N_1 and N_2 is then defined as the number of nodes in a common γ-quasi-clique C of N_1 and N_2 if the number of nodes in C is higher than some predefined threshold, or 0 otherwise.
- The length of the shortest path connecting a and b or some other function derived from all paths connecting a and b [15, 35].
- The probability that a random walker starting from a reaches b in a fixed number of random walk steps [41].

Previously mentioned graph-based similarity metrics can be used only to identify potential name synonyms. On the other hand, potential name homonyms can be identified by clustering ego-networks without ego-nodes considering both topological (e.g. cluster cohesiveness) and temporal (e.g. temporal coupling between ego-nodes and clusters) features [62] or by detecting non-overlapping cycles in co-authorship networks [67].

Learning a reference similarity function requires a training dataset which contains labeled vectors reflecting similarities between both coreferent and non-coreferent reference pairs.

Definition 6.2 (*Coreferent references*) Two references are coreferent if they are authored by the same person.

Generally speaking, each reference can be represented by a vector of n reference features $(r_1, r_2, ..., r_n)$ where features correspond to different reference attributes (authors, affiliations, title, publication venue, the year of publication, and so on). Reference features can be numerical, strings or vectors (e.g. strings represented by TF-IDF vectors). Then, the similarity between two references P and Q can be expressed by a similarity vector $S(P, Q) = (s_1, s_2, ...s_m)$, $m \geq n$, where s_i denotes the similarity between the k-th feature of P and the k-the feature of Q. More than one feature in S may reflect the similarity between two reference attributes. For example, the similarity between two publication titles can be represented by two features in S – the number of common terms in publication titles and the edit distance between publication titles. Reference similarity vectors may also include additional features such as the similarity or correlation between latent topics of corresponding papers [68, 71, 83], the similarity between cited articles [36, 43, 66, 73, 82], and the presence of two references on the same web page [28, 82, 83].

The similarity between two reference attributes is computed by some prede-fined similarity function. Then, a learned reference similarity function is actually a classifier C trained from a set of labeled similarity vectors $T = \{(S_i, L_i)\}$, where $L_i \in \{0, 1\}$ is the class attribute. $L_i = 1$ if S_i is a similarity vector of two coreferent references (the positive class), or 0 otherwise (the negative class). Finally, the similar-ity of two references is either the classifier's prediction of L for their similarity vector or the classifier's confidence in predicting the positive class. Various classification models, such as random forests [31, 70, 77], decision trees [68, 80], logistic regres-sion [20, 28], k-nearest neighbors classification [70] and support vector machines [25, 68, 83], are used in existing supervised author grouping methods to learn reference similarity functions. If predicted values of L are used as the similarity between ref-erences then the identification of reference clusters can be reduced to a simple graph problem. Namely, we can build an undirected graph Q whose nodes are references, while two nodes are connected if the classifier for those two references predicts the positive class. Then, references clusters correspond to connected components of Q which means that they can be determined using basic graph traversal algorithms (DFS or BFS).

The similarity between two references can be also expressed in probabilistic terms. Torvik et al. [75, 76] proposed the Authority model for author name disambiguation. In the Authority model, the similarity between two references is defined as the ratio $r(x)$ between $P_M(x)$ and $P_N(x)$, where x is a reference similarity vector, $P_M(x)$ is the probability of observing x for coreferent references, and $P_N(x)$ is the probability of observing x for non-coreferent references. The r function can be computed using statistical smoothing and interpolation of the training set and saved in a lookup table for later coreference classifications. If a reference similarity vector computed for two references is not in the lookup table then its r value is extrapolated from labeled reference similarity vectors contained in the lookup table. Soler proposed a reference similarity measure defined in the terms of the probability that two randomly selected references in a set of references would have as many coincidences (matching words or n-grams) as two compared references [69].

Finally, Culotta et al. [12] proposed an author grouping method based on a greedy agglomerative clustering technique which does not use a reference similarity measure, but relies on a learned cluster scoring function. The cluster scoring function reflects how likely is that all references in a cluster are authored by the same person. The authors proposed an error-driven algorithm that utilizes a ranking loss function to learn the best cluster scoring function from a training dataset.

6.6.2 Author Assignment Methods

In contrast to author grouping methods, author assignment name disambiguation approaches directly assign authors to references.

The first author grouping methods were introduced by Han et al. [21]. The authors proposed two methods based on two standard classification techniques, naive Bayes and support vector machines (SVM), which are used to decide whether a person given by its canonical name (a disambiguated name label that uniquely determines a person) is an author or co-author of a reference. Both methods require a training set D constructed from references whose authors are uniquely identified. Each instance in D represents one reference and consists of features derived from co-author names, publication title and the title of publication venue along with a class variable that denotes the canonical name of the author.

In the naive Bayes based method of Han et al., the Bayes rule is used to compute the posterior probability $P(X_i \mid C)$ that canonical name X_i ($1 \leq i \leq N$, where N is the number of canonical names in D) is an author of C. Then, C is assigned to the canonical name that has the maximal posterior probability. According to the Bayes rule, we have that

$$\max_i P(X_i \mid C) = \max_i \frac{P(C \mid X_i)\, P(X_i)}{P(C)} \tag{6.12}$$

where $P(X_i)$ is the prior probability of X_i authoring papers (this probability can be estimated as the fraction of papers in D authored by X_i) and $P(C)$ denotes the probability of C which can be omitted because it does not depend on X_i. Thus, X_i is determined by maximizing $P(C \mid X_i)P(X_i)$. The probability $P(C \mid X_i)$ is decomposed according to reference features:

$$P(C \mid X_i) = P_a(C \mid X_i)\, P_t(C \mid X_i)\, P_v(C \mid X_i) \tag{6.13}$$

where

1. $P_a(C \mid X_i)$ is the probability of X_i co-authoring C with other authors of C,
2. $P_t(C \mid X_i)$ is the probability of X_i authoring paper whose title is the title of C, and
3. $P_v(C \mid X_i)$ is the probability of X_i authoring paper published in the publication venue of C.

$P_a(C \mid X_i)$ is further hierarchically decomposed into several conditional probabilities which can be easily estimated from D: the probability of X_i producing solo-authored papers, the probability of X_i writing multi-authored papers, the probability of X_i writing papers with previous collaborators, the probability of X_i writing papers with new collaborators, and the probability of X_i writing papers with a particular co-author X_j. On the other hand, $P_t(C \mid X_i)$ and $P_v(C \mid X_i)$ are derived from conditional probabilities that a token appears in the publication title and the title of publication venue, respectively, in a paper authored by X_i (those conditional probabilities can be also trivially estimated from D). In their subsequent work [22], the authors presented an unsupervised version of the Naive Bayes approach to the name disambiguation problem. The main idea of the unsupervised approach is that each reference C_m in a set of references C corresponding to some ambiguous name can be viewed as a mixture of K components corresponding to K different individuals, i.e.

$$P(C_m) = \sum_{i=1}^{K} \big(P(X_i) \, P(C_m \mid X_i) \big) \tag{6.14}$$

As in the previous work, each of K persons is modeled by a hierarchical Naive Bayes model. In the first step of the algorithm, references in C are randomly assigned to K components. Then, the expectation maximization (EM) algorithm is used to iteratively reassign references to components by maximizing the likelihood of C which is equal to $\sum_{m \in C} P(C_m)$. The main disadvantage of this approach is that K has to be specified in advance. Other approaches based on mixture models were proposed by Bhattacharya and Getoor [1] who presented the LDA-ER model based on latent Dirichlet allocation and Tang et al. [72] who formalized the name disambiguation problem using hidden Markov random fields.

Regarding the SVM-based method of Han et al. [21], a binary SVM classifier is trained for each canonical name appearing in the training set D. For each X_i, D is converted into another training set $D(X_i)$ such that instances of $D(X_i)$ have a binary class variable ("yes" or "no") denoting whether X_i authored a reference or not. This means that the SVM is trained using the "one class versus all others" approach. For a two-class classification problem, an SVM receives training vectors $x_i \in R^n$, where n is the number of features, and finds an optimal hyperplane which maximally separates positive from negative training examples. The SVM classifier used by Han et al. relies on training vectors of the form $x_i = (w_1, w_2, ..., w_n)$ where n is the number of distinct tokens in D and w_i is the TF-IDF score of token i.

Veloso, Ferreira and colleagues proposed three author assignment methods called EAND (Eager Associative Name Disambiguation), LAND (Lazy Associative Name Disambiguation) and SLAND (Self-training LAND) [18, 78]. The methods are based on associations between reference features and authors. Such associations can be expressed by rules of the form $S \rightarrow a_k$ where a_k is an author name label uniquely identifying one person. S represents a set of propositions which are in the form $f = c$ where f is an arbitrary reference feature and c is a constant term. For example, the following rule

{co-author = "J. Doe", title = "classification", venue = "artificial"} $\rightarrow$ AlanSalton3

associates canonical author name "AlanSalton3" with publications in which "J. Doe" is a co-author, the term "classification" is contained in the publication title, and the term "artificial" is present in the title of publication venue. All three proposed methods can use any existing association rule mining algorithm to extract association rules from references in a training set. Let R denote the set of rules extracted from a training set of references D. A rule $r = S \rightarrow a_k$ matches a reference X if all propositions in S are satisfied for X. The extracted rules are not equally important when deciding the authorship of X. The confidence of the rule r can be expressed as the conditional probability of a_k being an author of a reference that matches $S \rightarrow a_k$. This probability can be estimated from D as $|P|/|Q|$ where Q is the subset of references in D which match r and P is the subset of Q encompassing all references authored by a_k. Let $R(a_k, X)$ denote the subset of rules in R that match reference X and predict that a_k authored X. Then, the strength of association between X and a_k, denoted by $W(a_k, X)$, can be expressed as the average confidence of rules in $R(a_k, X)$. Finally, the probability that a_k is the author of X can be obtained by normalizing $W(a_k, X)$:

$$P(a_k \mid X) = \frac{W(a_k, X)}{\sum_j W(a_j, X)} \qquad (6.15)$$

The main differences between EAND, LAND and SLAND are:

1. EAND uses the whole training set D to extract association rules and employs a simple pruning strategy to eliminate low-frequency rules. The frequency of a rule r is equal to the number of references in D that match r. All extracted rules having frequency below some predefined threshold are ignored.
2. LAND performs on-demand rule extraction when assigning an author to a reference r. The extraction of rules is not done on the whole training set D, but on its projection that contains just those references having the same features as r.
3. SLAND is an extension of LAND in which reliable predictions are included in D as new training examples. Additionally, if a reference X cannot be assigned to an author in D then SLAND creates a new, unseen author U and adds (X, U) to D.

6.7 Name Disambiguation Approach Based on Reference Similarity Network Clustering

We propose a novel supervised approach to the name disambiguation problem which handles both name synonyms and name homonyms. The main idea is to represent a set of bibliographic references by a network that reflects similarities between references and then determine unique authors by identifying cohesive sub-graphs in the network. The approach is based on the notion of a reference similarity network which is defined as follows.

Definition 6.3 (*Reference Similarity Network*) Let S be a set of bibliographic references. Let M denote a reference similarity metric, i.e. $M : S \times S \rightarrow [0, 1]$ where higher values of M indicate higher similarity between references ($M(s_1, s_2) = 1$ implies that references s_1 and s_2 are identical). The reference similarity network corresponding to S is an undirected weighted graph N constructed in the following way:

1. Each reference in S is represented by one node in N.
2. Two nodes P and Q are connected by an undirected link if $M(P, Q) > w$ where w is a reference similarity threshold. The weight of the link connecting P and Q is equal to $M(P, Q)$.

Let A and B be two different individuals having the same or similar name labels in a set of bibliographic references S. Let N denote the reference similarity network constructed from S. Then, it is highly probable that the references authored by A form a highly cohesive subgraph in N that is loosely coupled to the references authored by B which form another highly cohesive subgraph in the same network. To cluster reference similarity networks we use different community detection algorithms which identify non-overlapping cohesive subgraphs [19]. Compared to traditional graph partitioning approaches, community detection techniques are computationally feasible for extremely large networks and they do not require the number of clusters to be specified before clustering.

Our RSNC method (the abbreviation for **R**eference **S**imilarity **N**etwork **C**lustering) consists of the following steps (see Fig. 6.4):

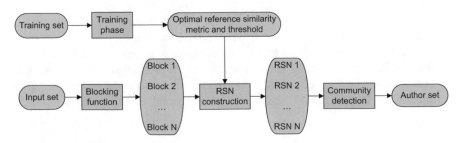

Fig. 6.4 The components and the data flow in the RSNC method

1. In the training phase, the best reference similarity metric is selected from a set of string similarity metrics and the optimal value of the reference similarity threshold is estimated by a greedy algorithm. The training is performed on a training dataset that contains pairs of both coreferent and non-coreferent bibliographic references (see Definition 6.2).
2. An input set of bibliographic references is partitioned into overlapping blocks of references. Each block encompasses references containing identical or highly similar author name labels. Highly similar author name labels are determined using string similarity metrics.
3. For each block B obtained in the previous step the following actions are performed:

   a. The reference similarity network corresponding to B is constructed using the optimal reference similarity metric and reference similarity threshold determined in the training phase.
   b. The reference similarity network is clustered using a community detection technique. The obtained clusters correspond to different authors.

After the name disambiguation procedure is performed, the co-authorship network corresponding to the input set of bibliographic references can be constructed in a straightforward manner:

1. The set of nodes in the co-authorship network corresponds to the set of clusters obtained by the name disambiguation procedure.
2. Two authors A and B are connected in the co-authorship network if $C(A) \cap C(B) \neq \emptyset$ where $C(X)$ denotes the cluster of references of an author X. The weight of the link is determined by a chosen link weighting scheme which is computed taking into account the intersection of $C(A)$ and $C(B)$.

The RSNC method has the training phase to determine the best reference similarity metric in a set of string similarity metrics and the reference similarity threshold used to construct reference similarity networks of reference blocks. Let M be a set of string similarity metrics, m one arbitrarily selected metric from M, and let D denote the training set. An optimal selection of the reference similarity threshold w for m implies that the reference similarity network N constructed from D has the strongest possible community structure. Since clusters in N are known (they are labeled in the training set D), we are in the position to classify links in N into two categories:

- intra-cluster links connecting nodes representing coreferent references, and
- inter-cluster links connecting nodes representing non-coreferent references.

Let $W(\text{intra})$ and $W(\text{inter})$ denote the total weight of intra-cluster and inter-cluster links in N, respectively, i.e.

$$W(\text{intra}) = \sum_{(a,b) \in \text{Intra}} m(a, b) \qquad (6.16)$$

$$W(\text{inter}) = \sum_{(a,b)\,\in\,\text{Inter}} m(a,b) \tag{6.17}$$

where (a, b) denotes the link connecting nodes a and b, Intra is the set of intra-cluster links and Inter is the set of inter-cluster links in N. The strongest possible community structure is obtained when

$$\Delta W = W(\text{intra}) - W(\text{inter}) \quad \text{is maximal possible} \tag{6.18}$$

Consequently, the optimal value of w corresponds to the reference similarity network constructed from D for which ΔW is maximized. This value can be determined by a simple greedy algorithm shown in Algorithm 6.2 in which similarities of all pairs of references in D are processed from the largest to the lowest one for each reference similarity metric.

Algorithm 6.2: The training phase of the RSNC method.

Input : D - training set, M - the set of reference similarity metrics
Output: $optM$ - the best metric in M, $optW$ - the optimal value of the reference similarity
          threshold for $optM$

$max\Delta = -\infty$
foreach $m \in M$ **do**
    L = an empty list of triplets
    /* compute similarity for each pair of references in D and add it to L */
    foreach $r_1, r_2 \in D,\ r_1 \neq r_2$ **do**
        L.insert($[m(r_1, r_2), r_1, r_2]$)

    sort L in the non-increasing order of the first component (reference similarity)

    /* compute ΔW and the optimal threshold for m */
    $W(\text{inter}) = 0,\ W(\text{intra}) = 0,\ m\Delta = -\infty,\ oW = 1.0$
    foreach $[w, r_1, r_2] \in L$ **do**
        if r_1 and r_2 are coreferent **then**
            $W(\text{intra}) = W(\text{intra}) + w$

        else
            $W(\text{inter}) = W(\text{inter}) + w$
        $\Delta W = W(\text{intra}) - W(\text{inter})$
        if $\Delta W > m\Delta$ **then**
            $m\Delta = \Delta W$
            $oW = w;$

    if $m\Delta > max\Delta$ **then**
        $max\Delta = m\Delta$
        $optM = m$
        $optW = oW$

Table 6.4 Characteristics of the experimental dataset

Ambiguous name label	Individuals	References
JMartin	16	112
MBrown	13	153
JRobinson	12	171
AKumar	14	244
MJones	13	260
KTanaka	10	280
DJohnson	15	368
MMiller	12	412
Total	105	2000

The RSCN method uses the following procedure to form blocks of references. Let L denote the set of name labels appearing in the input set of references S, let NG denote the set of n-grams in L, and let I be an inverted index which maps name labels to references such that $I(k)$ returns all references from S in which a name label k appears. Firstly, the set L is partitioned into overlapping blocks $NB = \{NB_i\}_{i \in NG}$ such that NB_i encompasses all name labels which contain n-gram i. Then, an undirected graph G depicting similarities between name labels is formed according to the following rule: two name labels A and B belonging to the same NB_i set are connected in G if the Jaccard n-gram similarity between A and B is higher than some predefined threshold (the default value is 0.6, see Sect. 6.5). This means that the string similarity function is computed for each pair of name labels having at least one n-gram in common. Then, the groups of highly similar name labels are obtained by identifying connected components of G. Finally, the set of blocks of references is formed according to the following rules:

- Each connected component in G corresponds to one block of references,
- The block B_k corresponding to the k-th connected component in G is equal to $B_k = \bigcup_{i \in C_k} I(k)$ where C_k is the set of nodes belonging to the k-th connected component.

6.7.1 Experimental Evaluation

To evaluate the RSNC method, we used a dataset encompassing 2000 bibliographic references authored by 8 ambiguous name labels (8 blocks) that correspond to 105 different individuals. The dataset was formed by C. Lee Giles from the CiteSeer metadata.[2] The basic characteristics of the dataset are given in Table 6.4.

---

[2]The dataset can be downloaded from http://clgiles.ist.psu.edu/data/nameset_author-disamb.tar.zip.

We employed four different community detection techniques provided by the iGraph library [11] to cluster reference similarity networks:

1. The Greedy Modularity Optimization (GMO) algorithm [8]. The algorithm relies a greedy hierarchical agglomeration strategy to maximize the weighted modularity measure. GMO starts with the partitioning in which each node is assigned to a singleton community. In each iteration of the algorithm, the variation in weighted modularity obtained by merging any two adjacent communities is computed. The merge operation that maximally increases (or minimally decreases) the modularity score is chosen and the merge of corresponding communities is performed.

2. The Louvain algorithm [4]. This method is an improved variant of the GMO method. The main idea of the algorithm is to apply a greedy multi-resolution strategy to maximize the weighted modularity metric starting from the partition in which nodes are put in singleton communities. When the modularity metric is optimized locally by moving nodes to neighboring communities, the algorithm creates a network of communities and then repeats the previous steps on that network.

3. The Walktrap algorithm [57]. This algorithm relies on a node distance measure reflecting the probability that a random walker moves from a node A to a node B in exactly k steps, where k is the only parameter of the algorithm. In our experiments we used the default value of k that is equal to 4. The clustering dendrogram is constructed by the Ward's agglomerative clustering technique and the partition maximizing weighted modularity is selected as the output of the algorithm.

4. The Infomap algorithm [61]. This method reveals communities by optimally compressing a description of information flows on the network. The algorithm applies a greedy strategy to minimize the map equation which reflects the expected description length of a random walk on a partitioned network.

The half of the dataset containing randomly selected reference pairs is used to train the RSNC method. In the training phase, the best reference similarity metric was selected among the following string similarity metrics: the Jaccard similarity at the token level, the Jaccard similarity at the n-gram level ($n = 3$), the TF-IDF similarity at the token level, and the TF-IDF similarity at the n-gram level ($n = 3$). String similarity metrics were computed on normalized references in which stop words were removed, while remaining tokens were stemmed using the Porter's stemming algorithm [58].

The results of the training phase are presented in Table 6.5. It can be seen that for each considered string similarity metric the corresponding reference similarity network exhibits a strong community structure (W(intra) $\gg W$(inter)) at the optimal value of the reference similarity threshold. The maximal value of ΔW is achieved using the TF-IDF distance at the token level for the reference similarity threshold equal to 0.184, so this metric is selected to form reference similarity networks that are later clustered by various community detection techniques.

To evaluate the accuracy of the RSNC method for different community detection techniques we rely on the following name disambiguation accuracy metrics:

Table 6.5 The results of the training phase. w – the optimal reference similarity threshold, L – the number of links in the reference similarity network for the optimal value of the reference similarity threshold, $W(\text{intra})$ – the total weight of intra-cluster links, $W(\text{inter})$ – the total weight of inter-cluster links. The bold value of ΔW ($\Delta W = W(\text{intra}) - W(\text{inter})$) indicates the best reference similarity metric

Distance	w	L	$W(\text{intra})$	$W(\text{inter})$	ΔW
Jaccard token	0.138	18910	2872.249	838.584	2033.664
Jaccard n-gram	0.138	13740	2162.476	584.856	1577.620
TF-IDF token	0.184	19441	4074.622	1076.181	**2998.441**
TF-IDF n-gram	0.224	11221	2882.541	761.694	2120.847

1. Pair-wise precision (denoted by P) – the fraction of reference pairs in the same cluster being coreferent [25]. This measure is computed by the following formula

$$P = \sum_{c \in C} \sum_{i,j \in c} \delta(i, j) \Big/ \sum_{c \in C} \frac{|c|(|c| - 1)}{2}, \qquad (6.19)$$

where C denotes the set of obtained clusters, c is a cluster in C, i and j are two different references in c, and $\delta(i, j) = 1$ if i and j are coreferent, or 0 otherwise. P is a measure of cluster purity or correctness – high values of P indicate that obtained clusters dominantly contain references of same individuals.

2. Pair-wise recall (denoted by R) – the fraction of coreferent reference pairs put in the same cluster [25]. R is computed by the following equation

$$R = \sum_{g \in G} \sum_{i,j \in g} \sigma(i, j) \Big/ \sum_{g \in G} \frac{|g|(|g| - 1)}{2}, \qquad (6.20)$$

where G denotes the set of real clusters (clusters of coreferent reference pairs in the input set of references), g is a cluster in G, i and j are two different coreferent references in g, $\sigma(i, j) = 1$ if i and j are located in the same cluster obtained by a clustering procedure, or 0 otherwise. R is a measure of cluster completeness – high values of R indicate that each obtained cluster encompasses almost all references of an individual.

3. Pair-wise F_1 which is an aggregated measure of accuracy defined as the harmonic mean of P and R.

4. Normalized mutual information (denoted by NMI) which is a measure of the similarity between real and obtained clusters [13]. NMI ranges from 0 to 1 and a higher value of NMI indicates that the partitioning obtained by a clustering technique is more similar to the real partitioning.

Table 6.6 The accuracy of the RSNC method. Bold values indicate the best community detection method. P – precision, R – recall, F_1 – the F_1 metric, NMI – normalized mutual information

	GMO				Infomap			
	P	R	F_1	NMI	P	R	F_1	NMI
JMartin	0.499	0.924	0.648	0.795	0.745	0.816	0.779	0.885
MBrown	0.749	0.926	0.828	0.833	0.827	0.845	0.836	0.866
JRobinson	0.400	0.694	0.507	0.610	0.614	0.528	0.568	0.711
AKumar	0.559	0.485	0.520	0.609	0.779	0.346	0.480	0.648
MJones	0.623	0.837	0.715	0.704	0.855	0.811	0.832	0.817
KTanaka	0.822	0.639	0.719	0.715	0.897	0.523	0.661	0.773
DJohnson	0.511	0.464	0.486	0.435	0.511	0.941	0.662	0.656
MMiller	0.950	0.672	0.787	0.797	0.983	0.819	0.894	0.849
Average	0.639	**0.705**	0.651	0.687	**0.776**	0.704	**0.714**	**0.776**
	Louvain				Walktrap			
	P	R	F_1	NMI	P	R	F_1	NMI
JMartin	0.499	0.924	0.648	0.795	0.533	0.816	0.645	0.801
MBrown	0.634	0.739	0.683	0.777	0.728	0.910	0.809	0.849
JRobinson	0.532	0.572	0.551	0.672	0.524	0.615	0.566	0.704
AKumar	0.530	0.453	0.489	0.563	0.778	0.361	0.493	0.654
MJones	0.684	0.830	0.750	0.728	0.612	0.807	0.696	0.728
KTanaka	0.861	0.567	0.683	0.751	0.883	0.449	0.595	0.742
DJohnson	0.660	0.461	0.543	0.593	0.758	0.322	0.452	0.612
MMiller	0.944	0.538	0.685	0.746	0.952	0.828	0.886	0.802
Average	0.668	0.635	0.629	0.703	0.721	0.638	0.643	0.736

The results of the evaluation are summarized in Table 6.6. It can be seen that the Infomap method exhibits the best name disambiguation performance achieving more than 70% pairwise F_1 and NMI equal to 0.776. This means that reference clusters obtained by Infomap are highly similar to real clusters in the dataset. Therefore, we can conclude that the RFNC method relying on the Infomap community detection algorithm achieves a good accuracy in solving the name disambiguation problem. Finally, it should be emphasized that the current level of accuracy of the RSNC method is achieved using basic string similarity metrics. In our future work we will investigate the performance of the RSNC method relying on learned reference similarity functions derived from numerical vectors expressing similarities between reference attributes (author names, title, publication venue, year, and so on).

6.8 Author Identification in Massive Bibliography Databases

The identification of authors based on the strict matching of name labels can be safely used when co-authorship networks are extracted from people-article centered bibliography databases. Our literature review showed that the strict name label matching was used to form co-authorship networks extracted from two massive bibliography databases – Mathematical Reviews and DBLP. Therefore, in this section we will review how authors are identified and disambiguated in those two databases.

Mathematical Reviews (MR, MathSciNet) is a subscription-based bibliography database maintained by the American Mathematical Society. The MR database indexes mathematical books and articles published in peer-reviewed journals. This database covers more than 1800 journals and contains bibliography records for more than 2 million publications authored by more than 800000 mathematicians. For each author, MR maintains an author profile – the record that contains all known name variants of the author, institutional affiliation(s), subject classifications contained in his/her papers, links to MR author profiles of co-authors, and references to cited articles that are indexed in the database.[3]

In the beginning of MR, the identification of authors was done entirely manually [74]. After 1985, the process of author identification was semi-automated using author matching algorithms. Namely, for each author of an article that is being added to the database a multi-criterion matching against existing author profiles is performed. The matching procedure takes into account three features: the name of an author, its institutional affiliation and subject classifications of the article assigned by the MR editors. Unfortunately, the details of the procedure cannot be found in the MR documentation, nor they are documented in relevant scientific articles. As reported in [74], in roughly eighty percent of cases an uniquely matching author profile can be found when indexing a new article. In the rest of cases, the matching algorithm ranks retrieved candidates that are later manually inspected in order to determine whether a new author profile should be created.

DBLP is a computer-science bibliography database developed by a research group from University of Trier led by professor Michael Lay. Currently, this database contains bibliographic information about more than 3 million publications written by more than 1.8 million researchers in more than 1500 journals and 5000 conference proceedings. DBLP is both people-centered and publication-centered database, but DBLP does not guarantee that an author profile corresponds to exactly one individual nor that one individual is represented by exactly one author profile [37]. For example, it can be easily verified that the DBLP profile of author "Miloš Savić" actually contains aggregated publication data of two distinct researchers (one affiliated to the University of Novi Sad, Serbia and another to the University of Oklahoma, USA). As an example on the opposite side, Djordje Herceg (full professor at the University of Novi Sad, Serbia) has two DBLP profiles (one as "Djordje Herceg" and the another

---

[3]http://www.ams.org/publications/math-reviews/mr-authors.

one as "Dorde Herceg"). In other words, both name homonymy and name synonymy problems are present in the DBLP database.

To increase the quality of data (decrease the number of appearances of name homonyms and synonyms) DBLP employs two strategies [37, 60]:

- potential name synonyms are identified using author name similarity functions, and
- potential name homonyms are detected using a heuristic based on ego co-authorship networks.

Different author name similarity measures are applied on blocks of authors in order to detect potential name synonyms. Due to a huge amount of authors in the database it is computationally unmanageable to compute author similarities considering the whole set of DBLP author profiles. Therefore, similarity measures are computed on subsets of DBLP profiles obtained by blocking functions. DBLP uses the following blocking functions:

1. the geodesic distance in the DBLP co-authorship network (the length of the shortest path between two nodes in the network),
2. the presence of authors in the same publication venue,
3. the same rare keywords in publication titles.

Author similarity measures are computed for authors that have distance smaller than or equal to two in the DBLP co-authorship network, authors who have published in the same journal or conference series and authors who have articles that contain the same rare word in their publication titles. The DBLP maintainers implemented more than 20 author similarity functions that are either classic string similarity functions/string edit distances, graph-based similarity functions (such as the number of co-authors in common) and hybrid measures that are combinations of the previous two types of author similarity measures.

The disconnected co-authors heuristic is applied on ego-networks of DBLP authors in order to detect potential name homonyms. More specifically, if there are two distinct persons that are represented by one node in DBLP the co-authorship network then it is highly probable that such node is an articulation point in its ego-network (network induced by the node and the nearest neighbors). In other words, the removal of a node representing two or more distinct persons from its ego-network will probably cause a fragmentation of the ego-network into several disjoint connected components representing unrelated groups of researchers. If the node is not an articulation point in its ego-network then we can be quite confident that the node represents a single person. When an author name homonymy is verified then the DBLP author profile is split into two or more DBLP author profiles and "mystical" numbers are added to author names in order to make authors with the same name distinguishable.

6.9 Impact of Name Disambiguation on Co-authorship Network Structure: A Case Study

In one of our previous publications we studied the co-authorship network of researchers who published their results in Serbian mathematical journals indexed in the eLib digital library [64]. As already mentioned, eLib belongs to the class of article-centered bibliography databases in which authors are not uniquely identified. The name disambiguation procedure used to identify authors in the eLib bibliographic records is described in Sect. 6.4.

In this section, we investigate the impact of name disambiguation on the structure of the eLib co-authorship network. Namely, we constructed two co-authorship networks from eLib bibliographical records:

- the first network, denoted by N', was constructed after author names were disambiguated (this network was analyzed in [64]), and
- the second network, denoted by N, was obtained without disambiguating author names. This means that N was constructed using the strict name label matching where identical author name labels denote the same author, while different author name labels indicate different authors.

Then, we compared basic structural characteristics of N and N'. For both networks we computed the number of nodes, the number of links, the average node degree, the number of isolated nodes, the number of connected components (without isolated nodes), the size of the largest connected component, the average clustering coefficient, the diameter of the largest connected component, and the small-world coefficient. The results of the comparison are summarized in Table 6.7. It can be seen that N' has a drastically smaller number of nodes (authors), isolated nodes (authors without collaborators) and connected components compared to N. Additionally, N' has a considerably higher characteristic path length and diameter in comparison to N. Moreover, the largest connected component in N' is drastically larger than the largest component in N. The large differences in previously mentioned structural quantities imply that the network obtained after disambiguating author names is significantly less fragmented and better reflects collaborative behavior of the eLib scientific community.

The structure of a co-authorship network can also be characterized by distributions of various node, link and component metrics. In order to provide a statistically sound comparison of the structure of N and N' we conducted several two-sample Kolmogorov–Smirnov (KS) tests. Namely, we investigated if there are statistically significant differences in distributions of basic node, link and component structural metrics when they are computed from N and N'. The results of the statistical testing are shown in Table 6.8.

It can be seen that there are no statistically significant differences in the distributions of link and component metrics: for each of those metrics obtained p-value of the KS test statistic (D) is higher than 0.05 which means that the null hypothesis of the KS test (no statistically significant differences between two compared

Table 6.7 Macroscopic view of name disambiguation impacts: N' is the eLib co-authorship network in which author names were disambiguated, while N is the same network obtained without name disambiguation

Property	N'	N
The number of nodes (authors)	3597	4247
The number of links (collaborations)	2766	2890
The average node degree (avg. number of co-authors)	2.31	2.21
The number of isolated nodes (authors without co-authors)	1202	1634
The number of connected components (without isolated nodes)	625	734
The size of the largest connected component	249	160
The average clustering coefficient	0.44	0.43
The diameter of the largest connected component	19	14
The small-world coefficient	2.11	1.64

Table 6.8 The results of two sample Kolmogorov–Smirnov tests for metric distributions obtained from the two variants of the eLib co-authorship network

	Metric	D	p
Node metrics	The number of papers per author	0.049	<0.001
	Author timespan	0.049	<0.001
	Degree centrality	0.051	<0.001
	Clustering coefficient	0.042	0.001
	Betweenness centrality	0.035	0.015
Link metrics	Link weight	0.027	0.259
	Link timespan	0.023	0.446
Component metrics	Component size	0.036	0.767
	The number of papers per component	0.054	0.268

distributions) is accepted. Although name disambiguation resulted in a less fragmented network, name disambiguation does not affect overall statistical properties of connected components. On the other hand, statistically significant differences are present for all examined node metrics. This result indicates that name disambiguation has a big impact when evaluating researchers contained in the network.

Fig. 6.5 The correlation between author rankings obtained from the eLib networks constructed after and without name disambiguation

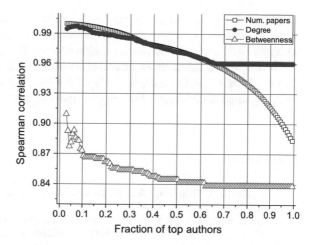

In other words, an accurate characterization of productivity, long-term presence and collaboration of authors present in the network cannot be obtained relying on the co-authorship network constructed without name disambiguation.

Ranking authors by research evaluation metrics is perhaps the most important aspect of co-authorship network analysis. Therefore, we examined the impact of name disambiguation to the stability of author rankings when authors are ranked by the number of published papers, degree and betweenness centrality. Let $T_{M,f}$ denote the set of top f-percent authors according to a researcher evaluation metric M in the network constructed after name disambiguation. For each author from $T_{M,f}$ we computed the value of M in the network constructed without name disambiguation. In other words, for a fixed f we obtained two ordered sequences of metric values:

- S^1 - the values of M for authors from $T_{M,f}$. This sequence is sorted in the non-increasing order of M values, i.e. S_i^1 is the value of M for i-th top ranked author when M is computed using the network constructed after name disambiguation.
- S^2 - the values of M for authors from $T_{M,f}$ when M is computed using the network constructed without name disambiguation. S_i^2 is the value of M for i-th author in S^1.

Then, we computed the Spearman rank correlation between S^1 and S^2 for various values of f (see Fig. 6.5). A higher value of the Spearman correlation between S^1 and S^2 indicates a higher similarity between author rankings. The value of the Spearman correlation that is equal to 1 implies that compared rankings are identical.

It can be observed from Fig. 6.5 that the Spearman correlation between author rankings is higher than 0.99 for the 25% top-ranked authors by the number of published papers and the 20% top-ranked authors by degree centrality. This means that the most productive and the best connected eLib authors can be determined using the network constructed without name disambiguation. On the other hand, the correlation between betweenness centrality rankings is always smaller than 0.93 which indicates that betweenness centrality rankings obtained from the network constructed without name disambiguation are less accurate. Also, the increase of f results in lower correlations between rankings which means that ranked sequences become less similar with more authors included in the ranking process. For centrality measures (degree and betweenness) the decrease of correlations in author rankings stops for f higher than 0.67 since isolated authors (33% of the total number of authors in the eLib co-authorship network) have both degree and betweenness centrality equal to zero.

References

1. Bhattacharya, I., Getoor, L.: A latent Dirichlet model for unsupervised entity resolution. In: Proceedings of the Sixth SIAM International Conference on Data Mining, April 20-22, 2006, Bethesda, MD, USA, pp. 47–58 (2006). https://doi.org/10.1137/1.9781611972764.5
2. Bhattacharya, I., Getoor, L.: Collective entity resolution in relational data. ACM Trans. Knowl. Discov. Data **1**(1), (2007). https://doi.org/10.1145/1217299.1217304
3. Bird, C., Barr, E., Nash, A., Devanbu, P., Filkov, V., Su, Z.: Structure and dynamics of research collaboration in computer science. In: Proceedings of the Ninth SIAM International Conference on Data Mining, p. 826837. SIAM (2009)
4. Blondel, V.D., Guillaume, J.L., Lambiotte, R., Lefebvre, E.: Fast unfolding of communities in large networks. Journal of Statistical Mechanics: Theory and Experiment **2008**(10), P10,008 (2008)
5. Bordes, A., Ertekin, S., Weston, J., Bottou, L.: Fast kernel classifiers with online and active learning. J. Mach. Learn. Res. **6**, 1579–1619 (2005)
6. Cen, L., Dragut, E.C., Si, L., Ouzzani, M.: Author disambiguation by hierarchical agglomerative clustering with adaptive stopping criterion. In: Proceedings of the 36th International ACM SIGIR Conference on Research and Development in Information Retrieval, SIGIR '13, pp. 741–744. ACM, USA (2013). https://doi.org/10.1145/2484028.2484157
7. Chen, Y., Brner, K., Fang, S.: Evolving collaboration networks in Scientometrics in 1978–2010: a micromacro analysis. Scientometrics **95**(3), 1051–1070 (2013). https://doi.org/10.1007/s11192-012-0895-2
8. Clauset, A., Newman, M.E.J., Moore, C.: Finding community structure in very large networks. Phys. Rev. E **70**, 066111 (2004). https://doi.org/10.1103/PhysRevE.70.066111
9. Cohen, W.W., Ravikumar, P., Fienberg, S.E.: A comparison of string distance metrics for name-matching tasks. In: Proceedings of IJCAI-03 Workshop on Information Integration, pp. 73–78 (2003)
10. Cota, R.G., Ferreira, A.A., Nascimento, C., Gonçalves, M.A., Laender, A.H.F.: An unsupervised heuristic-based hierarchical method for name disambiguation in bibliographic citations. J. Am. Soc. Inf. Sci. Tech. **61**(9), 1853–1870 (2010). https://doi.org/10.1002/asi.v61:9
11. Csardi, G., Nepusz, T.: The igraph software package for complex network research. InterJ. Complex Syst. p. 1695 (2006)
12. Culotta, A., Kanani, P., Hall, R., Wick, M., McCallum, A.: Author disambiguation using error-driven machine learning with a ranking loss function. In: Sixth International Workshop on Information Integration on the Web (IIWeb-07). Vancouver, Canada (2007)
13. Danon, L., Daz-Guilera, A., Duch, J., Arenas, A.: Comparing community structure identification. J. Stat. Mech. Theory Exp. **2005**(09), P09008 (2005)
14. Ester, M., Kriegel, H., Sander, J., Xu, X.: A density-based algorithm for discovering clusters in large spatial databases with noise. In: Proceedings of the Second International Conference on Knowledge Discovery and Data Mining (KDD-96), pp. 226–231. Portland, Oregon, USA (1996)
15. Fan, X., Wang, J., Pu, X., Zhou, L., Lv, B.: On graph-based name disambiguation. J. Data Inf. Qual. **2**(2), 10:1–10:23 (2011). https://doi.org/10.1145/1891879.1891883

16. Fegley, B.D., Torvik, V.I.: Has large-scale named-entity network analysis been resting on a flawed assumption?. PLoS ONE **8**(7), e70299 (2013). https://doi.org/10.1371/journal.pone. 0070299
17. Ferreira, A.A., Gonçalves, M.A., Laender, A.H.: A brief survey of automatic methods for author name disambiguation. SIGMOD Rec. **41**(2), 15–26 (2012). https://doi.org/10.1145/2350036. 2350040
18. Ferreira, A.A., Veloso, A., Gonçalves, M.A., Laender, A.H.: Effective self-training author name disambiguation in scholarly digital libraries. In: Proceedings of the 10th Annual Joint Conference on Digital Libraries, JCDL '10, pp. 39–48. ACM, New York, USA (2010). https:// doi.org/10.1145/1816123.1816130
19. Fortunato, S.: Community detection in graphs. Phys. Rep. **486**(35), 75–174 (2010). https://doi. org/10.1016/j.physrep.2009.11.002
20. Gurney, T., Horlings, E., Van Den Besselaar, P.: Author disambiguation using multi-aspect similarity indicators. Scientometrics **91**(2), 435–449 (2012). https://doi.org/10.1007/s11192- 011-0589-1
21. Han, H., Giles, L., Zha, H., Li, C., Tsioutsiouliklis, K.: Two supervised learning approaches for name disambiguation in author citations. In: Proceedings of the 4th ACM/IEEE-CS Joint Conference on Digital Libraries, JCDL '04, pp. 296–305. ACM, New York, USA (2004). https://doi.org/10.1145/996350.996419
22. Han, H., Xu, W., Zha, H., Giles, C.L.: A hierarchical naive bayes mixture model for name disambiguation in author citations. In: Proceedings of the 2005 ACM Symposium on Applied Computing, SAC '05, pp. 1065–1069. ACM, New York, USA (2005). https://doi.org/10.1145/ 1066677.1066920
23. Han, H., Zha, H., Giles, C.L.: A model-based k-means algorithm for name disambiguation. In: ISWC 2003 Workshop on Semantic Web Technologies for Searching and Retrieving Scientific Data. CEUR-WS (2003). http://ceur-ws.org/Vol-83/int_2.pdf
24. Han, H., Zha, H., Giles, C.L.: Name disambiguation in author citations using a k-way spectral clustering method. In: Proceedings of the 5th ACM/IEEE-CS Joint Conference on Digital Libraries, JCDL '05, pp. 334–343. ACM, New York, USA (2005). https://doi.org/10.1145/ 1065385.1065462
25. Huang, J., Ertekin, S., Giles, C.L.: Efficient name disambiguation for large-scale databases. In: Proceedings of the 10th European Conference on Principle and Practice of Knowledge Discovery in Databases, PKDD'06, pp. 536–544. Springer, Berlin (2006). https://doi.org/10. 1007/11871637_53
26. Huang, J., Zhuang, Z., Li, J., Giles, C.L.: Collaboration over time: Characterizing and modeling network evolution. In: Proceedings of the 2008 International Conference on Web Search and Data Mining, WSDM '08, pp. 107–116. ACM, New York, USA (2008). https://doi.org/10. 1145/1341531.1341548
27. Jaro, M.A.: Advances in record-linkage methodology as applied to matching the 1985 census of Tampa, Florida. J. Am. Stat. Assoc. **84**(406), 414–420 (1989)
28. Kanani, P., McCallum, A., Pal, C.: Improving author coreference by resource-bounded information gathering from the web. In: Proceedings of the 20th International Joint Conference on Artifical Intelligence, IJCAI'07, pp. 429–434. Morgan Kaufmann Publishers Inc., San Francisco, USA (2007)
29. Kang, I.S., Na, S.H., Lee, S., Jung, H., Kim, P., Sung, W.K., Lee, J.H.: On co-authorship for author disambiguation. Inf. Proces. Manag. **45**(1), 84–97 (2009). https://doi.org/10.1016/ j.ipm.2008.06.006
30. Khabsa, M., Treeratpituk, P., Giles, C.L.: Large scale author name disambiguation in digital libraries. In: 2014 IEEE International Conference on Big Data, pp. 41–42 (2014). https://doi. org/10.1109/BigData.2014.7004487
31. Khabsa, M., Treeratpituk, P., Giles, C.L.: Online person name disambiguation with constraints. In: Proceedings of the 15th ACM/IEEE-CS Joint Conference on Digital Libraries, JCDL '15, pp. 37–46. ACM, New York, USA (2015). https://doi.org/10.1145/2756406.2756915

32. Kim, J., Diesner, J.: The effect of data pre-processing on understanding the evolution of collaboration networks. J. Inf. **9**(1), 226 – 236 (2015). https://doi.org/10.1016/j.joi.2015.01.002
33. Kim, J., Diesner, J.: Distortive effects of initial-based name disambiguation on measurements of large-scale coauthorship networks. J. Assoc. Inf. Sci. Tech. **67**(6), 1446–1461 (2016). https://doi.org/10.1002/asi.23489
34. Laender, A.H., Gonçalves, M.A., Cota, R.G., Ferreira, A.A., Santos, R.L.T., Silva, A.J.: Keeping a digital library clean: new solutions to old problems. In: Proceedings of the Eighth ACM Symposium on Document Engineering, DocEng '08, pp. 257–262. ACM, New York, USA (2008). https://doi.org/10.1145/1410140.1410195
35. Levin, F.H., Heuser, C.A.: Evaluating the use of social networks in author name disambiguation in digital libraries. J. Inf. Data Manag. **1**(2), 183–198 (2010)
36. Levin, M., Krawczyk, S., Bethard, S., Jurafsky, D.: Citation-based bootstrapping for large-scale author disambiguation. J. Am. Soc. Inf. Sci. Tech. **63**(5), 1030–1047 (2012). https://doi.org/10.1002/asi.22621
37. Ley, M.: DBLP: Some lessons learned. Proc. VLDB Endow. **2**(2), 1493–1500 (2009). https://doi.org/10.14778/1687553.1687577
38. Lindsey, D.: Production and citation measures in the sociology of science: the problem of multiple authorship. Soc. Stud. Sci. **10**(2), 145–162 (1980)
39. Liu, W., Islamaj Dogan, R., Kim, S., Comeau, D.C., Kim, W., Yeganova, L., Lu, Z., Wilbur, W.J.: Author name disambiguation for pubmed. J. Assoc. Inf. Sci. Tech. **65**(4), 765–781 (2014). https://doi.org/10.1002/asi.23063
40. Liu, Y., Li, W., Huang, Z., Fang, Q.: A fast method based on multiple clustering for name disambiguation in bibliographic citations. J. Assoc. Inf. Sci. Tech. **66**(3), 634–644 (2015). https://doi.org/10.1002/asi.23183
41. Malin, B.: Unsupervised name disambiguation via social network similarity. In: Proceedings of the Workshop on Link Analysis, Counterterrorism, and Security, in conjunction with the SIAM International Conference on Data Mining, pp. 93–102 (2005)
42. Martin, T., Ball, B., Karrer, B., Newman, M.E.J.: Coauthorship and citation patterns in the Physical Review. Phys. Rev. E **88**, 012814 (2013). https://doi.org/10.1103/PhysRevE.88.012814
43. McRae-Spencer, D.M., Shadbolt, N.R.: Also by the same author: Aktiveauthor, a citation graph approach to name disambiguation. In: Proceedings of the 6th ACM/IEEE-CS Joint Conference on Digital Libraries, JCDL '06, pp. 53–54. ACM, New York, USA (2006). https://doi.org/10.1145/1141753.1141762
44. Mena-Chalco, J.P., Digiampietri, L.A., Lopes, F.M., Cesar, R.M.: Brazilian bibliometric coauthorship networks. J. Assoc. Inf. Sci. Tech. **65**(7), 1424–1445 (2014). https://doi.org/10.1002/asi.23010
45. Mijajlović, Z., Ognjanovic, Z., Pejovic, A.: Digitization of mathematical editions in Serbia. Math. Comput. Sci. **3**(3), 251–263 (2010). https://doi.org/10.1007/s11786-010-0021-x
46. Milojević, S.: Accuracy of simple, initials-based methods for author name disambiguation. J. Informetr. **7**(4), 767–773 (2013). https://doi.org/10.1016/j.joi.2013.06.006
47. Moody, J.: The structure of a social science collaboration network: disciplinary cohesion from 1963 to 1999. Am.Sociol. Rev. **69**(2), 213–238 (2004)
48. Newman, M.E.J.: Scientific collaboration networks I: network construction and fundamental results. Phys. Rev. E **64**, 016131 (2001). https://doi.org/10.1103/PhysRevE.64.016131
49. Newman, M.E.J.: Scientific collaboration networks II: shortest paths, weighted networks, and centrality. Phys. Rev. E **64**, 016132 (2001). https://doi.org/10.1103/PhysRevE.64.016132
50. Newman, M.E.J.: The structure of scientific collaboration networks. Proc. Natl. Acad. Sci. **98**(2), 404–409 (2001). https://doi.org/10.1073/pnas.98.2.404
51. Newman, M.E.J.: Who is the best connected scientist? A study of scientific coauthorship networks. In: Ben-Naim E., Frauenfelder H., Toroczkai Z. (eds.) Complex Networks. Lecture Notes in Physics, vol. 650, pp. 337–370. Springer, Berlin (2004). https://doi.org/10.1007/978-3-540-44485-5_16
52. Ochoa, X., Mndez, G., Duval, E.: Who we are: Analysis of 10 years of the ED-MEDIA conference. In: Siemens G., Fulford C. (eds.) Proceedings of World Conference on Educational Multimedia, Hypermedia and Telecommunications 2009, pp. 189–200. AACE (2009)

53. On, B., Elmacioglu, E., Lee, D., Kang, J., Pei, J.: Improving grouped-entity resolution using quasi-cliques. In: Proceedings of the 6th IEEE International Conference on Data Mining (ICDM 2006), pp. 1008–1015 (2006). 18–22 December 2006, Hong Kong, China. https://doi.org/10.1109/ICDM.2006.85

54. On, B.W., Lee, D.: Scalable name disambiguation using multi-level graph partition. In: Proceedings of the 2007 SIAM International Conference on Data Mining, pp. 575–580 (2007). https://doi.org/10.1137/1.9781611972771.64

55. On, B.W., Lee, D., Kang, J., Mitra, P.: Comparative study of name disambiguation problem using a scalable blocking-based framework. In: Proceedings of the 5th ACM/IEEE-CS Joint Conference on Digital Libraries, JCDL '05, pp. 344–353. ACM, New York, USA (2005). https://doi.org/10.1145/1065385.1065463

56. On, B.W., Lee, I., Lee, D.: Scalable clustering methods for the name disambiguation problem. Knowl. Inf. Syst. **31**(1), 129–151 (2012). https://doi.org/10.1007/s10115-011-0397-1

57. Pons, P., Latapy, M.: Computing communities in large networks using random walks. J. Graph Algorithms Appl. **10**(2), 191–218 (2006)

58. Porter, M.F.: An algorithm for suffix stripping. In: Sparck Jones K., Willett P (eds.) Readings in Information Retrieval, pp. 313–316. Morgan Kaufmann Publishers Inc., San Francisco, USA (1997)

59. Radovanović, M., Ferlež, J., Mladenić, D., Grobelnik, M., Ivanović, M.: Mining and visualizing scientific publication data from Vojvodina. Novi Sad J. Math. **37**(2), 161–180 (2007)

60. Reuther, P., Walter, B., Ley, M., Weber, A., Klink, S.: Managing the quality of person names in dblp. In: Gonzalo J., Thanos C., Verdejo M., Carrasco R. (eds.) Research and Advanced Technology for Digital Libraries. Lecture Notes in Computer Science, vol. 4172, pp. 508–511. Springer, Berlin (2006). https://doi.org/10.1007/11863878_55

61. Rosvall, M., Bergstrom, C.T.: Maps of information flow reveal community structure in complex networks. **105**(4), 1118–1123 (2007). https://doi.org/10.1073/pnas.0706851105

62. Saha, T.K., Zhang, B., Hasan, M.A.: Name disambiguation from link data in a collaboration graph using temporal and topological features. Soc. Netw. Anal. Min. **5**(1), 11 (2015). https://doi.org/10.1007/s13278-015-0249-1

63. Savić, M., Ivanović, M., Dimić Surla, B.: Analysis of intra-institutional research collaboration: a case of a Serbian faculty of sciences. Scientometrics pp. 1–22 (2016). https://doi.org/10.1007/s11192-016-2167-z

64. Savić, M., Ivanović, M., Radovanović, M., Ognjanović, Z., Pejović, A., Jakšić Kruger, T.: The structure and evolution of scientific collaboration in serbian mathematical journals. Scientometrics **101**(3), 1805–1830 (2014). https://doi.org/10.1007/s11192-014-1295-6

65. Savić, M., Ivanović, M., Radovanović, M., Surla, B.D.: Towards culture-sensitive extensions of CRISs: Gender-based researcher evaluation. In: Model and Data Engineering: 6th International Conference, MEDI 2016, Almería, Spain, 21-23 Sept 2016, pp. 332–345. Springer International Publishing, New York (2016). https://doi.org/10.1007/978-3-319-45547-1_26

66. Schulz, C., Mazloumian, A., Petersen, A.M., Penner, O., Helbing, D.: Exploiting citation networks for large-scale author name disambiguation. EPJ Data Sci. **3**(1), 11 (2014). https://doi.org/10.1140/epjds/s13688-014-0011-3

67. Shin, D., Kim, T., Choi, J., Kim, J.: Author name disambiguation using a graph model with node splitting and merging based on bibliographic information. Scientometrics **100**(1), 15–50 (2014). https://doi.org/10.1007/s11192-014-1289-4

68. Shu, L., Long, B., Meng, W.: A latent topic model for complete entity resolution. In: Proceedings of the 2009 IEEE International Conference on Data Engineering, ICDE '09, pp. 880–891. IEEE Computer Society, Washington, USA (2009). https://doi.org/10.1109/ICDE.2009.29

69. Soler, J.M.: Separating the articles of authors with the same name. Scientometrics **72**(2), 281–290 (2007). https://doi.org/10.1007/s11192-007-1730-z

70. Song, M., Kim, E.H.J., Kim, H.J.: Exploring author name disambiguation on PubMed-scale. J. Informetr. **9**(4), 924–941 (2015). https://doi.org/10.1016/j.joi.2015.08.004

71. Song, Y., Huang, J., Councill, I.G., Li, J., Giles, C.L.: Efficient topic-based unsupervised name disambiguation. In: Proceedings of the 7th ACM/IEEE-CS Joint Conference on Digital Libraries, JCDL '07, pp. 342–351. ACM, USA (2007). https://doi.org/10.1145/1255175. 1255243
72. Tang, J., Fong, A.C.M., Wang, B., Zhang, J.: A unified probabilistic framework for name disambiguation in digital library. IEEE Trans. Knowl. Data Eng. **24**(6), 975–987 (2012). https:// doi.org/10.1109/TKDE.2011.13
73. Tang, L., Walsh, J.P.: Bibliometric fingerprints: name disambiguation based on approximate structure equivalence of cognitive maps. Scientometrics **84**(3), 763–784 (2010). https://doi. org/10.1007/s11192-010-0196-6
74. TePaske-King, B., Richert, N.: The identification of authors in the Mathematical Reviews database. Issues in Science and Technology Librarianship (31) (2001). https://doi.org/10.5062/ F4KH0K9M
75. Torvik, V.I., Smalheiser, N.R.: Author name disambiguation in MEDLINE. ACM Trans. Knowl. Discov. Data **3**(3), 11:1–11:29 (2009). https://doi.org/10.1145/1552303.1552304
76. Torvik, V.I., Weeber, M., Swanson, D.R., Smalheiser, N.R.: A probabilistic similarity metric for medline records: a model for author name disambiguation. J. Am. Soc Inf. Sci. Tech. **56**(2), 140–158 (2005). https://doi.org/10.1002/asi.v56:2
77. Treeratpituk, P., Giles, C.L.: Disambiguating authors in academic publications using random forests. In: Proceedings of the 9th ACM/IEEE-CS Joint Conference on Digital Libraries, JCDL '09, pp. 39–48. ACM, USA (2009). https://doi.org/10.1145/1555400.1555408
78. Veloso, A., Ferreira, A.A., Gonçalves, M.A., Laender, A.H.F., Meira Jr., W.: Cost-effective on-demand associative author name disambiguation. Inf. Process. Manag. **48**(4), 680–697 (2012). https://doi.org/10.1016/j.ipm.2011.08.005
79. Wang, F., Li, J., Tang, J., Zhang, J., Wang, K.: Name disambiguation using atomic clusters. In: The Ninth International Conference on Web-Age Information Management, pp. 357–364 (2008). https://doi.org/10.1109/WAIM.2008.96
80. Wang, J., Berzins, K., Hicks, D., Melkers, J., Xiao, F., Pinheiro, D.: A boosted-trees method for name disambiguation. Scientometrics **93**(2), 391–411 (2012). https://doi.org/10.1007/s11192-012-0681-1
81. Winkler, W.E.: Overview of record linkage and current research directions. Tech. Rep. RR2006/02, US Bureau of the Census (2006)
82. Wu, H., Li, B., Pei, Y., He, J.: Unsupervised author disambiguation using dempster—shafer theory. Scientometrics **101**(3), 1955–1972 (2014). https://doi.org/10.1007/s11192-014-1283-x
83. Yang, K.H., Peng, H.T., Jiang, J.Y., Lee, H.M., Ho, J.M.: Author Name Disambiguation for Citations Using Topic and Web Correlation, pp. 185–196. Springer, Berlin (2008). https://doi. org/10.1007/978-3-540-87599-4_19
84. Zhu, J., Yang, Y., Xie, Q., Wang, L., Hassan, S.U.: Robust hybrid name disambiguation framework for large databases. Scientometrics **98**(3), 2255–2274 (2014). https://doi.org/10.1007/ s11192-013-1151-0

Chapter 7
Analysis of Co-authorship Networks

Abstract Scientific collaboration can be quantitatively studied by analyzing the structure and evolution of co-authorship networks. In this chapter we present a comprehensive overview of research studies focused on empirical analysis of co-authorship networks. Typical structural and evolutionary characteristics of co-authorship networks are identified by an aggregate analysis of examined studies.

According to the scope-based classification of co-authorship networks (Chap. 5, Sect. 5.4) we can also group existing empirical studies dealing with their analysis. Existing empirical studies of field co-authorship networks cover a wide range of scientific fields including physics [10, 91–93, 95, 101, 108], mathematics [8, 11, 19, 20, 56, 57], computer science [13, 14, 32, 36, 47, 64, 91–93, 95, 114, 116], biomedicine [91–93, 95], economy [54], management [3], library and information science [1, 128], sociology [87], as well as more narrower sub-disciplines such as genetic programming [81, 119, 120], evolutionary computation [28, 29], information visualization [18], social network analysis [99], econophysics [41] and steel structures [2, 121, 122], to mention a few. A general overview of studies investigating field co-authorship networks is given in Sect. 7.1. The next two sections provide a more detailed overview of studies focused on co-authorship networks of computer scientists and mathematicians (Sects. 7.2 and 7.3, respectively). Section 7.4 presents an overview of research works analyzing journal co-authorship networks. An overview of studies investigating national co-authorship networks is presented in Sect. 7.5. In the last section we give a summary of frequently observed structural and evolutionary characteristics of co-authorship networks.

© Springer International Publishing AG, part of Springer Nature 2019
M. Savić et al., *Complex Networks in Software, Knowledge,*
and Social Systems, Intelligent Systems Reference Library 148,
https://doi.org/10.1007/978-3-319-91196-0_7

7.1 Empirical Studies of Field Co-authorship Networks

Perhaps the most influential studies investigating the structure of field co-authorship networks are those authored by Mark E. Newman [91–93, 95]. In his seminal papers, Newman established general methodological guidelines for empirical studies of scientific collaboration that are based on metrics and methods of complex network theory. Relying on publication metadata from four digital libraries (Los Alamos e-Print Archive, MEDLINE, SPIRES and NCSTRL), Newman extracted and studied co-authorship networks representing research collaboration in physics (Los Alamos Archive, SPIRES) biomedicine (MEDLINE) and computer science (NCSTRL). Additionally, Newman divided bibliographic records from the Los Alamos e-Print Archive into three sub-disciplines within physics (astrophysics, physics of condensed matter and theory of high energy physics), thus investigating properties of seven co-authorship networks in total. At the time of the study, available bibliographic records covered 9 years of submissions to the Los Alamos electronic preprint archive (nowadays known as arXiv), 40 years in the case of MEDLINE, 27 years for SPIRES and 10 years for NCSTRL. However, Newman decided to construct co-authorship networks considering only data from an identical five-year period, from 1995 to 1999 inclusive, for two reasons:

- publication records of papers published before 1995 were less complete than publication records of papers published after than, and
- the same time period ensures a valid comparison of collaboration patterns in different scientific fields.

Firstly, Newman observed that the distribution of the number of papers per author in each of examined co-authorship networks closely follows either a pure power-law or truncated power-law (a power-law with an exponential cut-off). This implies that there are big variations in the scientific productivity of researchers in examined research fields: a large majority of researchers publish a relatively small number of papers (close to the average number of papers per author), but there is also a small fraction of researchers whose scientific productivity is significantly higher than the average productivity. For example, the average number of papers per author in the astrophysics network is 4.8, while each of the top ten most productive authors in this network authored more than 66 papers. It is also important to emphasize that power-laws in the number of papers per author were observed (by hand) way back in the early 20th century by Lotka [80] (so-called the Lotka's law of scientific productivity) and confirmed by subsequent computerized studies [102, 125]. Newman noticed that distributions of the number of authors per paper can also be very well approximated by power-laws. This result Newman explained by relatively large experimental collaborations in scientific fields covered in his study.

Secondly, Newman investigated the structure of extracted co-authorship networks. Similarly to empirically observed distributions of the number of papers per author and distributions of the number of authors per paper, empirically observed degree distributions of co-authorship networks do not have a characteristic scale, i.e. they are heavy-tailed in the sense that their tails are not exponentially bounded. However, only the degree distribution of the SPIRES co-authorship network closely follows a power-law. For the rest of networks investigated by Newman, degree distributions obey either a power-law with an exponential cut-off or a two regime power-law where a degree distribution initially follows a power-law with one scaling exponent and then in the tail changes to a power-law with a different scaling exponent.

Newman also found that each examined co-authorship network has a giant connected component. Except for the co-authorship networks of researchers in high-energy theory and computer science, the size of the largest connected component is higher than 80% of the total number of authors. The largest connected components in high-energy theory and computer science encompass 71.4% and 57.2% of the total number of authors, respectively. Newman explained that the size of the largest connected component in those two networks can be considered as an "anomaly" due to a poorer coverage of the corresponding bibliography databases. Newman stressed out the importance of giant connected components in co-authorship networks. Collaboration between two researchers usually implies a personal contact. Consequently, a vast majority of researchers in a co-authorship network having a giant connected component are either directly or indirectly connected. Therefore, the presence of the giant connected component enables more effective dissemination of research results. Also, the existence of a giant connected component in a field co-authorship network indicates that scientific achievements in the research field are products of many joint rather than many independent and isolated efforts. On the opposite side, the absence of a giant connected component suggests a poorly cohesive community of researchers and an immaturity of the research field.

Thirdly, Newman observed that examined co-authorship networks possess the small-world property [126] that is reflected by short distances between randomly selected researchers. Moreover, Newman noticed that there is a trend towards shorter distances to other researchers as the number of co-authors increases. The existence of giant connected components and short distances between researchers are phenomena that can be explained by the Erdős-Rényi (ER) model of random graphs [38, 39, 48]. However, another typical characteristic of co-authorship networks, which can not be explained by the ER model, is a high degree of local clustering. Clustering coefficients of co-authorship networks are much higher than those for comparable random graphs. The existence of high local clustering in co-authorship networks Newman explained by three factors:

- Common co-authors: if two researchers who have not collaborated in the past have common co-authors then (1) they share similar research interests, and (2) there is a high chance that one of common co-authors will introduce them to each other (this chance increases with the number of common co-authors).
- Presence in the same scientific circle: if researchers who have no collaborated in the past revolve in the same scientific circle (e.g. attend the same conferences, read and publish in the same journals) then they may start to collaborate due to similar research interests.

- Geographical proximity: this is a special case of the previous possibility when researchers work at the same institution.

Newman also studied the centrality of researchers in co-authorship networks relying on the betweenness centrality measure. He showed that examined networks exhibit so-called *funnelling effect*. For most researchers only a few collaborators lie on a majority of the shortest paths. Such collaborators are also called "sociometric superstars". This means that co-authors of a researcher are not equally important to his/her connectivity with other researchers. Additionally, research collaboration with just one or two famous or influential members of a scientific community would result in a drastically shorter distances to other researchers. In his subsequent study [90], Newman also investigated properties of the co-authorship network of mathematicians extracted from the "Mathematical Reviews" database confirming the conclusions from his previous studies. However, in [90] Newman made two additional important contributions. Namely, he showed that

- The co-authorship networks he previously studied exhibit assortatitive mixing patterns, i.e. highly connected authors (authors having a high number of collaborators) tend to be directly connected among themselves.
- Newman applied the Girvan–Newman community detection algorithm [49] to the co-authorship network of the Santa Fe Institute showing that the network has a community structure. Moreover, each detected community corresponds to one research division at the institute and encompasses researchers having similar research interests. Many later studies showed that community structure is a typical feature of co-authorship networks [46, 77, 130]. Additionally, authors proposing new community detection techniques usually test them on co-authorship networks [33, 34, 42, 55, 94, 106, 107].

The study by Barabási et al. [8] is the first study dealing with the evolution of large-scale field co-authorship networks. The authors investigated the evolution of two co-authorship networks reflecting research collaboration in mathematics and neuroscience in the period from 1991 to 1998 (inclusive). The main empirical findings of the study are:

1. The degree distributions of investigated networks follow power-laws,
2. The average separation (the length of the shortest path connecting two nodes) between researchers decreases in time implying that the networks evolve into a more compact state (a small-world gets smaller over time),
3. The relative size of the largest connected component increases in time leading to the emergence of giant connected components in two investigated networks.
4. The average node degree increases in time implying that the average degree of research collaboration increases as the networks evolve.
5. The clustering coefficient of both investigated networks decreases in time. The observed evolutionary decrease of local clustering can by explained the observed evolutionary increase of the average node degree in sparse networks.

As the most important contribution made in [8], Barabási et al. showed that the integration of new researchers in co-authorship networks can be explained by the preferential attachment principle, which is the founding principle of the scale-free model of complex networks [7]. Obviously, a co-authorship network evolves by addition of new authors and new collaboration links. Barabási et al. showed that the probability that a new researcher establishes a link to an "old" researcher (a researcher that is already present in the network) is proportional to the degree of the old researcher. Additionally, they showed that the probability of creating a new link between two old researchers is proportional to the product of their degrees. Based on the evidence of the preferential attachment for both external links (links between new and old researchers) and internal links (links between old researchers), Barabási et al. proposed a simple model of complex networks that can explain both heavy-tailed degree distributions and observed dynamics of co-authorship networks. The model is based on the following rules:

- The evolution of a network starts with a small random graph $G = (V, E)$.
- In each iteration of the algorithm β new nodes are created (added to V) and integrated into the network, where β is a parameter of the model. Each new node establishes k links to old nodes, where k is another parameter of the model. The probability that a new node A connects to an old node B is equal to

$$P(A \leftrightarrow B) = \frac{d(B)}{\sum_{X \in V} d(X)} \qquad (7.1)$$

where $d(X)$ stands for the degree of a node X.
- In each iteration of the algorithm τ new links connecting old nodes are created. The probability that a link between two old nodes O_1 and O_2 is created is equal to

$$P(O_1 \leftrightarrow O_2) = \frac{d(O_1)\, d(O_2)}{\sum_{X,Y \in V} d(X)\, d(Y)} \qquad (7.2)$$

An empirical evidence that the preferential attachment principle governs the evolution of large-scale field co-authorship networks was also given by Newman [89]. Using co-authorship networks he previously investigated in [93], Newman also showed that:

- the probability that two researchers establish collaboration increases with the number of their common co-authors, and
- the probability that a researcher acquires a new collaborator increases with the number of his/her co-authors.

Goh et al. [51] investigated properties of the distribution of betweenness centrality in various complex networks from different domains, including also the co-authorship network in the field of neuroscience whose evolution was previously studied by Barabási et al. [8]. The authors showed that betweenness centrality, similarly to degree centrality, displays a power-law behavior in large-scale co-authorship networks. The authors also investigated correlations between betweenness centralities of collaborators in co-authorship networks of neuroscience and mathematics [52]. The obtained results show that examined co-authorship networks do not exhibit neither assortative nor disassortative betweenness centrality mixing patterns.

The co-authorship network of Los Alamos e-Print Archive examined by Newman was also empirically studied by Barrat et al. [10]. The authors showed that the distribution of collaboration strength (i.e. the distribution of link weights), similarly to the degree distribution of the network, is heavy-tailed, implying that the network contains extremely frequent collaborators. The authors also proposed modifications of two basic structural measures to take into account the weight of links (weighted clustering coefficient and weighted average nearest-neighbors degree) concluding that "the study of correlations between weights and topology provide a complementary perspective on the structural organization of the network that might be undetected by quantities based only on topological information". In their subsequent work [9], Barrat et al. proposed a modification of the Barabási-Albert model of scale-free networks with a weight-driven, instead of degree-driven, preferential attachment rule. The weight-driven preferential attachment rule says that the probability that a new node establishes a link of a prespecified weight with an old node is proportional to the sum of weights of links incident with the old node. The authors showed that their model can explain power-law distributions of both link weights and node strengths.

Giant connected components and heavy-tailed degree distributions were also reported by Guimera at al. [60] for co-authorship networks in social psychology, economics, ecology and astronomy. Newman [90, 91] emphasized that structural properties of co-authorship networks vary across scientific disciplines and that peculiarities, customs and practices within a scientific discipline are visible through variations in the corresponding co-authorship network. The article by Bettencourt et al. [12] expresses such differences by mapping the evolution of collaboration networks of eight scientific fields (superstring theory, cosmic strings, cosmological inflation, carbon nanotubes, quantum computing, prions and scarpie, H5N1 influenza and cold fusion which is marked as an example of a "pathological" scientific field) over time, from their inception to maturity. The main premise the authors started with is that "the creation and spread of new discoveries through a scientific community creates qualitative, measurable changes in its social structure". The authors investigated changes in the structure of examined co-authorship networks by analyzing the evolution of several structural measures such as node-edge ratio, diameter and the size of the largest connected component. The major conclusions of the study are:

- The investigated co-authorship networks "densify" in time in the sense that the average number of links per node increases during their evolution. The authors showed that the densification of examined co-authorship networks can be described by a simple law previously empirically observed by Leskovec et. al [75, 76] for large-scale networks from other domains:

$$E(t) = c \cdot N(t)^\alpha \qquad (7.3)$$

where $E(t)$ and $N(t)$ are the number of links and nodes in the network at time t, c is a normalization constant, and α is the scaling exponent of the power law. For each examined field, except for cold fusion ("pathological" field), $\alpha > 1$ which means that the number of links (collaborations) grows at a faster rate than the number of nodes (researchers).

- There is a topological transition from a fragmented network structure, which is characterized by several independent and small connected components in an early stage of the evolution, to the emergence of a giant connected component in all examined scientific fields except for cold fusion ("pathological" field).

The authors make a parallel between the Kuhn's theory of scientific revolutions [72] and observed topological transitions in successful, non-pathological scientific fields. According to Kuhn, successful scientific fields arise from discovery (the novelty of facts, "anomalies" that cannot be explained by currently available theories) and invention (the novelty of a theory that successfully explains previously observed "anomalies"). A successful scientific field typically starts with discoveries made by small and independent research groups. Therefore, in early phases of the field development the corresponding co-authorship network has a fragmented structure without a giant connected component. Several incompatible and incomplete theories regarding present discoveries are also characteristic for immature research fields. If the fragmented community reaches consensus about the explanatory potential of one among several competitive theories then a process of a large-scale adoption of that theory starts, inevitably leading to a widespread collaboration in the field. This widespread collaboration oriented towards common goals causes the emergence of a giant connected component in the corresponding co-authorship network.

Ramasco et al. [108] studied collaboration networks as evolving, self-organizing bipartite graphs. Namely, they represented research collaboration in a field as a bipartite graph with two types of nodes: nodes representing researchers and nodes representing *acts of collaboration* (papers). As a part of their study, the authors also analyzed one scientific collaboration network previously investigated by Newman – the network extracted from Los Alamos e-Print Archive that represents scientific collaboration in condensed matter physics. They confirmed the power-law distribution of the number of co-authors per author. In contrast to the Newman's findings, Ramasco et al. indicated that the distribution of the number of authors per paper is exponential, not a power-law. The study also provided the evidence of strong assortative mixing patterns in the network. As the main contribution, the authors proposed a minimal model of evolving, self-organizing collaboration networks that combines preferential attachment with the bipartite graph structure.

Yoshikane and Kageura [131] compared co-authorship networks of four different scientific fields: electrical engineering, information processing, polymer science and biochemistry. The authors used the Gini index to quantify inequalities in the strength of co-authorship links. The obtained results show that there are considerable differences between scientific fields: (1) researchers in information processing are collaborating with a relatively small number of partners compared to other three fields, and (2) the inequality in the strength of research collaboration is much higher in biochemistry and polymer science compared to information processing and electrical engineering.

Scientific collaboration trends and co-authorship patterns in sociology were studied by Moody [87]. The examined co-authorship network was constructed taking into account all journal articles indexed in the Sociological Abstracts database that were published between 1963 and 1999. Moody observed that co-authorship in sociology increases both in volume (the number of multi-authored papers) and extent (the average number of authors per multi-authored papers). Every article indexed in Sociological Abstract is assigned to one of 36 broad specialty areas, so Moody also compared co-authorship trends across different sociological sub-disciplines. The degree distribution of the investigated network does not conform to a strict power-law. However, it is highly skewed to the right indicating the presence of highly collaborative sociologists. Moody noticed that the network has a giant connected component and analyzed its sensitivity. The obtained results show that a relatively large number of highly collaborative researchers have to be removed from the giant component in order to obtain a fragmented community of researchers. Consequently, he concluded that "while the network contains clear star actors – people with a disproportionate number of ties, and such actors are likely very influential within a local region of the network, information diffusion through the network does not depend on such actors". Moody divided the set of nodes in the network according to structural embeddedness into three disjoint subsets:

1. nodes from the largest biconnected component,
2. nodes from the largest connected component which are not located in the largest biconnected component, and
3. nodes from other connected components.

Then, he applied ordered logistic regression to examine whether structural embeddednes can be predicted from researcher evaluation metrics and attributes (productivity, timespan, gender and co-authorship diversity). The obtained results show that the position of a researcher within the co-authorship network can be predicted from his/her individual and publication characteristics.

Goyal et al. [54] examined the evolution of a co-authorship network of economists publishing their research in journals indexed in EconLit (a bibliographic database maintained by the American Economic Association). The authors constructed three evolutionary snapshots of the network corresponding to 1970–1979, 1980–1989 and 1990–1999 periods. They computed basic network statistics (the size of the largest connected component, the number of isolated nodes, the average node degree, the average distance between nodes in the largest component and the clustering coefficient) for each of those evolutionary snapshots. The obtained results show that the largest connected component has grown substantially and, simultaneously, the average distance between nodes become significantly smaller. This means that the cohesiveness of the economics research community increased significantly in later years. Additionally, the clustering coefficient of each evolutionary snapshot has a very high value implying that the network possesses the small-world property in the Watts-Strogatz sense. The authors also observed that there has been a significant increase in the average node degree which implies that economy has evolved into a substantially more collaborative research field.

An exploratory analysis of co-authorship in the field of management was carried out by Acedo et al. [3]. The co-authorship network was formed according to more than 11000 papers authored by approximately 10000 researchers in the field. The authors observed that the network differs considerably from co-authorship networks reflecting scientific collaboration in natural sciences with respect to the basic structural descriptors such as density, neighborhood size, the size of the largest connected component and the clustering coefficient. Various centrality metrics were employed to identify the most prominent researchers in the field. The obtained results show that the most central positions within the network are occupied by authors from a few prestigious universities playing important roles in professional associations and/or authors serving as journal editors.

Palla et al. [100] and Pollner et al. [105] studied the evolution of communities in the co-authorship network of the Los Alamos cond-mat e-print archive (one of the networks previously analyzed by Mark E. Newman). In both works, the communities were extracted with the clique percolation method. Palla et al. [100] investigated lifetime and evolutionary stability of research communities observing that large communities persist longer if they are capable to continually change members, while small communities live longer if they undergo minor membership changes. Secondly, they observed that the probability that a researcher will abandon community is proportional to the total weight of links connecting him/her to researchers from other communities. On the other hand, Pollner et al. [105] investigated evolutionary characteristics of co-authorship links connecting researchers from different communities concluding that community development is driven by the preferential attachment principles, i.e. the probability that a new node joins a community is proportional to the size of the community and the strength of co-authorship links within the community.

The degree distribution of a co-authorship network encompassing nanoscience researchers was examined by Milojević [86]. The network was extracted from the NanoBank database taking into account articles published between 2000 and 2004. The obtained results show that the preferential attachment principle does not hold for authors having a relatively small number of collaborators (less than 20), neither for researchers having an extremely large number of collaborators. Namely, nanoscience researchers with more than 250 co-authors are more frequent than expected by a fitted power-law model due to hyper-authorship practices in certain subfields of nanoscience.

Pan and Saramäki [101] investigated correlations between the strength of co-authorship links and other topological characteristics of two co-authorship networks of physicists. The first network was extracted taking into account all articled posted on arXiv till March 2010 and the second network was formed considering all articles published in Physical Review (PR) journals between 1893 and 2009. The authors observed that the overlap of neighborhoods of adjacent nodes decreases with link weight for a vast majority of links. This decrease is followed by an increase but only for a extremely small number of links having the highest weight. Therefore, it can be concluded that weak co-authorship links mainly reside inside dense ego-networks.

Velden et al. [123] examined the structure of three co-authorship networks corresponding to three sub-fields within chemistry. The authors showed that each examined network has a giant connected component exhibiting a strong community structure revealed by the InfoMap algorithm. Co-authorship links connecting different communities were classified into two categories: transfer links which connect loosely coupled communities (two communities are considered loosely coupled if they are no longer connected after removing one or two nodes) and collaboration links which connect communities that are not loosely coupled. The authors showed that a vast majority of co-authorship links connecting researchers from different communities belong to the category of transfer links.

The structure and evolution of co-authorship networks in the field of steel structures were investigated by Abbasi et al. [2] and Uddin et al. [121, 122]. Abbasi et al. [2] examined the extent to which the fundamental centrality measures (degree, closeness and betweenness) are associated with research collaboration with newcomers in the field. The results of the analysis showed that newcomers prefer to attach to researchers who have a high betweenness centrality rather than to those with a high degree or closeness centrality. Uddin et al. [121] investigated evolutionary trends of a co-authorship network formed considering papers published from 1990 to 2009 in the most prominent journals in the field. The authors observed tendencies towards a lower number of single-authored papers in the field, a higher number of papers having three or more authors, a lower percentage of isolated nodes in the network and a higher average node degree. The authors also induced an international collaboration network (ICN) of the field from the co-authorship network. The nodes in the ICN represent different countries. Two countries A and B are connected by an undirected link if there are two researchers a and b connected in the co-authorship network such that a is from A and b is from B. The authors investigated associations between the centrality of countries in the ICN and the average efficiency of papers (the number of citations normalized by the number of years passed after the paper was published) from a country showing that there is a strong correlation between those two measures. In their subsequent work [122], Uddin et al. examined how the citation count of a publication is influenced by the centrality of its authors in the co-authorship network. They additionally analyzed how the centrality of researchers is related to the strength of co-authorship links. The obtained results revealed that

1. the number of citations of a research paper is positively correlated with the values of centrality measures of its authors.
2. the strength of co-authorship links is positively correlated with centralities of directly connected researchers, i.e. frequent co-authors tend to occupy central positions in the network.

Abbasi et al. [1] performed an analysis of ego-networks in a co-authorship network reflecting research collaboration in information and library sciences. The co-authorship network was formed considering papers published between 2000 and 2009 in the top 9 journals in the field. The authors investigated how structural characteristics of researchers' ego-networks correlate with their research performance measured by the g-index. The obtained results indicate that the number of co-authors (node degree) and the betweenness centrality within ego networks exhibit significant positive correlations with the g-index. On the other hand, the density of ego-networks negatively correlates with the g-index suggesting that a significant brokerage role of a researcher within his/her ego-network positively influences his/her research performance. The study by Ortega [97] also examined relationships between research impact of authors and structural characteristics of their ego-networks. Ortega used the Microsoft Academic Search to extract ego-networks of 500 researchers from 11 different scientific disciplines and collect the number of citations to their papers. The main conclusion of the study is that researchers having sparse ego networks in which ego nodes have a high betweenness centrality tend to have a higher number of citations per paper compared to researchers whose ego networks are dense and compact.

7.2 Co-authorship Networks of Computer Science Authors

Large-scale networks reflecting collaboration in computer science were investigated by Elmacioglu and Lee [36] in 2005, Huang et al. [64] in 2008, Bird et al. [13] in 2009, Biryukov and Dong [14] in 2010, Shi et al. [114] and Franceschet [47] in 2011, Staudt et al. [116] in 2012, and Divakarmurthy and Menezes [32] in 2013. Huang et al. [64] relied on bibliographical records contained in the *CiteSeer* digital library to extract a co-authorship network of computer scientists. The study of Divakarmurthy and Menezes [32] is based on bibliographic records in the *ACM* digital library. In other studies, the *DBLP* computer science bibliography database was used as the source of relevant bibliographic records. Researchers also investigated the structure and evolution of scientific collaboration in a more narrower computer science communities working on specific research topics such as genetic programming [81, 119, 120], evolutionary computation [28, 29], computational geometry [65], information retrieval [31], information systems [132], information fusion [66] and intelligence in computer games [73]. There are also studies that examined properties of co-authorship networks of computer science conferences [24, 25, 61, 79, 88, 96, 104, 109, 115, 124, 127].

Elmacioglu and Lee [36] explored statistical properties of a co-authorship network extracted from a subset of DBLP data. More specifically, the authors studied the structure and evolution of the co-authorship network that emerges from articles published in 19 journals and 81 conferences in the database research field in the period from 1968 to 2013. The main findings of the study are:

1. The network is a small-world network with six degrees of separation between researchers. Additionally, the small-world coefficient exhibits an evolutionary stable value.

2. The distribution of the number of papers per author follows a power-law with scaling exponent $\alpha = 2$. The degree distribution of the network also exhibits a power-law in the tail.
3. The average number of co-authors per author and the average number of authors per paper steadily increase in time implying an increasing collaboration in the database research field.
4. The network has a giant connected component encompassing approximately 57% of the total number of nodes. The clustering coefficient of the giant connected component also steadily increase in time reaching value of 0.63 in 2003. Such high value and the observed growth of the clustering coefficient can be explained by the fact that the study covers researchers from only one, mature research field in computer science.

Huang et al. [64] investigated properties of the co-authorship network formed from bibliographic data contained in the CiteSeer database. The network covers the period from 1980 to 2005 and 451305 papers authored by 283174 distinct researchers in computer science. The authors observed that the network has a giant connected component and exhibits the scale-free and small-world properties. The small-world coefficient of the investigated network gradually decreases from 1994 to 2005 suggesting that the small-world of computer science researchers gets smaller over time. The network also exhibits a significant assortativity. The authors also investigated assortativity and co-authorship reciprocity over time, where co-authorship reciprocity is defined as the tendency for a pair of co-authors to exchange their position in the list of authors, finding that both quantities decay over time. To study scientific collaboration in a more grained manner, the authors mapped bibliographical records to six computer science disciplines (artificial intelligence, applications, architectures, databases, systems and theory). Such division enabled the authors to identify and compare co-authorship patterns in different computer science disciplines. For example, the artificial intelligence community has the highest assortativity but the lowest reciprocity, while the database community has the highest number of papers per author, the highest number of co-authors per researcher and the highest reciprocity. The authors also proposed a model to predict increments in the strength of research collaboration.

Similarly to Huang et. al [64], Bird et. al [13] also constructed several co-authorship networks that reflect scientific collaboration in various computer science disciplines. The used bibliography data contained 83587 papers authored by 76598 distinct researchers (the information about the time frame of publications is not given in the article). The authors investigated 14 co-authorship networks corresponding to different disciplines ranging from artificial intelligence to WWW. All examined networks exhibit the scale-free property and have a strong community structure. Namely, the authors employed the Girvan–Newman algorithm [49] to detect communities in analyzed networks. Obtained community partitions have very high modularity scores (higher than 0.6 for each examined network). Secondly, the authors measured assortativity with respect to the number of publications and career length (the time passed from the first to the last publication of an author). One of interesting observations is that the cryptography research community has very high values of degree, publication and career assortativity indices. Such result can be explained by the fact that cryptography is a very technical field requiring a solid mathematical background. Therefore, it imposes a high entrance barrier for junior researchers. The authors also expressed their worries on the dynamism of the field suggesting that senior researchers in cryptography should put some extra effort to attract and train newcomers. By investigating the evolution of the betweenness centralization of examined networks, the authors were able to identify periods characterized by the dominance of a small number of researchers in terms of their vital role to information diffusion. Finally, Bird et al. investigated overlaps among different computer science disciplines (in terms of the number of authors that publish in different disciplines) and the migration of researchers across disciplines. They observed that there are highly interdisciplinary computer science fields (e.g. data mining and software engineering), as well as fields with closed research communities (e.g. cryptography).

In the study of Birykov and Dong [14], which was also based on the DBLP data, each author was classified either as a "short-time" researcher (a researcher whose career does not exceed 5 years) or as an "experienced" researcher. Birykov and Dong constructed a transition graph between computer science disciplines considering only "experienced" researchers. The transition graph contains computer science disciplines as nodes. Two disciplines A and B are connected by a directed link $A \rightarrow B$ if there is a researcher who worked in a discipline A and later moved to a discipline B. The weight of $A \rightarrow B$ represents the probability that a researcher working in A later moves to work in B. The authors emphasized that various kind of asymmetries between computer science disciplines can be observed in the transition graph. For example, there are researchers who initially started to work on algorithms and theory which later also investigated topics related to distributed computing. However, researchers which initially worked on distributed computing do not tend to explore topics related to algorithms and theory. Finally, the authors compared research communities corresponding to different computer science disciplines with respect to their productivity (publication growth rate), collaboration trends (the average number of authors per paper, the number of isolated nodes and the clustering coefficient) and population stability (changes in research population in terms of the number of "newcomers" and "leavers").

Franceschet [47] also studied co-authorship networks extracted from bibliographic records contained in the DBLP database. He examined structural and evolutionary properties of three co-authorship networks:

1. the network of research collaboration among all DBLP authors present from 1936 to 2008 inclusive (the DBLP network),
2. the network determined by co-authorship in journal articles indexed in DBLP in the same period (the DBLP journal network), and
3. the network determined by co-authorship in conference articles indexed in DBLP in the same period (the DBLP conference network).

Similarly to previously mentioned studies, for each analyzed network Franceschet observed a heavy-tailed distribution of researcher productivity, the small-world property, the existence of a giant connected component and assortative mixing patterns (the strongest for the DBLP journal network). An interesting observation is that single authored articles more frequently appear in journals than in conference proceedings. Franceschet used the power-law fitting technique introduced by Clauset et al. [27] to examine whether investigated networks possess the scale-free property. The results showed that the DBLP network and the DBLP journal network are scale-free. On the other hand, power-law is not a plausible statistical model for the degree distribution of the DBLP conference network. Franceschet also investigated the stability of giant connected components by iteratively removing nodes according to three different removal schemes: (1) the removal of a randomly selected node in each iteration, (2) the removal of the highest degree node, and (3) the removal of the node with the highest eigenvector centrality. The results showed that the degree-driven node removal disintegrates giant connected components in the shortest time. Approximately 15% of the top connected researchers have to be removed from the DBLP network in order to break apart its giant connected component and such researchers have degree centrality higher than 11. The obtained small value of the critical degree centrality required to disintegrate the network suggests that "the network is not glued together by star collaborators and, while such actors are likely very influential within their local communities, they do not control information diffusion on the whole computer science collaboration network". Regarding the evolution of the DBLP network, Franceschet observed two characteristic periods: the period before 1983 characterized by the absence of a giant connected component, and the period after 1983 when the giant connected component emerged and at the same year the small-world, clustering and assortativity coefficients of the network started to decrease.

Shi et al. [114] analyzed diversity of ties in a co-authorship network determined by publications from top-rated conferences indexed in DBLP. Intuitively speaking, the connections of a researcher can be considered diverse if his/her co-authors belong to many different research communities. The authors assigned a label to each researcher in the network, where one label represents one research field according to the Microsoft Academic Search categorization of computer science conferences. Namely, a researcher A is labeled by a research field F if a majority of publications of A were published in conference proceedings belonging to F. The authors introduced a measure to quantify the diversity of ties of an author that is defined as the number of different labels of his/her collaborators. The main empirical result of the study is that authors with a high degree of diversity of ties tend to be connected among themselves.

Staudt et al. [116] also analyzed the DBLP co-authorship network. They confirmed the previously reported scale-free property of the network. The authors applied the Louvain community detection method [15] to identify research communities in the network. The obtained results show that the network possesses a strong community structure (the obtained value of modularity is equal to 0.89). On the other hand, there is a weak correspondence between automatically recovered communities and topical clusters determined by computer science conferences (authors attending the same conference are in one topical cluster). The authors also investigated the impact of scientific seminars organized to stimulate collaboration on "hot" research topics on the strength of research collaboration. More specifically, the authors examined whether participating at *Dagstuhl Seminars* makes significant collaboration changes manifested by an increased collaboration among researchers that were at the same event. The obtained results suggest that that single events such as *Dagstuhl Seminars* are not influential enough to cause significant changes in the co-authorship network structure.

Divakarmurthy and Menezes [32] investigated the impact of citations to the structure of co-authorship networks. They used the ACM digital library to form two co-authorship networks: the first one extracted from all publications indexed in ACM (223464 papers authored by 62758 researchers and published in the period from 1951 to 2011) and the second one formed considering only those publications that are cited at least once. The authors then compared basic structural characteristics (the average node degree, the power-law scaling exponent of the degree distribution, the clustering and small-world coefficients) of those two networks showing that there are slight structural differences. The authors also compared rankings of authors determined by their centrality in the two co-authorship networks showing that the "citation filter" (the removal of uncited papers when forming a co-authorship network) causes significant author ranking shifts.

7.2.1   Co-authorship Networks of Topical Computer Science Communities

The structure of the co-authorship network reflecting collaboration among researchers working on genetic programming was investigated by Tomassini et. al [120]. The network was formed using the genetic programming bibliography database created and maintained by William B. Langdon and Steven Gustafson. This bibliography database covers the majority of publications in the field since its inception. Similarly to other co-authorship networks, the genetic programming co-authorship network exhibits the scale-free and small-world properties, the clustering coefficient of the network is significantly higher than the clustering coefficient of a comparable random graph and the network is assortative. Tomassini and Luthi also investigated the evolution of the network [119] providing the evidence that the preferential attachment principle of the Barabási-Albert model can explain how new researchers integrate into the network. Finally, Luthi et. al [81] used the greedy modularity maximization algorithm [94] to identify cohesive research groups in the network. Since the authors were familiar with the genetic programming research community, they verified that automatically identified communities actually correspond to institutional research groups.

The co-authorship network of evolutionary computation (EC) researchers was analyzed by Cotta and Merelo [29]. The authors made a web crawler to collect bibliographic information about papers published by researchers who attended at least once one of four large evolutionary computation conferences indexed in DBLP

(GEECO, PPSN, EuroGP and EvoWorkshops). The results of the analysis showed that the network has a giant connected component and possesses the scale-free and small-world properties. In a subsequent study [28], the authors demonstrated that the network has a community structure and investigated the evolution of communities. The conclusion of the analysis is that the co-authorship network and the network of research communities evolve towards higher connectivity and lower centralization. The authors also pointed out that the most central community in the network of communities is theoretically oriented suggesting that the theory of evolutionary computation "drives" the whole research field.

Ding [31] examined both co-authorship and citation network of researchers in the information retrieval (IR) area. The data for the study was collected using the Web of Science indexing service and covers 15367 articles published in the period from 1956 to 2008. Ding identified 5 major research topics in information retrieval (multimedia IR, databases and query processing, medical IR, Web IR and IR theory) by applying the Author-Conference-Topic model of Tang et al. [118]. For each major research topic, the top 20 most productive authors were identified and then co-authorship and citation practices among them were investigated. The study revealed that the most productive IR authors do not tend to collaborate with productive colleagues exploring different research topics, but they tend to cite them (directly or indirectly through a citation chain shorter than three papers). Additionally, highly cited IR authors do not tend to collaborate among themselves.

The structure of a network that represents collaboration in computational geometry was investigated by Hui et al. [65]. The network was built from GeomBib, a BibTEX database of papers in computational geometry maintained under the supervision of Bill Jones. The authors showed that the network exhibits the scale-free and small-world properties and a strong community structure. The study also provides information about top ranked authors in the field according to different centrality measures. The small-world property and a strong community structure was also observed for the co-authorship network of information fusion research community [66].

The collaboration among scholars in the information systems (IS) research field was studied by Zhai et. al [132]. The authors make collaboration networks at the individual and country level using publication data from eight mainstream IS journals. The main findings of the study are:

- The number of authors and the number of publications per journal increase linearly in time.
- The distribution of the number of articles per author and the number of articles per country are both heavy-tailed implying that there are highly productive authors and highly productive countries.
- The average node degree in both collaboration networks increases steadily over time, indicating an evolutionary increase of research collaboration in the field at both levels.
- United States, England, Canada and Australia are the top 4 ranked countries according to the Schubert's partnership ability index [113]. The same countries are also in the top 5 ranked countries in the country collaboration network according to

the degree centrality measure. The strongest cross-country collaboration in the field occurs between scientist from the United States and Canada. Additionally, the United States has been in the core of the country collaboration network from the beginning of its evolution.

The co-authorship networks of researchers working in the field of computer supported cooperative work (CSCW) and a more broader field of human-computer interaction (HCI) were analyzed by Horn et al. [62]. The networks were extracted considering articles contained in the HCI bibliography database published between 1982 and 2004. The obtained results show that the HCI co-authorship network has a giant connected component with the small-world property. The authors investigated characteristics of ties between the CSWC community and the larger HCI community, as well as the visibility of CSCW researchers within the HCI community, showing that (1) CSCW researchers have roughly equal numbers of CSCW co-authors and HCI co-authors from non-CSCW disciplines, and (2) the most central researchers in the CSWC co-authorship network tend also to occupy central positions in the HCI co-authorship network.

Lara-Cabrera et al. [73] presented an analysis of the structure and evolution of the co-authorship network of researchers working on computer intelligence in games (CIG). The authors made a program for keyword, journal and conference based queering of the DBLP database in order to collect data about relevant articles and corresponding authors. Two types of evolutionary network snapshots were constructed in order to investigate the evolution of co-authorship patterns: *cumulative* co-authorship network snapshots formed considering all publications before a certain year, and *effective* co-authorship network snapshots formed considering only publications in the previous 5 years with respect to a certain year. Firstly, the authors noticed that the largest connected component of the CIG cumulative co-authorship network is of relatively modest size (18.6% authors) and much smaller than the largest components of co-authorship networks of genetic programming [120] and evolutionary computation [29]. This results indicates the CIG field is still in an early stage of the development. Similarly as for the genetic programming co-authorship network [119], the growth of the cumulative CIG network is driven by the preferential attachment principle. The authors also observed that the probability that two authors establish research collaboration increases with the number of common co-authors. This phenomenon is also manifested by an evolutionary increase of the clustering coefficient in both cumulative and effective CIG network. The probability that two CIG authors renew their collaboration increases with the number of joint papers. The authors also used the greedy modularity maximization algorithm to detect communities in the largest connected component and investigate their evolution. The number of communities and the average size of communities increase in time implying a decreasing degree of centralization of the largest connected component.

7.2.2   Co-authorship Networks of Computer Science Conferences

Nascimento et. al [88] analyzed the structure of the co-authorship network obtained considering all papers published in the SIGMOD (Special Interest Group on Management of Data) conference proceedings between 1975 and 2002. The bibliographic records of relevant publications were retrieved from the DBLP database. The authors computed the distribution of the number of papers per author showing that 90% of SIGMOD authors published less than 3 SIGMOD articles, but that there are SIGMOD authors having a high number of SIGMOD papers (up to 32 papers). A similar asymmetry is also observed for the number of collaborators per author. Moreover, there is a strong correlation between the number of published papers and the number of collaborators. The authors also observed that the network has a giant connected component and that it possesses the small-world property.

As a part of an content analysis of papers published in the proceedings of SIGIR (Special Interest Group on Information Retrieval) conferences, Smeaton et al. [115] also briefly reported on basic structural characteristics of the SIGIR co-authorship network. The authors noticed that the network is very fragmented (it has one relatively large and several small disjoint connected components). As an interesting fact, the authors pointed out that there is an usually large distance in the network between the current and previous SIGIR chair at the time of the study (Sue Dumais and Bruce Croft, respectively). Chris Buckley is the researcher with the shortest average distance to other researchers in the network. The announcement of this fact was made at the SIGIR2002 conference dinner and Chris Buckley was awarded with a certificate and a Christopher Lee DVD.[1]

The co-authorship network of WCRE (Working Conference on Reverse Engineering) was examined by Hasan et al. [61]. The network was extracted from the DBLP database. The authors showed that the WCRE co-authorship graph possesses the small-world property. They also compared the WCRE network to other two co-authorship networks: the first one representing collaboration in several software maintenance and software reengineering conferences, and the second one representing collaboration in general software engineering conferences. The authors observed similar structural characteristics (the size of the largest connected component, the average node degree, the small-world and clustering coefficients) and evolutionary trends (an initially slow growth of the largest connected component followed by a more rapid growth) for all three examined networks.

Liu et. al [79] analyzed the structure of the co-authorship network of digital library researchers presenting their research at ACM, IEEE, and joint ACM/IEEE digital library conferences. The network was formed using bibliographic records available on DBLP. The authors noticed that the largest connected component in

---

[1]At that time, Christopher Lee had the lowest closeness centrality in the collaboration network of Hollywood actors (a network frequently analyzed in complex network analysis literature).

the network is considerably smaller compared to the largest connected components of co-authorship networks studied by Newman [91–93, 95]. The authors indicated several reasons for the previous observation: a relative immaturity of the field, the multidisciplinary nature of digital library conferences, a low international collaboration in the field (the authors also pointed out that only about 7% of the total number of collaboration are cross-country collaborations) and the fact that many digital library projects are institutionally oriented and, consequently, not particularly suitable for cross-institutional collaboration. The authors applied a hierarchical bottom-up clustering algorithm to cluster nodes in the network. The results showed that the network possesses a community structure where obtained clusters correspond to different research institutions. The authors also proposed a network-based measure of author importance called *AuthorRank*. AuthorRank is a modification of the PageRank metric that takes into account the weight of co-authorship links. Assuming that program committee members can be regarded as prestigious actors in the field, the authors showed that AuthorRank is a better indicator of author importance compared to basic node centrality measures (degree, closeness and betweenness).

The co-authorship network of the ICIS conference (International Conference of Information Systems) was examined by Xu and Chau [127]. The authors showed that the network contains a giant connected component, possesses the small-world property and its degree distribution follows a power-law with an exponential cutoff. Similarly to most of previous studies of co-authorship networks, the authors identified the most eminent researchers and the most productive institutions in the ICIS research community using various productivity and centrality measures. From the co-authorship network the authors induced the corresponding cross-institutional collaboration network and studied its structure indicating that geographical proximity of institutions may not always result in cross-institutional collaboration.

From an Endnote database containing all publications from the European conference on information systems (ECIS) in the period from 1993 to 2005, Vidgen et al. [124] built and investigated two types of scientific collaboration networks: the co-authorship network of the ECIS research community and a panel network determined by the participation at the conference panel discussions. The authors observed that the co-authorship network is more fragmented than the panel network. The panel network is more denser and has a smaller diameter. Additionally, the panel network has a clear core-periphery structure (a dense and cohesive core of highly connected nodes, and a sparse periphery of loosely connected nodes). The results of the analysis enabled authors to propose a range of possible interventions that might be taken to strengthen the ECIS community of researchers and increase their collaboration.

Cheong and Corbitt investigated the structure of the co-authorship network of the Australasian conference on information systems [24] and the Pasific Asia conference on information systems [25]. The authors showed that both networks possess the small-world property and identified important researchers in the community relying on various node centrality measures.

Reinhardt et al. [109] explored the co-authorship network of researchers that participated in European Conference on Technology Enhanced Learning (ECTEL) from

2006 to 2010 inclusive. They found that the ECTEL community is very fragmented having a small number of very prolific researchers.

The analysis of co-authorship, citation and co-citation practices of researchers participated at the ED-MEDIA conferences was conducted by Ochoa et al. [96]. The authors observed that the distribution of author productivity and the distribution of author impact (measured by the number of citations) are heavy-tailed distributions following power-laws. Secondly, they found that the co-authorship network is composed out of three layers: the first layer is actually the largest connected component of the network and encompasses regular participants of the ED-MEDIA conferences, the second layer is made of connected components of a moderate size organized around one or two researchers that are well known in the ED-MEDIA community, and the third layer consists of very small components encompassing authors from the same institution that always publish together and never cross institutional borders. The authors also investigated the cross-country collaboration network obtained from the co-authorship network observing that:

- the strongest collaboration occurs among the most productive countries, and
- collaborations between researchers from different less developed countries are rare.

The ED-MEDIA authors are also related through co-citations which further implies that they form a strongly cohesive community of researchers. The authors also designed and implemented a personalized recommendation system based on the co-authorship and citation networks that recommends relevant papers and potential collaborators to ED-MEDIA participants.

Pham et al. [104] studied the evolution of the co-authorship and citation networks corresponding to the five major technology enhanced learning conferences (ICALT, ECTEL, ICWL, ITS, and AIED). The DBLP bibliography database was used to collect relevant references, while the CiteSeer database was used to gather information about citations. The authors noticed that densification laws [75, 76] can explain the evolution of all examined networks. The diameters and small-world coefficients of examined co-authorship networks steadily increase in time suggesting that the corresponding research communities are still in an early stage of the development. The authors observed that the size of largest connected component for ITS, AIED and ECTEL initially increased faster than the same quantity for ICALT and ICWL concluding that the former three conferences have developed faster. The absence of a giant connected component in the ICWL co-authorship network was explained by a large number of ICWL authors that participated just once at the conference and the existence of a large number of ICWL authors who tend to collaborate only with their previous collaborators (thus making no new connections that cross their ICWL sub-community). The authors also compared evolutionary trends observed for TEL conferences to evolutionary trends in scientific collaboration in four established conferences on databases concluding that "TEL conferences exhibit a pattern typical of young communities".

7.3   Co-authorship Networks of Mathematicians

Research of co-authorship networks in mathematics has been influenced by the existence of one prominent mathematician, Paul Erdős (1913–1996), whose unique work ethic and lifestyle led to the publication of over 1500 papers with a great number of collaborators [58]. The notion of the Erdős number was formally defined by Goffman in 1969 [50]. In response to the Goffman's paper, several mathematicians indicated some interesting facts about the Erdős number, even Erdős himself in a sarcastic tone speculated on the properties of the Erdős collaboration network [37]. More serious studies of the Erdős co-authorship network were conducted by Grossman and Ion [59] in 1995 and Batagelj and Mravar in 2000 [11]. More general analysis of mathematics collaboration networks were performed by Grossman in 2002 [56, 57], who examined statistical properties of the network derived from Mathematical Reviews (MR), and Brunson et al. [19] in 2014, who studied the evolution of the MR network, identifying two points of drastic reorganization of the network, as well as an increased collaboration between mathematicians in more recent times. Cerinšek and Batagelj [20] in 2014, as a part of a more general study of various two-mode bibliography networks, investigated the co-authorship network derived from bibliographic records in the Zentralblatt database.

Grossman and Ion [59] examined basic properties of the co-authorship network encompassing Erdős and his co-authors. They observed that a small fraction of Erdős's co-authors (9% of the total number) have never collaborated with a mathematician having Erdős number equal to 1. Secondly, the authors investigated evolutionary trends in mathematical production using the MR database. The results showed that the number of solo-authored papers is decaying in time (from over 90% in 1940 to less than 60% in 1995). Consequently, there is a steady increase in the number of multi-authored publications in the same time frame.

Batagelj and Mravar [11] used a truncated Erdős co-authorship network to illustrate different techniques for identifying important researchers and cohesive groups in co-authorship networks. The truncated Erdős co-authorship network was obtained by removing Paul Erdős from the co-authorship network that encompasses mathematicians having Erdős number smaller than or equal to 2. Batagelj and Mravar observed that the truncated Erdős co-authorship network contains a giant connected component encompassing 99% of the total number of mathematicians in the network. Various clustering techniques (prespecified block-modeling and the Ward's hierarchical clustering) were applied to the core of the giant connected component in order to identify cliques and cohesive groups in the network. The authors also introduced a measure of collaboration that quantifies the openness of an author towards authors located at the periphery of a co-authorship network and employed it to identify the most collaborative co-authors of Paul Erdős.

Grossman investigated the structure [57] and evolution [56] of the co-authorship network that was constructed from MR bibliography records. The network covers articles published in the period from 1940 to 2002. He found that (1) the network has a giant connected component encompassing more than 60% of the total number

of authors, (2) it is a small-world network with eight degrees of separation, and
(3) it exhibits local clustering extremely stronger than in a comparable Erdős-Renyi
random graph (more than 10000 time stronger). Secondly, Grosmann observed that
the network has the scale-free property where the degree distribution of the network
follows a power-law with an exponential cut-off. Additionally, the distribution of
component sizes can also be characterized by a power-law. Finally, Grossman noticed
that the average node degree increases in time (from 0.49 in 1940 to 2.94 in 2002)
suggesting that research in mathematics has transformed from being dominantly
individual to being dominantly collaborative.

Brunson et al. [19] also investigated the evolution of the co-authorship network associated to the MR database. The authors investigated long-term trends
and observed shifting behavior in mathematical collaboration. In contrast to Grosmann [56] who investigated the evolution of the network cumulatively, the authors
examined the evolution of the network using a fixed-duration sliding window. For
each year in the examined time frame (1990–2009) only publications from the previous 5 years were taken into account when constructing an evolutionary snapshot of the
network. Additionally, the authors studied two sub-networks of the network: the first
one constructed from publications that can be considered as research works in "pure"
mathematics, and the second one built from publications that can be regarded as more
"applied" (the classification of papers into pure and applied was done using the MSC
classification scheme). The authors found that mathematicians from pure and applied
mathematics are differently positioned in the aggregated network: researchers in
pure mathematics tend to be located in the core of the network, while researchers in
applied are mostly located on the periphery. Secondly, the co-authorship network of
researchers in applied mathematics is less cohesive than the network of researchers
in pure mathematics. The authors noticed two points of a drastic reorganization of
the aggregated network:

- The first point occurred in the middle of 1990s and its main characteristic is a
  noticeable increase of the average degree and clustering coefficient, especially for
  researchers working on applied mathematics. At the same time there is an decrease
  in cross-disciplinary research (collaboration between researchers from pure and
  applied mathematics).
- The second point is from early 2000s when the average number of publications per
  author and the average strength of collaborative ties started to abruptly decrease.
  The authors explained this event by a large influx of mathematicians from the
  former Soviet Union into the MR database.

Secondly, the authors noticed that the small-world coefficients of all three analyzed
networks (aggregated, pure, applied) decrease in time, much faster than predicted
by several random graph models. Simultaneously, the average clustering coefficients
increase as examined networks evolve. However, clustering trends in pure and applied
mathematics are of quite different nature: the increase of the clustering coefficient in
pure mathematics is due to an increase in the average number of collaborators, while
the increase in applied mathematics is caused by an increase of highly cooperative
publications (publications that have a relatively large number of authors).

Cerinšek and Batagelj [20] performed various network-based analyses of bibliographic records in the Zentralblatt database. The authors also investigated properties of the Zentralblatt co-authorship network observing that a sub-network induced by highly collaborative authors is dominated by "tandems" – connected components of exactly two nodes. Similarly to previous studies of co-authorship networks, the authors exploited the co-authorship network to identify the most productive and collaborative authors present in the database.

In our previous works, we investigated the structure and evolution of a co-authorship network formed from bibliographical records in eLib – the electronic library of the Mathematical Institute of the Serbian Academy of Sciences and Arts [111, 112]. This particular electronic library enabled us to investigate the structure of scientific collaboration among authors that publish papers in the most prestigious Serbian mathematical journals, where a vast majority of present authors are Serbian (Yugoslav) mathematicians. The bibliographic records also cover a wide time range: the first research article indexed in eLib was published in 1932. Thus, we also investigated the evolution of the eLib co-authorship network in an 80 year period, from 1932 to 2011, with a yearly resolution in order to observe general trends in the evolution of collaboration among researchers from the eLib community. The main findings of the study can be summarized as follows:

- There are several time periods in which the number of papers per year is steadily increasing, as well as several periods when this quantity shows a decreasing trend. The time points at which publication trends change correspond to historically significant political events in Serbia (Yugoslavia). The evolution of the number of authors per year closely follows the evolution of the number of papers per year, i.e. there is a strong correlation between those two variables. Additionally, in each year the number of male authors is significantly higher than the number of female authors indicating a strong gender gap in Serbian mathematical research.

- Nearly 75% of articles published in the journals indexed in eLib are single-authored papers. However, the average number of authors per paper increases in time, while the fraction of single-authored papers decreases indicating that Serbian mathematical scene is constantly transforming towards more intense cooperation. A similar evolutionary trend was also observed by Grossman at the world-wide level for articles indexed in "Mathematical Reviews" in the period from 1940 to 2000 [56]. The distribution of the number of papers per researcher follows a truncated power-law with a faster decay for more than 25 papers. The power-law nature of the distribution implies that a majority of Serbian mathematician publish a relatively small number of papers in Serbian mathematical journals, but that there are also researchers whose contribution to domestic journals is extremely high. For example, the most productive eLib author published 75 papers in eLib journals, while the average number of publications per author is more than 28 times lower. The observed cut-off in the distribution has a suitable theoretical explanation taking into account a wide time range of analyzed publications (80 years). Cut-offs in power-law distributions appear when time or capacity constraints are incorporated into the principle of cumulative advantage [4]. Even extremely prolific authors after some time stop publishing papers (due to retirement or death) thus introducing time constraints into the principle of the cumulative advantage.

- In contrast to a majority of co-authorship networks studied in the literature, the eLib co-authorship network is extremely fragmented: the network contains a large number of connected components whose sizes obey a power-law, a vast majority of components are isolated researchers or small-size trivial components, but there is also a small number of relatively large, non-trivial components of connected researchers. To identify highly cohesive research groups within non-trivial connected components we used five different community detection methods (the Girvan–Newman algorithm based on edge betweenness, Walktrap, Infomap, Label propagation and the Louvain method) showing that (1) the largest connected components of the eLib network possess a strong community structure, and (2) the Louvain method exhibits the best performance regarding the quality of obtained clusters (this method identified the largest fraction of Radicchi strong clusters compared to other methods).
- The sub-networks corresponding to individual journals indexed in eLib have similar structural properties as the whole eLib network: they are sparse, fragmented, do not have a giant connected component and contain a significant amount of isolated nodes. Additionally, we used structural characteristics of those sub-networks to rank eLib journals according to the degree of collaborative behavior of their authors.
- Evolutionary analysis of the network revealed that there are six different time periods characterized by different intensity and type of collaborative behavior of researchers publishing in Serbian mathematical journals. The intensity of collaboration in the last two periods (from 1975 to 2011) exhibits a growing trend and non-trivial connected components within the network evolve in a way to become significantly larger and more cohesive.
- The analysis of correlations between different types of researcher evaluation metrics (metrics of productivity, collaboration and centrality) showed that betweenness centrality is a better indicator of productivity and long-term presence in the eLib journals than degree centrality. Additionally, the strength of correlations between productivity metrics and betweenness centrality increases as the network grows suggesting that even more stronger correlations can be expected in the future.

7.4   Journal Co-authorship Networks

Co-authorship patterns in individual publication venues can be obtained by analyzing corresponding journal co-authorship networks [17, 23, 43, 45, 63, 78, 84].

Borner et al. [17] studied the co-authorship and citation networks of the PNAS journal (*Proceedings of the National Academy of Sciences of the United States of America*) covering 45120 articles published in the period from 1982 to 2001. The authors observed that both networks have the scale-free and small-world properties. Secondly, the number of PNAS authors increases faster than the number of

PNAS publications due to an increase of the average number of authors per paper. The authors also proposed a model of simultaneous evolution of co-authorship and citation networks that is validated on the PNAS dataset.

The co-authorship network of *Scientometrics* was studied by Hou et al. [63] in 2008 and Chen et al. [23] in 2013. Hou et al. [63] examined the structure of the network using a methodology that combines social network analysis metrics, cluster analysis and frequency analysis of words appearing in paper titles. More specifically, the authors employed

1. the clustering algorithm provided by the Bibexcel software tool to identify cohesive research communities in the network,
2. word frequency analysis to identify research themes within research communities, and
3. node centrality metrics to identify the most prominent authors in each community.

Such integrated approach enabled the authors to construct a coarse-grained visually comprehensible description of the network, i.e. the authors formed the network of communities in which nodes are augmented with the most central authors and dominant research themes within communities. At the micro-level, Hou et al. investigated correlations between centrality metrics and the productivity of authors (measured by the number of papers) observing that there are positive and significant correlations between measured quantities. On the other hand, Chen et al. [23] conducted a multi-level analysis of the evolution of the co-authorship network of authors publishing in Scientometrics. The authors investigated evolutionary trends in collaboration among individual researchers (micro-level), institutions (meso-level) and countries (macro-level). Similarly to Bettencourt et al. [12], the authors observed that densification laws [75, 76] combined with evolutionary stable diameters can explain the evolution of the network at all three levels.

The co-authorship and citation networks of *Physical Review* journals were studied by Martin et al. [84]. The study encompassed nearly half a million articles published since the inception of the journal in 1893. The main findings of the study are:

- The productivity of authors roughly follows the Pareto principle (the 80-20 rule): 80% of the total number of publications are authored by 20% of the most prolific authors.
- The number of papers and the number of unique authors per year grow exponentially.
- There are increasing trends in the average number of authors per paper and the average number of co-authors per author.
- There is a strong tendency towards reciprocal citations which are especially strong among co-authors.
- Similarly to the findings reported by Newman [89], the probability that two researchers, who have never collaborated previously, establish collaboration increases with the number of joint co-authors, but the upper bound of the probability is not especially high (it is lower than 0.2).

The structure of research collaboration in *Journal of Finance* was investigated by Fatt et al [43]. The authors observed that the network has a giant connected component which obeys scale-free and small-world characteristics. Several node centrality measures were employed to identify the most important researchers in the giant component.

The scale-free property was also observed for the co-authorship network of the *IEEE Transactions on Intelligent Transportation Systems* journal [78] at different levels of research collaboration (individual researchers, institutions, countries). The authors of the study also identified dominant research topics of papers published in the journal.

Fischbach et al. [45] examined the properties of co-authorship network of the *Electronic Markets* journal. The authors extended each node in the network with the impact of the corresponding researcher measured by the number of citations. They then investigated associations between various node centrality metrics and the impact of authors. The obtained results indicate that authors having high degree, betweenness and closeness centralities in the network tend to be cited more often compared to the rest of researchers in the network.

7.5 National Co-authorship Networks

Several studies investigated properties of co-authorship networks at the national level [6, 16, 21, 22, 26, 30, 35, 40, 44, 53, 67–71, 74, 82, 83, 85, 98, 103, 110, 117, 129, 133]. The main characteristic of national co-authorship networks is that they depict collaboration among researchers from one country, considering either all researchers from the country or a subset of them working in a particular scientific field.

Perc [103] studied the evolution of the co-authorship network of Slovenian researchers from 1960 to 2010 with a yearly resolution. The nodes of the network represent 7380 Slovenian researchers registered in SICRIS (Slovenian Current Research Information System) that have at least one publication indexed in the Web of Science. Perc observed that there is a point in time (1985) when a giant connected component emerged. After that point, the mean distance between researchers and the diameter of the network decrease yielding to a national research collaboration structure with the small-world property. Additionally, Perc showed that the growth of the network is governed by the preferential attachment principle with a nearly linear attachment rate. The observed preferential attachment rate causes a log-normal degree distribution of the network. The evolution of the Slovenian co-authorship network was also studied by Kastrin et al. [68] who confirmed the findings reported by Perc. The authors also showed that there are strong evolutionary correlations between productivity/interdisciplinarity and collaboration for Slovenian researchers.

Kronegger et al. [71] investigated the structure of four co-authorship networks of Slovenian researchers corresponding to four different scientific fields (physics, mathematics, biotechnology and sociology). The authors found that all examined networks are small-worlds. The authors also investigated the stability of the networks employing the degree-driven removal scheme and showing that a very small number of highly connected researchers have a significant role to the overall connectedness of Slovenian mathematicians and biotechnologists. Ferligoj et al. [44] employed stochastic actor-oriented modeling to test the small-world and preferential attachment hypothesis for disciplinary co-authorship networks of Slovenian researchers. The obtained results show that examined networks have a high clustering driven by transitive collaboration closure processes. The authors also showed that the preferential attachment principle does not hold at the national level, i.e. Slovenian researchers do not tend to form new collaboration links with those Slovenian researchers who have a high degree of domestic collaboration. On the other side, a high degree of international collaboration has a positive impact on co-authorship link formation in natural, technical, medical and biotechnical sciences, but a negative impact in social sciences and humanities. It was also shown that disciplinary co-authorship networks of Slovenian researchers sooner or later consolidate into core-periphery structures [30, 70].

Karlovčec et al. [67] analyzed transition dynamics of individuals migrating between the core and periphery of the co-authorship network of Slovenian researchers over 44 years. The authors applied the Lip algorithm to each of 44 evolutionary snapshots of the network in order to obtain core-periphery partitions. They then described each researcher in the network by a feature vector reflecting the stability, balance and strength of retention in the core and applied the K-means clustering algorithm to identify groups of researchers having similar characteristics of core-periphery transition dynamics. Finally, the groups obtained by the clustering procedure were used to train decision-tree classification models in order to produce human-interpretative descriptions of the core-periphery transition dynamics for individual researchers.

Lužar et al. [82] studied the interdisciplinarity of the co-authorship network of Slovenian researchers (the same network that was previously examined by Perc [103]) using a methodology based on community detection techniques. The main finding of the study is that the number of communities in the network increases as the network grows, but the average level of interdisciplinarity within communities remains constant. From this result the authors concluded that "a healthy and flourishing interdisciplinary research environment in Slovenia is in need of additional and stronger stimulation than it has received thus far".

Çavuşoğlu and Türker [21] constructed the co-authorship network of Turkish scientists relying on bibliographic records in the Web of Science for publications coming from Turkey between 1980 and 2010. The authors studied the structure and evolution of the network observing several evolutionary trends in co-authorship practices of Turkish researchers. Namely, the authors identified increasing trends in the average degree of the network, the average number of papers per author, and the average number of authors per paper. The authors also observed that the network exhibits scale-free and small-world characteristics. In a subsequent study, the authors also examined four Turkish co-authorship networks representing collaboration in engineering, mathematics, physics and surgery [22] and identified similarities and differences in the structure and evolution of scientific collaboration among Turkish researchers from different scientific disciplines. The co-authorship network of Turkish social scientists was also analyzed by Gossart and Özman [53]. The connected component analysis revealed that the network has a large number of isolated nodes and small-size components. Furthermore, 99% node pairs are not mutually reachable in the network via a path in the network implying the need to strengthen the domestic research collaboration of social scientists from Turkey.

Mena-Chalco et al. [85] examined properties of eight Brazilian co-authorship networks corresponding to the following research areas: agricultural sciences, biological sciences, earth sciences, humanities, applied social sciences, health sciences, engineering, linguistic and arts. The analyzed networks were constructed using bibliographic records stored in *Brazilian Lattes Platform* – an information system maintained by the Brazilian Government to manage information about scientific production of individual researchers and research institutions from Brazil. The authors used several network based measures (the average node degree, density, the clustering coefficient, the average path length, diameter, the assortativity index and the size of the largest component) to compare co-authorship patterns in different research areas. The general conclusions of the study are that all examined networks are small-worlds (the average path length varies from 4.6 in agricultural sciences to 8.1 in linguistics), possess the scale-free property characterized by a power-law in the tail of a degree distribution, and that the majority of the networks exhibit positive medium to strong assortativity (ranging from 0.02 for engineering to 0.37 in humanities). The analysis of the evolution of the networks revealed evolutionary patterns similar to those previously reported in studies dealing with the evolution of large-scale co-authorship networks: there is an increase of the average node degree and the size of the largest connected component followed by a decrease of the small-world coefficient as examined co-authorship networks evolve. The co-authorship network of Brazilian researchers extracted from the Lattes Platform was also investigated by Araujo et al. [5] whose results confirm the findings reported by Mena-Chalco et al. [85].

The structure and evolution of the national co-authorship network of Korea was investigated by Kim et al. [69]. The network was extracted from a dataset built by the Korea Institute of Science and Technology Information that covers more than 700000 papers authored by approximately 415000 researchers in the period between 1948 and 2011. The authors observed that the number of papers and the number of authors exponentially grow in time. Secondly, the number of single-authored papers has decreased constantly from 67.85% in 1964 to 24.48% in 2011 implying an increasing tendency of Korean researchers to collaborate domestically, that is also reflected by an evolutionary increase in the average node degree. The characteristic path length of the network started to decrease in 1978 when a giant connected component of the network has emerged. The giant connected component in the last examined year (2011) encompasses approximately 90% of nodes and exhibits the small-world property with 5 degrees of separation and a highly skewed degree distribution. The authors used the Gini coefficient to quantify inequalities in scientific productivity and collaboration of Korean researchers. Temporal changes of the Gini coefficient show an increase over time indicating deep inequalities in productivity and collaboration at the domestic level.

Bordons et al. [16] explored relationships between research performance indicators and the position of researchers in co-authorship networks of three different scientific fields (nanoscience, pharmacology and statistics) in Spain. More specifically, the authors used a Poisson regression model to investigate how the g-index is related to different co-authorship network measures. The obtained results show that Spanish researchers with a high degree centrality and strong collaborative ties tend to have a higher g-index in all three fields. The authors also observed that co-authorship networks of experimental fields (nanoscience and pharmacology) are more denser compared to the co-authorship network of Spanish statisticians which is less connected and more fragmented.

The study by Chinchilla-Rodríguez et al. [26] presents an analysis of an evolutionary sequence of three co-authorship networks of researchers from Argentina active in library and information sciences. The pre-specified blockmodeling technique based on the notion of structural equivalence was used to detect clusters in the networks. The authors observed that the networks exhibit a core-periphery structure with small-size cores.

The collaboration of Chinese researchers in particular research fields was investigated by Yan et al. [129] in 2010 and Ma et al. [83] and Li et al. [133] in 2014. Yan et al. [129] analyzed the co-authorship network of library and information science (LIS) researchers from China. The network was constructed using bibliographic data from 18 core LIS journals indexed in Chinese Social Science Citation Index. The analysis of the network revealed its scale-free and small-world characteristics. The authors investigated correlations between centrality measures and the impact of researchers estimated by the number of citations. The obtained results show that there are strong Spearman's correlations between centrality measures and the number of citations, the strongest being for betweenness centrality. This result indicates that Chinese LIS researchers having a central position in the domestic co-authorship network also have a high research impact at the international level. The same results is also obtained for an international co-authorship network of library and information science extracted from articles published in 16 top rated LIS journals [128]. Similar findings were also reported by Abbasi et al. [1] who investigated correlations between centrality measures and research performance measured using the g-index. Ma et al. [83] studied scientific collaboration among Chinese researchers in the field of humanities and social sciences. The authors constructed 23 co-authorship networks corresponding to 23 different disciplines (literature, archeology, sociology, psychology, and so on). Similarly to the study by Mena-Chalco et al. [85], several network based measures are employed to observe similarities and differences in co-authorship patterns between different research areas. Li et al. [133] studied the evolution of the co-authorship network of Chinese researchers in management. The authors observed that the number of nodes in the network grows at a faster rate than the number of nodes in the largest connected component and, at the same time, the average distance between authors has an increasing trend indicating that the collaboration network has not yet reached a mature and steady stage of the development.

7.6 Summary

From the previously presented review of empirical studies of co-authorship networks we can derive some general, frequently observed characteristics of their structure and evolution.

Heavy-tailed distributions of research productivity and collaboration [8, 13, 17, 21, 29, 32, 36, 43, 47, 57, 64, 65, 78, 79, 84, 85, 88, 91–93, 95, 96, 108, 120, 127, 129]. Generally speaking, an average researcher from a scientific field/community produces a relatively small number of papers and collaborates with a relatively small number of other researchers. However, within scientific fields/communities we can also find a small fraction of researchers having an extremely high degree of research productivity and collaboration. Empirically observed distributions of productivity and collaboration frequently obey (truncated) power-laws. This means that corresponding co-authorship networks possess the scale-free property and that inequalities in research productivity can be described by the Lotka's law of scientific productivity. The scale-free property of a complex network can be explained by the preferential attachment principle (also known as "rich get richer", "the Matthew effect" and the principle of cumulative advantage). There is a body of empirical evidence [8, 73, 89, 103, 119] showing that the preferential attachment principle indeed governs the evolution of large-scale co-authorship networks, i.e. new researchers tend to integrate into a co-authorship network by establishing research collaboration with highly collaborative researchers. As the consequence, the degree of connectedness of highly collaborative researchers increases faster compared to loosely connected researchers making the gap between highly connected and loosely connected researchers even more bigger.

The small-world property [17, 21, 24, 25, 29, 32, 36, 43, 47, 57, 61, 64–66, 71, 79, 85, 88, 91–93, 95, 103, 120, 127, 129]. Co-authorship networks exhibit the small-world property in the Watts-Strogatz sense [126] meaning that the distance between two randomly selected researchers in a co-authorship network is a small value (much smaller than the size of the network) and that the network exhibits a high degree of local clustering (much higher than for a comparable random graph). Table 7.1 presents the values of the small-world and clustering coefficients for various co-authorship networks studied in the literature. The table also shows the values of the small-world and clustering coefficients for comparable random graphs. It can be observed that the small-world coefficients of co-authorship networks are either much smaller or close to the small-world coefficients of comparable random graphs (it is important to recall that random graphs are also small-world networks, but not in the Watts-Strogatz sense). On the other hand, the clustering coefficients of co-authorship networks are much higher than the clustering coefficients of comparable random graphs (at least 92.5 times higher). In other words, the probability that two collaborators of a researcher are collaborators among themselves is significantly higher than expected by random chance alone.

Giant connected components [8, 11, 12, 29, 36, 43, 47, 57, 60, 64, 88, 91–93, 95, 103, 127]. One of the typical features of large-scale co-authorship networks is

Table 7.1 Characteristics of the small-world property for various co-authorship networks analyzed in the literature: l – the small-world coefficient, l_r – the small-world coefficient of a comparable random graph, C – the clustering coefficient, C_r – the clustering coefficient of a comparable random graph

Network	Ref.	l	l_r	$l - l_r$	C	C_r	C/C_r
EconLit	[54]	9.47	21.99	−12.52	0.16	0.000041	3813.1
Mathematical reviews	[57]	8.00	11.80	−3.80	0.15	0.000009	17202.0
Brazil	[85]	5.83	9.35	−3.52	0.14	0.000004	36455.3
DBLP-DP	[36]	6.00	8.30	−2.30	0.63	0.000107	5883.8
Computational geometry	[65]	5.31	7.57	−2.26	0.41	0.000441	928.9
LIS	[128]	9.68	11.49	−1.81	0.58	0.000212	2739.1
Mathematical reviews	[90]	7.60	9.14	−1.54	0.15	0.000015	9743.8
WCRE	[61]	4.26	5.27	−1.01	0.76	0.008213	92.5
LIS China	[129]	8.84	9.71	−0.86	0.43	0.000236	1798.4
ICIS	[127]	5.40	6.26	−0.86	0.60	0.001788	335.6
Genetic programming	[120]	4.74	5.56	−0.82	0.67	0.001484	448.1
DBLP	[47]	6.41	7.11	−0.70	0.75	0.000010	77869.3
DBLP conferences	[47]	6.54	7.14	−0.60	0.75	0.000012	60035.5
SIGMOD	[88]	5.65	6.14	−0.49	0.69	0.001484	464.8
MEDLINE	[93]	4.60	4.92	−0.32	0.07	0.000012	5543.5
CiteSeer	[64]	7.10	7.32	−0.22	0.63	0.000020	32289.9
DBLP Journals	[47]	7.26	7.48	−0.22	0.77	0.000016	49661.5
Evolutionary computation	[29]	6.10	6.00	0.10	0.80	0.000765	1043.3
MR	[61]	6.23	6.10	0.13	0.68	0.002296	296.2
ArXiv-hep-tx	[93]	6.91	6.67	0.24	0.33	0.000463	706.4
ACM	[32]	4.99	4.63	0.36	0.60	0.000173	3465.3
SE	[61]	7.70	7.26	0.44	0.67	0.000377	1778.1
PNAS	[17]	5.89	5.27	0.62	0.40	0.000084	4735.9
Turkey	[21]	4.00	3.35	0.65	0.75	0.000462	1624.4
Slovenia	[103]	4.60	3.76	0.84	0.22	0.001450	151.7
ArXiv-cond-mat	[93]	6.40	5.50	0.90	0.35	0.000350	993.2
ArXiv-astro-ph	[93]	4.66	3.58	1.08	0.41	0.000904	458.0
ArXiv	[93]	5.90	4.79	1.11	0.43	0.000183	2345.4
JCDL	[79]	6.58	5.01	1.57	0.89	0.002772	321.1
SPIRES	[93]	4.00	2.12	1.88	0.73	0.003055	237.6
NCSTRL	[93]	9.70	7.35	2.35	0.50	0.000299	1657.0

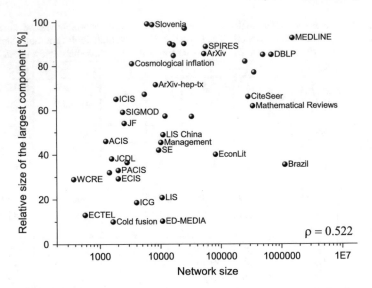

Fig. 7.1 The relative size of the largest connected component versus network size for various co-authorship networks analyzed in the literature. Each point in the plot represents one co-authorship network

the existence of a giant connected component which encompasses a large majority of nodes present in a co-authorship network and whose size is much larger than the size of the second largest connected component. The size of the largest connected component in a co-authorship network is an indicator of overall cohesiveness of the corresponding research field/community. The existence of a giant connected component suggests a widespread research collaboration in the scientific field/community oriented towards common goals. On the other hand, the absence of a giant connected component implies a poorly cohesive community of researchers and in the case of field co-authorship networks suggests that the research field is still in an early phase of development. Figure 7.1 shows the relative size of the largest connected component versus the number of nodes for various co-authorship networks studied in the literature. It can be observed that the size of the largest connected component is moderately correlated with the number of nodes (the value of the Spearman correlation is $\rho = 0.522$), i.e. larger co-authorship networks moderately tend to be more globally cohesive. More stronger correlation can be observed between the size of the largest connected component and the average node degree (see Fig. 7.2, $\rho = 0.669$) suggesting that a widespread increase of the research collaboration at the individual level leads to more globally cohesive research communities. In several studies investigating the evolution of co-authorship networks [8, 12, 36, 47, 54, 61, 85, 103, 119] it was observed that the size of the largest connected component steadily increases in time leading to the emergence of the giant connected component.

Assortative mixing [13, 47, 64, 85, 90, 108, 120] Although the assortativity of co-authorship networks is less investigated compared to other structural characteristics, the existing body of empirical evidence suggests that co-authorship

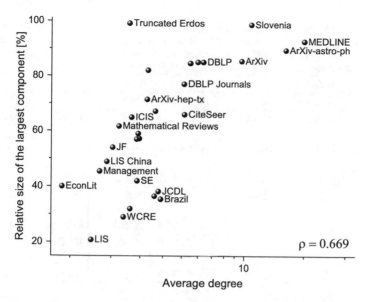

Fig. 7.2 The relative size of the largest connected component versus the average node degree for various co-authorship networks analyzed in the literature. Each point in the plot represents one co-authorship network

networks tend to exhibit assortative degree mixing patterns (see Table 7.2). This property implies that the best connected researchers in a co-authorship network (researchers having a high number of collaborators) tend to be collaborators among themselves and form a densely connected core of the network.

Community structure [11, 13, 28, 33, 34, 42, 46, 55, 65, 66, 73, 77, 79, 81, 82, 94, 100, 105–107, 112, 116, 123, 130]. Another typical and frequently observed feature of co-authorship networks is that they have a strong community structure. This means that co-authorship networks, in spite of being globally sparse, contain dense and highly cohesive subgraphs (clusters) which usually correspond to different institutional research groups or encompass researchers dealing with the same or similar research topics.

Collaboration increases in time [12, 21, 23, 36, 47, 54, 56, 84, 85, 104, 119, 132]. One of frequently observed trends in the evolution of co-authorship networks is an evolutionary increase of the average node degree. In other words, the average number of co-authors of an average researcher increases in time. In some cases this tendency can be expressed by a densification law [12, 23, 104].

Average separation decreases in time [8, 21, 47, 54, 64, 85, 103]. As already mentioned, large-scale co-authorship networks are small-world networks. One of the main characteristics of small-world random graphs is that their small-world coefficient increases very slowly with the number of nodes. However, in several studies it was observed that the small-world coefficient of co-authorship networks either decreases in time or has an evolutionary stable value. In other words, co-authorship

Table 7.2 The value of the assortativity index for various co-authorship networks investigated in the literature P – the number of papers used to form the co-authorship network, T – the time frame of publications, N – the number of nodes in the network, L – the number links in the network, A – the value of the assortativity index. N/A denotes missing (not reported) values.

Network	Ref.	P	T	N	L	A
ArXiv	[90]	98502	1995–1999	52909	256608	0.36
ArXiv-cond-mat	[108]	17828	N/A	16258	47555	0.31
DBLP journals	[47]	N/A	1936–2008	356822	987059	0.3
CiteSeer	[64]	451305	1980–2005	283174	787224	0.28
Genetic programming	[120]	4564	1986–2007	2809	5853	0.27
DBLP	[47]	N/A	1936–2008	688642	2283764	0.17
DBLP conferences	[47]	N/A	1936–2008	503595	1584108	0.16
Brazil	[85]	7351957	1990–2010	1131912	2512872	0.15
MEDLINE	[90]	2163923	1995–1999	1520251	13758272	0.13
Mathematical reviews	[90]	N/A	1940–2004	253339	494011	0.12

networks tend to evolve into a state which does not decrease the efficiency of spreading processes, i.e. the ability of co-authorship networks to quickly propagate information either gradually improves or stays constant during evolution.

Correlations between research impact and centrality. Several studies indicate that there are strong correlations between the centrality of researchers in co-authorship networks and their scientific impact [1, 45, 122, 128, 129]. This means that researchers having a high number of collaborators or researchers located in the core of a co-authorship network tend to be cited more frequently.

References

1. Abbasi, A., Chung, K.S.K., Hossain, L.: Egocentric analysis of co-authorship network structure, position and performance. Inf. Process. Manag. **48**(4), 671–679 (2012). https://doi.org/10.1016/j.ipm.2011.09.001
2. Abbasi, A., Hossain, L., Leydesdorff, L.: Betweenness centrality as a driver of preferential attachment in the evolution of research collaboration networks. J. Informetr. **6**(3), 403–412 (2012). https://doi.org/10.1016/j.joi.2012.01.002
3. Acedo, F.J., Barroso, C., Casanueva, C., Galn, J.L.: Co-authorship in management and organizational studies: an empirical and network analysis. J. Manag. Stud. **43**(5), 957–983 (2006). https://doi.org/10.1111/j.1467-6486.2006.00625.x
4. Amaral, L.A.N., Scala, A., Barthlmy, M., Stanley, H.E.: Classes of small-world networks. Proc. Natl. Acad. Sci. **97**(21), 11149–11152 (2000). https://doi.org/10.1073/pnas.200327197

5. Arajo, E.B., Moreira, A.A., Furtado, V., Pequeno, T.H.C., Andrade Jr., J.S.: Collaboration networks from a large CV database: dynamics, topology and bonus impact. PLOS ONE **9**(3), 1–7 (2014). https://doi.org/10.1371/journal.pone.0090537

6. Badar, K., Hite, J.M., Badir, Y.F.: Examining the relationship of co-authorship network centrality and gender on academic research performance: the case of chemistry researchers in Pakistan. Scientometrics **94**(2), 755–775 (2013). https://doi.org/10.1007/s11192-012-0764-z

7. Barabási, A.L., Albert, R.: Emergence of scaling in random networks. Science **286**(5439), 509–512 (1999). https://doi.org/10.1126/science.286.5439.509

8. Barabási, A.L., Jeong, H., Néda, Z., Ravasz, E., Schubert, A., Vicsek, T.: Evolution of the social network of scientific collaborations. Phys. A **311**, 590–614 (2002). https://doi.org/10.1016/S0378-4371(02)00736-7

9. Barrat, A., Barthélemy, M., Vespignani, A.: Weighted evolving networks: coupling topology and weight dynamics. Phys. Rev. Lett. **92**, 228701 (2004). https://doi.org/10.1103/PhysRevLett.92.228701

10. Barrat, A., Barthlemy, M., Pastor-Satorras, R., Vespignani, A.: The architecture of complex weighted networks. Proc. Natl. Acad. Sci. U. S. A. **101**(11), 3747–3752 (2004). https://doi.org/10.1073/pnas.0400087101

11. Batagelj, V., Mrvar, A.: Some analyses of Erdos collaboration graph. Soc. Netw. **22**(2), 173–186 (2000). https://doi.org/10.1016/s0378-8733(00)00023-x

12. Bettencourt, L.M.A., Kaiser, D.I., Kaur, J.: Scientific discovery and topological transitions in collaboration networks. J. Informetr. **3**(3), 210–221 (2009). https://doi.org/10.1016/j.joi.2009.03.001

13. Bird, C., Barr, E., Nash, A., Devanbu, P., Filkov, V., Su, Z.: Structure and dynamics of research collaboration in computer science. In: Proceedings of the Ninth SIAM International Conference on Data Mining, pp. 826–837. SIAM (2009)

14. Biryukov, M., Dong, C.: Analysis of computer science communities based on DBLP. In: Proceedings of the 14th European Conference on Research and Advanced Technology for Digital Libraries, ECDL'10, pp. 228–235. Springer, Berlin (2010)

15. Blondel, V.D., Guillaume, J.L., Lambiotte, R., Lefebvre, E.: Fast unfolding of communities in large networks. J. Stat. Mech. Theory Exp. **2008**(10), P10008 (2008)

16. Bordons, M., Aparicio, J., González-Albo, B., Daz-Faes, A.A.: The relationship between the research performance of scientists and their position in co-authorship networks in three fields. J. Informetr. **9**(1), 135–144 (2015). https://doi.org/10.1016/j.joi.2014.12.001

17. Borner, K., Maru, J.T., Goldstone, R.L.: The simultaneous evolution of author and paper networks. Proc. Natl. Acad. Sci. U. S. A. **101**(Suppl 1), 5266–5273 (2004). https://doi.org/10.1073/pnas.0307625100

18. Brner, K., Dall'Asta, L., Ke, W., Vespignani, A.: Studying the emerging global brain: analyzing and visualizing the impact of co-authorship teams. Complexity **10**(4), 57–67 (2005). https://doi.org/10.1002/cplx.20078

19. Brunson, J.C., Fassino, S., McInnes, A., Narayan, M., Richardson, B., Franck, C., Ion, P., Laubenbacher, R.: Evolutionary events in a mathematical sciences research collaboration network. Scientometrics **99**(3), 973–998 (2014). https://doi.org/10.1007/s11192-013-1209-z

20. Cerinek, M., Batagelj, V.: Network analysis of Zentralblatt MATH data. Scientometrics, 1–25 (2014). https://doi.org/10.1007/s11192-014-1419-z

21. Çavuşo glu, A., Türker, I.: Scientific collaboration network of Turkey. Chaos, Solitons and Fractals **57**(0), 9–18 (2013). https://doi.org/10.1016/j.chaos.2013.07.022

22. Çavuşo glu, A., Türker, I.: Patterns of collaboration in four scientific disciplines of the Turkish collaboration network. Phys. A Stat. Mech. Appl. **413**(0), 220–229 (2014). https://doi.org/10.1016/j.physa.2014.06.069

23. Chen, Y., Brner, K., Fang, S.: Evolving collaboration networks in scientometrics in 19782010: a micromacro analysis. Scientometrics **95**(3), 1051–1070 (2013). https://doi.org/10.1007/s11192-012-0895-2

24. Cheong, F., Corbitt, B.J.: A social network analysis of the co-authorship network of the Australasian conference of information systems from 1990 to 2006. In: 17th European Conference on Information Systems, ECIS 2009, Verona, Italy, pp. 292–303 (2009)
25. Cheong, F., Corbitt, B.J.: A social network analysis of the co-authorship network of the pacific Asia conference on information systems from 1993 to 2008. In: Pacific Asia Conference on Information Systems, PACIS 2009, Hyderabad, India, 10–12 July, p. 23 (2009)
26. Chinchilla-Rodríguez, Z., Ferligoj, A., Miguel, S., Kronegger, L., de Moya-Anegón, F.: Block-modeling of co-authorship networks in library and information science in argentina: a case study. Scientometrics 93(3), 699–717 (2012). https://doi.org/10.1007/s11192-012-0794-6
27. Clauset, A., Shalizi, C., Newman, M.: Power-law distributions in empirical data. SIAM Rev. 51(4), 661–703 (2009). https://doi.org/10.1137/070710111
28. Cotta, C., Guervós, J.J.M.: Where is evolutionary computation going? a temporal analysis of the EC community. Genet. Program. Evolvable Mach. 8(3), 239–253 (2007). https://doi.org/10.1007/s10710-007-9031-0
29. Cotta, C., Merelo, J.: The complex network of evolutionary computation authors: an initial study. Preprint available at http://arxiv.org/abs/physics/0507196 (2005)
30. Cugmas, M., Ferligoj, A., Kronegger, L.: The stability of co-authorship structures. Scientometrics 106(1), 163–186 (2016). https://doi.org/10.1007/s11192-015-1790-4
31. Ding, Y.: Scientific collaboration and endorsement: Network analysis of co-authorship and citation networks. J. Informetr. 5(1), 187–203 (2011). https://doi.org/10.1016/j.joi.2010.008
32. Divakarmurthy, P., Menezes, R.: The effect of citations to collaboration networks. In: Menezes, R., Evsukoff, A., Gonzlez, M.C. (eds.) Complex Networks. Studies in Computational Intelligence, vol. 424, pp. 177–185. Springer, Berlin (2013). https://doi.org/10.1007/978-3-642-30287-9_19
33. Donetti, L., Muoz, M.A.: Detecting network communities: a new systematic and efficient algorithm. J. Stat. Mcch. Theory Exp. 2004(10), P10012 (2004)
34. Duch, J., Arenas, A.: Community detection in complex networks using extremal optimization. Phys. Rev. E 72, 027104 (2005). https://doi.org/10.1103/PhysRevE.72.027104
35. Durbach, I.N., Naidoo, D., Mouton, J.: Co-authorship networks in South African chemistry and mathematics. S. Afr. J. Sci. 104(2), 487–492 (2008)
36. Elmacioglu, E., Lee, D.: On six degrees of separation in DBLP-DB and more. SIGMOD Rec. 34(2), 33–40 (2005). https://doi.org/10.1145/1083784.1083791
37. Erdős, P.: On the fundamental problem of mathematics. Am. Math. Mon. 79(2), 149 (1972)
38. Erdős, P., Rnyi, A.: On random graphs. I. Publ. Math. Debr. 6, 290–297 (1959)
39. Erdős, P., Rnyi, A.: On the evolution of random graphs. Publ. Math. Inst. Hung. Acad. Sci. 5, 17–61 (1960)
40. Eslami, H., Ebadi, A., Schiffauerova, A.: Effect of collaboration network structure on knowledge creation and technological performance: the case of biotechnology in Canada. Scientometrics 97(1), 99–119 (2013). https://doi.org/10.1007/s11192-013-1069-6
41. Fan, Y., Li, M., Chen, J., Gao, L., Di, Z., Wu, J.: Network of econophysicists: a weighted network to investigate the development of econophysics. Int. J. Mod. Phys. B 18(17n19), 2505–2511 (2004). https://doi.org/10.1142/S0217979204025579
42. Farkas, I., Ábel, D., Palla, G., Vicsek, T.: Weighted network modules. New J. Phys. 9(6), 180 (2007)
43. Fatt, C.K., Ujum, E., Ratnavelu, K.: The structure of collaboration in the journal of finance. Scientometrics 85(3), 849–860 (2010). https://doi.org/10.1007/s11192-010-0254-0
44. Ferligoj, A., Kronegger, L., Mali, F., Snijders, T.A., Doreian, P.: Scientific collaboration dynamics in a national scientific system. Scientometrics 104(3), 985–1012 (2015). https://doi.org/10.1007/s11192-015-1585-7
45. Fischbach, K., Putzke, J., Schoder, D.: Co-authorship networks in electronic markets research. Electron. Mark. 21(1), 19–40 (2011). https://doi.org/10.1007/s12525-011-0051-5
46. Fortunato, S.: Community detection in graphs. Phys. Rep. 486(35), 75–174 (2010). https://doi.org/10.1016/j.physrep.2009.11.002

47. Franceschet, M.: Collaboration in computer science: a network science approach. J. Am. Soc. Inf. Sci. Technol. **62**(10), 1992–2012 (2011). https://doi.org/10.1002/asi.21614
48. Gilbert, E.N.: Random graphs. Ann. Math. Stat. **30**(4), 1141–1144 (1959)
49. Girvan, M., Newman, M.E.J.: Community structure in social and biological networks. Proc. Natl. Acad. Sci. **99**(12), 7821–7826 (2002). https://doi.org/10.1073/pnas.122653799
50. Goffman, C.: And what is your Erdős number? Am. Math. Mon. **76**(7), 149 (1969)
51. Goh, K.I., Oh, E., Jeong, H., Kahng, B., Kim, D.: Classification of scale-free networks. Proc. Natl. Acad. Sci. U. S. A. **99**, 12583–12588 (2002). https://doi.org/10.1073/pnas.202301299
52. Goh, K.I., Oh, E., Kahng, B., Kim, D.: Betweenness centrality correlation in social networks. Phys. Rev. E **67**, 017101 (2003). https://doi.org/10.1103/PhysRevE.67.017101
53. Gossart, C., Özman, M.: Co-authorship networks in social sciences: the case of turkey. Scientometrics **78**(2), 323–345 (2009). https://doi.org/10.1007/s11192-007-1963-x
54. Goyal, S., van der Leij, M.J., Moraga-Gonzales, J.L.: Economics: an emerging small world. J. Polit. Econ. **114**(2), 403–412 (2006)
55. Gregory, S.: An algorithm to find overlapping community structure in networks. In: Kok, J.N., Koronacki, J., Lopez de Mantaras, R., Matwin, S., Mladenič, D., Skowron, A. (eds.) Knowledge Discovery in Databases: PKDD 2007. Lecture Notes in Computer Science, vol. 4702, pp. 91–102. Springer, Berlin (2007). https://doi.org/10.1007/978-3-540-74976-9_12
56. Grossman, J.: The evolution of the mathematical research collaboration graph. Congr. Numer. 201–212 (2002)
57. Grossman, J.: Patterns of collaboration in mathematical research. SIAM News **35**(9), 8–9 (2002)
58. Grossman, J.W.: In: Graham, R.L., Neetil, J., Butler, S. (eds.) Paul Erds: The Master of Collaboration. The Mathematics of Paul Erds II, pp. 489–496. Springer, New York (2013). https://doi.org/10.1007/978-1-4614-7254-4_27
59. Grossman, J.W., Ion, P.D.F.: On a portion of the well known collaboration graph. Congr. Numer. **108**, 129–131 (1995)
60. Guimera, R., Uzzi, B., Spiro, J., Amaral, L.: Team assembly mechanisms determine collaboration network structure and team performance. Science **308**(5722), 697–702 (2005). https://doi.org/10.1126/science.1106340
61. Hassan, A.E., Holt, R.C.: The small world of software reverse engineering. In: 2013 20th Working Conference on Reverse Engineering (WCRE), vol. 0, pp. 278–283 (2004). https://doi.org/10.1109/WCRE.2004.37
62. Horn, D.B., Finholt, T.A., Birnholtz, J.P., Motwani, D., Jayaraman, S.: Six degrees of jonathan grudin: a social network analysis of the evolution and impact of CSCW research. In: Proceedings of the 2004 ACM Conference on Computer Supported Cooperative Work, CSCW '04, pp. 582–591. ACM, New York, NY, USA (2004). https://doi.org/10.1145/1031607.1031707
63. Hou, H., Kretschmer, H., Liu, Z.: The structure of scientific collaboration networks in scientometrics. Scientometrics **75**(2), 189–202 (2008). https://doi.org/10.1007/s11192-007-1771-3
64. Huang, J., Zhuang, Z., Li, J., Giles, C.L.: Collaboration over time: characterizing and modeling network evolution. In: Proceedings of the 2008 International Conference on Web Search and Data Mining, WSDM '08, pp. 107–116. ACM, New York, NY, USA (2008). https://doi.org/10.1145/1341531.1341548
65. Hui, Z., Cai, X., Greneche, J.M., Wang, Q.: Structure and collaboration relationship analysis in a scientific collaboration network. Chin. Sci. Bull. **56**(34), 3702–3706 (2011). https://doi.org/10.1007/s11434-011-4756-9
66. Johansson, F., Martenson, C., Svenson, P.: A social network analysis of the information fusion community. In: 2011 Proceedings of the 14th International Conference on Information Fusion (FUSION), pp. 1–8 (2011)
67. Karlovčec, M., Lužar, B., Mladenić, D.: Core-periphery dynamics in collaboration networks: the case study of slovenia. Scientometrics **109**(3), 1561–1578 (2016). https://doi.org/10.1007/s11192-016-2154-4

68. Kastrin, A., Klisara, J., Lužar, B., Povh, J.: Analysis of slovenian research community through bibliographic networks. Scientometrics **110**(2), 791–813 (2017). https://doi.org/10. 1007/s11192-016-2203-z
69. Kim, J., Tao, L., Lee, S.H., Diesner, J.: Evolution and structure of scientific co-publishing network in korea between 1948–2011. Scientometrics **107**(1), 27–41 (2016). https://doi.org/ 10.1007/s11192-016-1878-5
70. Kronegger, L., Ferligoj, A., Doreian, P.: On the dynamics of national scientific systems. Qual. Quant. **45**(5), 989–1015 (2011). https://doi.org/10.1007/s11135-011-9484-3
71. Kronegger, L., Mali, F., Ferligoj, A., Doreian, P.: Collaboration structures in Slovenian scientific communities. Scientometrics **90**(2), 631–647 (2012). https://doi.org/10.1007/s11192-011-0493-8
72. Kuhn, T.S.: The Structure of Scientific Revolutions. University of Chicago Press, Chicago (1970)
73. Lara-Cabrera, R., Cotta, C., Fernndez-Leiva, A.: An analysis of the structure and evolution of the scientific collaboration network of computer intelligence in games. Phys. A Stat. Mech. Appl. **395**, 523–536 (2014). https://doi.org/10.1016/j.physa.2013.10.036
74. Larivire, V., Gingras, Y., Archambault, É.: Canadian collaboration networks: a comparative analysis of the natural sciences, social sciences and the humanities. Scientometrics **68**(3), 519–533 (2006). https://doi.org/10.1007/s11192-006-0127-8
75. Leskovec, J., Kleinberg, J., Faloutsos, C.: Graphs over time: densification laws, shrinking diameters and possible explanations. In: Proceedings of the Eleventh ACM SIGKDD International Conference on Knowledge Discovery in Data Mining, KDD '05, pp. 177–187. ACM, New York, NY, USA (2005). https://doi.org/10.1145/1081870.1081893
76. Leskovec, J., Kleinberg, J., Faloutsos, C.: Graph evolution: densification and shrinking diameters. ACM Trans. Knowl. Discov. Data **1**(1), (2007). https://doi.org/10.1145/1217299. 1217301
77. Leskovec, J., Lang, K.J., Dasgupta, A., Mahoney, M.W.: Community structure in large networks: natural cluster sizes and the absence of large well-defined clusters. Internet Math. **6**(1), 29–123 (2009). https://doi.org/10.1080/15427951.2009.10129177
78. Li, L., Li, X., Cheng, C., Chen, C., Ke, G., Zeng, D., Scherer, W.: Research collaboration and ITS topic evolution: 10 years at T-ITS. IEEE Trans. Intell. Transp. Syst. **11**(3), 517–523 (2010). https://doi.org/10.1109/TITS.2010.2059070
79. Liu, X., Bollen, J., Nelson, M.L., Van de Sompel, H.: Co-authorship networks in the digital library research community. Inf. Process. Manag. **41**(6), 1462–1480 (2005). https://doi.org/ 10.1016/j.ipm.2005.03.012
80. Lotka, A.J.: The frequency distribution of scientific production. J. Wash. Acad. Sci. **16**, 317–323 (1926)
81. Luthi, L., Tomassini, M., Giacobini, M., Langdon, W.B.: The genetic programming collaboration network and its communities. In: Proceedings of the 9th Annual Conference on Genetic and Evolutionary Computation, GECCO '07, pp. 1643–1650. ACM, New York, NY, USA (2007). https://doi.org/10.1145/1276958.1277284
82. Lužar, B., Levnajić, Z., Povh, J., Perc, M.: Community structure and the evolution of interdisciplinarity in Slovenia's scientific collaboration network. PLoS ONE **9**(4), e94429 (2014). https://doi.org/10.1371/journal.pone.0094429
83. Ma, F., Li, Y., Chen, B.: Study of the collaboration in the field of the Chinese humanities and social sciences. Scientometrics **100**(2), 439–458 (2014). https://doi.org/10.1007/s11192-014-1301-z
84. Martin, T., Ball, B., Karrer, B., Newman, M.E.J.: Co-authorship and citation patterns in the physical review. Phys. Rev. E **88**, 012814 (2013). https://doi.org/10.1103/PhysRevE.88. 012814
85. Mena-Chalco, J.P., Digiampietri, L.A., Lopes, F.M., Cesar, R.M.: Brazilian bibliometric co-authorship networks. J. Assoc. Inf. Sci. Technol. **65**(7), 1424–1445 (2014). https://doi.org/ 10.1002/asi.23010

86. Milojević, S.: Modes of collaboration in modern science: beyond power laws and preferential attachment. J. Am. Soc. Inf. Sci. Technol. **61**(7), 1410–1423 (2010). https://doi.org/10.1002/asi.v61:7

87. Moody, J.: The structure of a social science collaboration network: disciplinary cohesion from 1963 to 1999. Am. Sociol. Rev. **69**(2), 213–238 (2004). https://doi.org/10.1177/000312240406900204

88. Nascimento, M.A., Sander, J., Pound, J.: Analysis of SIGMOD's co-authorship graph. SIGMOD Rec. **32**(3), 8–10 (2003). https://doi.org/10.1145/945721.945722

89. Newman, M.: Clustering and preferential attachment in growing networks. Phys. Rev. E **64**(2), 025102 (2001)

90. Newman, M.: Co-authorship networks and patterns of scientific collaboration. Proc. Natl. Acad. Sci. **101**(1), 5200–5205 (2004)

91. Newman, M.E.J.: Scientific collaboration networks I: network construction and fundamental results. Phys. Rev. E **64**, 016131 (2001). https://doi.org/10.1103/PhysRevE.64.016131

92. Newman, M.E.J.: Scientific collaboration networks II: shortest paths, weighted networks, and centrality. Phys. Rev. E **64**, 016132 (2001). https://doi.org/10.1103/PhysRevE.64.016132

93. Newman, M.E.J.: The structure of scientific collaboration networks. Proc. Natl. Acad. Sci. **98**(2), 404–409 (2001). https://doi.org/10.1073/pnas.98.2.404

94. Newman, M.E.J.: Fast algorithm for detecting community structure in networks. Phys. Rev. E **69**, 066133 (2004). https://doi.org/10.1103/PhysRevE.69.066133

95. Newman, M.E.J.: Who is the best connected scientist? A study of scientific co-authorship networks. In: Ben-Naim, E., Frauenfelder, H., Toroczkai, Z. (eds.) Complex Networks. Lecture Notes in Physics, vol. 650, pp. 337–370. Springer, Berlin (2004). https://doi.org/10.1007/978-3-540-44485-5_16

96. Ochoa, X., Mndez, G., Duval, E.: Who we are: analysis of 10 years of the ED-MEDIA conference. In: Siemens, G., Fulford, C. (eds.) Proceedings of World Conference on Educational Multimedia, Hypermedia and Telecommunications, pp. 189–200. AACE (2009)

97. Ortega, J.L.: Influence of co-authorship networks in the research impact: Ego network analyses from microsoft academic search. J. Informetr. **8**(3), 728–737 (2014). https://doi.org/10.1016/j.joi.2014.07.001

98. Osca-Lluch, J., Velasco, E., Lpez, M., Haba, J.: Co-authorship and citation networks in Spanish history of science research. Scientometrics **80**(2), 373–383 (2009). https://doi.org/10.1007/s11192-008-2089-5

99. Otte, E., Rousseau, R.: Social network analysis: a powerful strategy, also for the information sciences. J. Inf. Sci. **28**(6), 441–453 (2002). https://doi.org/10.1177/016555150202800601

100. Palla, G., Barabasi, A.L., Vicsek, T.: Quantifying social group evolution. Nature **446**, 664–667 (2007). https://doi.org/10.1038/nature05670

101. Pan, R.K., Saramki, J.: The strength of strong ties in scientific collaboration networks. EPL (Europhys. Lett.) **97**(1), 18007 (2012)

102. Pao, M.L.: An empirical examination of Lotka's law. J. Am. Soc. Inf. Sci. **37**(1), 26–33 (1986). https://doi.org/10.1002/asi.4630370105

103. Perc, M.: Growth and structure of Slovenia's scientific collaboration network. J. Informetr. **4**(4), 475–482 (2010). https://doi.org/10.1016/j.joi.2010.04.003

104. Pham, M.C., Derntl, M., Klamma, R.: Development patterns of scientific communities in technology enhanced learning. Educ. Technol. Soc. **15**(3), 323–335 (2012)

105. Pollner, P., Palla, G., Vicsek, T.: Preferential attachment of communities: the same principle, but a higher level. EPL (Europhys. Lett.) **73**(3), 478 (2006). https://doi.org/10.1209/epl/i2005-10414-6

106. Pons, P., Latapy, M.: Computing communities in large networks using random walks. J. Graph Algorithms Appl. **10**(2), 191–218 (2006). https://doi.org/10.1007/11569596_31

107. Raghavan, U.N., Albert, R., Kumara, S.: Near linear time algorithm to detect community structures in large-scale networks. Phys. Rev. E **76**, 036106 (2007). https://doi.org/10.1103/PhysRevE.76.036106

108. Ramasco, J.J., Dorogovtsev, S.N., Pastor-Satorras, R.: Self-organization of collaboration networks. Phys. Rev. E **70**, 036106 (2004). https://doi.org/10.1103/PhysRevE.70.036106
109. Reinhardt, W., Meier, C., Drachsler, H., Sloep, P.: Analyzing 5 years of EC-TEL proceedings. In: Kloos, C.D., Gillet, D., Crespo Garca, R.M., Wild, F., Wolpers, M. (eds.) Towards Ubiquitous Learning. Lecture Notes in Computer Science, vol. 6964, pp. 531–536. Springer, Berlin (2011). https://doi.org/10.1007/978-3-642-23985-4_51
110. Rivellini, G., Rizzi, E., Zaccarin, S.: The science network in Italian population research: an analysis according to the social network perspective. Scientometrics **67**(3), 407–418 (2006). https://doi.org/10.1556/Scient.67.2006.3.5
111. Savić, M., Ivanović, M., Radovanović, M., Ognjanović, Z., Pejović, A., Kruger, T.J.: The structure and evolution of scientific collaboration in Serbian mathematical journals. Scientometrics **101**(3), 1805–1830 (2014). https://doi.org/10.1007/s11192-014-1295-6
112. Savić, M., Ivanović, M., Radovanović, M., Ognjanović, Z., Pejović, A., Kruger, T.J.: Exploratory analysis of communities in co-authorship networks: a case study. In: Bogdanova, A.M., Gjorgjevikj, D. (eds.) ICT Innovations 2014. Advances in Intelligent Systems and Computing, vol. 311, pp. 55–64. Springer International Publishing, Berlin (2015). https://doi.org/10.1007/978-3-319-09879-1_6
113. Schubert, A.: A Hirsch-type index of co-author partnership ability. Scientometrics **91**(1), 303–308 (2012). https://doi.org/10.1007/s11192-011-0559-7
114. Shi, Q., Xu, B., Xu, X., Xiao, Y., Wang, W., Wang, H.: Diversity of social ties in scientific collaboration networks. Phys. A Stat. Mech. Appl. **390**(2324), 4627–4635 (2011). https://doi.org/10.1016/j.physa.2011.06.072
115. Smeaton, A.F., Keogh, G., Gurrin, C., McDonald, K., Sødring, T.: Analysis of papers from twenty-five years of SIGIR conferences: What have we been doing for the last quarter of a century? SIGIR Forum **37**(1), 49–53 (2003). https://doi.org/10.1145/945546.945550
116. Staudt, C., Schumm, A., Meyerhenke, H., Görke, R., Wagner, D.: Static and dynamic aspects of scientific collaboration networks. In: International Conference on Advances in Social Networks Analysis and Mining, ASONAM 2012, Istanbul, Turkey, 26–29 August 2012, pp. 522–526 (2012). https://doi.org/10.1109/ASONAM.2012.90
117. Stefano, D.D., Fuccella, V., Vitale, M.P., Zaccarin, S.: The use of different data sources in the analysis of co-authorship networks and scientific performance. Soc. Netw. **35**(3), 370–381 (2013). https://doi.org/10.1016/j.socnet.2013.04.004
118. Tang, J., Jin, R., Zhang, J.: A topic modeling approach and its integration into the random walk framework for academic search. In: Eighth IEEE International Conference on Data Mining, ICDM '08, pp. 1055–1060 (2008). https://doi.org/10.1109/ICDM.2008.71
119. Tomasini, M., Luthi, L.: Empirical analysis of the evolution of a scientific collaboration network. Phys. A Stat. Mech. Appl. **385**(2), 750 – 764 (2007). https://doi.org/10.1016/j.physa.2007.07.028
120. Tomassini, M., Luthi, L., Giacobini, M., Langdon, W.: The structure of the genetic programming collaboration network. Genet. Program. Evolvable Mach. **8**(1), 97–103 (2007). https://doi.org/10.1007/s10710-006-9018-2
121. Uddin, S., Hossain, L., Abbasi, A., Rasmussen, K.: Trend and efficiency analysis of co-authorship network. Scientometrics **90**(2), 687–699 (2012). https://doi.org/10.1007/s11192-011-0511-x
122. Uddin, S., Hossain, L., Rasmussen, K.: Network effects on scientific collaborations. PLoS ONE **8**(2), e57546 (2013). https://doi.org/10.1371/journal.pone.0057546
123. Velden, T., Haque, A.u., Lagoze, C.: A new approach to analyzing patterns of collaboration in co-authorship networks: mesoscopic analysis and interpretation. Scientometrics **85**(1), 219–242 (2010). https://doi.org/10.1007/s11192-010-0224-6
124. Vidgen, R.T., Henneberg, S., Naudé, P.: What sort of community is the European conference on information systems? A social network analysis 1993–2005. Eur. J. Inf. Syst. **16**(1), 5–19 (2007). https://doi.org/10.1057/palgrave.ejis.3000661
125. Voos, H.: Lotka and information science. J. Am. Soc. Inf. Sci. **25**(4), 270–272 (1974). https://doi.org/10.1002/asi.4630250410

126. Watts, D.J., Strogatz, S.H.: Collective dynamics of 'small-world' networks. Nature **393**(6684), 409–10 (1998). https://doi.org/10.1038/30918
127. Xu, J.J., Chau, M.: The social identity of IS: analyzing the collaboration network of the ICIS conferences (1980-2005). In: Proceedings of the International Conference on Information Systems, ICIS 2006, Milwaukee, Wisconsin, USA, 10–13 December 2006, p. 39 (2006)
128. Yan, E., Ding, Y.: Applying centrality measures to impact analysis: a co-authorship network analysis. J. Am. Soc. Inf. Sci. Technol. **60**(10), 2107–2118 (2009). https://doi.org/10.1002/asi.21128
129. Yan, E., Ding, Y., Zhu, Q.: Mapping library and information science in China: a co-authorship network analysis. Scientometrics **83**(1), 115–131 (2010). https://doi.org/10.1007/s11192-009-0027-9
130. Yang, J., Leskovec, J.: Structure and overlaps of ground-truth communities in networks. ACM Trans. Intell. Syst. Technol. **5**(2), 26:1–26:35 (2014). https://doi.org/10.1145/2594454
131. Yoshikane, F., Kageura, K.: Comparative analysis of co-authorship networks of different domains: the growth and change of networks. Scientometrics **60**(3), 435–446 (2004). https://doi.org/10.1023/B:SCIE.0000034385.05897.46
132. Zhai, L., Li, X., Yan, X., Fan, W.: Evolutionary analysis of collaboration networks in the field of information systems. Scientometrics 1–21 (2014). https://doi.org/10.1007/s11192-014-1360-1
133. Zhai, L., Yan, X., Shibchurn, J., Song, X.: Evolutionary analysis of international collaboration network of Chinese scholars in management research. Scientometrics **98**(2), 1435–1454 (2014). https://doi.org/10.1007/s11192-013-1040-6

Chapter 8
Analysis of Enriched Co-authorship Networks: Methodology and a Case Study

Abstract The nodes of an enriched co-authorship network are annotated with various types of nominal and numeric attributes that provide additional information about researchers present in the network. In this chapter we propose a novel methodology to study the structure and evolution of enriched co-authorship networks. The proposed methodology is illustrated on an enriched co-authorship network encompassing researchers employed at one large faculty of sciences.

Enriched co-authorship networks are co-authorship networks in which nodes are annotated with

1. attributes indicating demographic characteristics of researchers (e.g. gender, age, academic position, and so on), and
2. researcher evaluation metrics that quantify various determinants of research performance.

In one of our previous publications [22], we introduced a hybrid methodology to analyze the structure of enriched intra-institutional co-authorship network extracted from CRIS databases. The main premise of the methodology proposed in [22] is that the structure of an institution determines research groups within the enriched co-authorship network. However, research communities are self-organizing social systems and research collaboration may transcend institutional boundaries in the sense that the members of a research group may belong to different institutional units.

In this chapter we propose a novel methodology to analyze the structure and evolution of enriched co-authorship networks in which research groups are not known in advance (Sect. 8.1). To demonstrate the usefulness of the proposed methodology we applied it on an enriched co-authorship network encompassing researchers employed at the Faculty of Sciences, University of Novi Sad, Serbia (FS-UNS). The background of the case study and the extraction of the examined network are

© Springer International Publishing AG, part of Springer Nature 2019
M. Savić et al., *Complex Networks in Software, Knowledge,*
and Social Systems, Intelligent Systems Reference Library 148,
https://doi.org/10.1007/978-3-319-91196-0_8

explained in Sect. 8.2. The next Sect. 8.3 presents and discusses the results of the analysis. The last section summarizes our main findings and concludes this chapter.

8.1   Methodology

Our methodology to study the structure and evolution of enriched co-authorship networks includes and combines

1. domain-independent metrics and methods used in complex network analysis,
2. domain-dependent researcher evaluation metrics,
3. non-parametric statistical tests applied to the sets of metric values of independent groups of nodes in an enriched co-authorship network, and
4. graph representations derived from enriched co-authorship networks.

We assume that the nodes in an investigated enriched co-authorship network are augmented with gender information and metrics quantifying productivity, collaboration and social capital of researchers. Additionally, we assume that each researcher belongs to exactly one research group. However, the methodology can be adapted in a straightforward manner for analysis of enriched co-authorship networks containing overlapping research groups.

The degree of a node in an enriched co-authorship network is equal to the number of co-authors of the corresponding researcher. This measure reflects the capacity of researchers to collaborate – a higher number of co-authors implies a more diverse collaborative behavior. We also use the weighted degree measure which is equal to the sum of links incident with a node. This measure reflects the strength of research collaboration of a researcher with his/her co-authors. Additionally, we also consider the (weighted) node degree with respect to a partition of the set of nodes. Let P denote a non-overlapping partition of the set of nodes of an enriched co-authorship network, let A be an arbitrary node in the network and let S_A be the group in P to which A belongs.

Definition 8.1   (*Intra-P (weighted) degree*) The intra-P (weighted) degree of A with respect to P is equal to the number (total weight) of links connecting A with nodes that are in S_A.

Definition 8.2   (*Inter-P (weighted) degree*) The inter-P (weighted) degree of A with respect to P is equal to the number (total weight) of links connecting A with nodes that do not belong to S_A.

For any partition P, we have that the (weighted) degree of a node is equal to the sum of its intra-P and inter-P (weighted) degrees. In our case study, we use the following partitions of the set of nodes:

1. The partition of researchers according to their institutional affiliations. For this partition, the intra-P degree reflects the degree of intra-institutional research

collaboration, while the inter-P degree quantifies the degree of inter-institutional research collaboration of a researcher. The intra-P weighted degree and inter-P weighted degree quantify the strength of intra-institutional and inter-institutional research collaboration, respectively.

2. The partition of researchers into research groups obtained after a community detection algorithm was applied to the network. In this case, the intra-P degree and intra-P weighted degree of a researcher reflect the degree and strength of research collaboration, respectively, with researchers from the same research group. On the other hand, the inter-P degree and inter-P weighted degree quantify the degree and strength of research collaboration with researchers belonging to other research groups.

The existence of giant connected components is one of the main structural properties of co-authorship networks (see Chap. 7). The absence of a giant connected component in a co-authorship network implies that the corresponding research community is poorly cohesive. Structural characteristics of connected components can be quantitatively expressed by several metrics proposed under the framework of complex network theory, e.g. the clustering coefficient, the characteristic path length and the Newman degree assortativity coefficient [4]. In the case of enriched co-authorship networks, the assortativity coefficient can be generalized with respect to numeric node attributes. More specifically, we propose the P-assortativity coefficient with respect to a research productivity metric P.

Definition 8.3 (*P-assortativity coefficient*) Let P be a metric quantifying the productivity of researchers and let G denote an enriched co-authorship network. The P-assortativity index of G is equal to the Person correlation coefficient between P values of researchers connected in G.

The P-assortativity coefficient takes a value in the range $[-1, 1]$ where high positive values indicate that highly productive researchers dominantly tend to collaborate among themselves, while high negative values suggest that highly productive researchers dominantly tend to collaborate with lowly productive researchers.

The scale-free property is another frequently observed characteristic of co-authorship networks. A network exhibits the scale-free property if its degree distribution $P(k)$ follows a power-law ($P(k) \sim k^{-\gamma}$, where $P(k)$ is the fraction of nodes having degree k and γ is a constant). In the context of our methodology, the power-law test proposed by Clauset et al. [6] is used to examine whether the degree distribution and the distribution of node strength (the sum of weights of links incident with a node) of an enriched co-authorship network follow power-laws. The emergence of power-laws in complex evolving networks can be explained by the principle of linear preferential attachment [2] – the probability that a new node A establishes a link with an old node B is directly proportional to the degree (strength) of B. To confirm the scale-free property of a network we also have to rule out two complementary principles that may govern the evolution of complex networks:

- *non-linear preferential attachment* – the attachment probability depends on node degree but not linearly, and

- *uniform attachment* – the attachment probability does not depend on node degree.

The uniform attachment principle is related to complex networks whose degree distributions are exponential [2]. On the other hand, nearly-linear preferential attachment leads to complex networks having log-normal degree distributions [17, 20]. Therefore, we consider two additional theoretical distributions, exponential and log-normal, to model empirically observed node degree/strength distributions. The fitted power-law model is compared to the fitted models of the alternative distributions using the log-likelihood ratio test [6].

We also rely on the k-core decomposition technique [23] to investigate structural characteristics of enriched co-authorship networks. The k-core of an undirected graph is a maximal subgraph in which the degree of each node is higher than or equal to k. The k-core can be obtained by recursively deleting all nodes whose degree is smaller than k. A node has shell index k if it belongs to the k-core but not to the $(k + 1)$-core of the network. The shell index enables us to distinguish between two types of researchers having a high number of collaborators: high-degree researchers dominantly connected to low-degree researchers have a low shell index, whereas high-degree researchers dominantly connected to other high-degree researchers have a high shell index.

A network has a core-periphery structure if the set of nodes can be divided into two groups: (1) a group of densely connected nodes which from the core of the network, and (2) a group of loosely connected nodes that are on the periphery of the network. An enriched co-authorship network has a nested core-periphery structure if its k-cores are either connected graphs or they have giant connected components. If the network has a nested-core periphery structure then we are able to divide its nodes into core nodes (researchers with a high shell index forming a dense and highly cohesive subgraph within the network) and periphery nodes (researchers with a low shell index sparsely connected to researchers from the core). We use the following rule to assign researchers either to the core group (C) or to the periphery group (P): the core group C is a minimal subset of nodes such that the sum of shell indexes of nodes in C is higher than then sum of shell indexes of nodes that are not in C (i.e. nodes belonging to P). The sets C and P can be determined by a simple greedy algorithm starting from nodes having the highest shell index.

The Mann-Whitney U (MWU) test [12] and accompanying probabilities of superiority [7] can be exploited to compare two independent groups of nodes in an enriched co-authorship network with respect to a numerical property of nodes. The MWU test is a non-parametric statistical procedure which means that the test does not assume any particular distribution of metric values. Let M be an arbitrary node metric quantifying some aspect of research performance, and let G_1 and G_2 be the sets of M values for two independent groups of nodes. The MWU test is a test of stochastic superiority, i.e. it can be applied to test the null hypothesis that

$$P(g_1 > g_2) = P(g_2 > g_1) \qquad (8.1)$$

where $P(g_1 > g_2)$ is the probability of superiority of G_1 over G_2, i.e. the probability that a randomly selected value from G_1 (denoted by g_1) is higher than a randomly selected value from G_2 (denoted by g_2), and $P(g_2 > g_1)$ is the probability of superiority of G_2 over G_1. In other words, the null hypothesis of the test is that metric values of one group do not tend to be significantly higher or smaller than metric values of another group. The test is based on the U statistic which is the number of times a value from G_2 precedes a value from G_1 in the ranked sequence of values from both groups. Under the null hypothesis U closely follows a normal distribution. It should be emphasized that $P(g_1 > g_2) + P(g_2 > g_1)$ is not necessarily equal to 1 since $P(g_1 > g_2) + P(g_2 > g_1) + P(g_1 = g_2) = 1$ and $P(g_1 = g_2)$ may be non-zero. We use the MWU test to examine differences between

- researchers from the core of an investigated enriched co-authorship network and researchers from the periphery if the network has a core-periphery structure.
- researchers belonging to different research groups,
- researchers collaborating only with researchers from their own research groups and researchers participating in inter-group research collaborations, and
- male and female researchers within a research group.

Community detection techniques are utilized in our methodological framework to identify research groups in enriched co-authorship networks. In our case study six different community detection techniques suitable for weighted graphs are employed to determine the best partition of the enriched co-authorship network into cohesive research groups:

1. The fast greedy modularity optimization algorithm (GMO) by Clauset et al. [5],
2. The Louvain algorithm (LV) by Blondel et al. [3],
3. The Walktrap algorithm (WT) by Pons and Latapy [18],
4. The Girvan-Newman community detection algorithm based on the edge betweenness measure [8] (abbreviated as EB)
5. The Infomap algorithm (IM) by Rosvall and Bergstrom [21], and
6. The algorithm proposed by Newman [16] that finds communities by computing the leading non-negative eigenvector of the modularity matrix of the network (abbreviated as SOM – spectral optimization of modularity).

The best community partition is selected considering the following two measures:

1. the weighted modularity metric [14], denoted by Q, and
2. the ratio between the total weight of inter-cluster links (co-authorship links connecting nodes that belong to different communities) and the total weight of intra-cluster links (co-authorship links connecting nodes that belong to the same community), denoted by r.

Let P_A and P_B denote two community partitions obtained by two community detection algorithms A and B. A good community detection algorithm should result in a community partition that has a high value of Q and a low value of r. Therefore, $Q(P_A) > Q(P_B)$ and $r(P_A) < r(P_B)$ implies that P_A is better than P_B in the sense that communities determined by P_A are more cohesive than communities determined by P_B or, equivalently, that algorithm A performs better than B. The quality of obtained communities is additionally checked using the Raddichi definitions of strong and weak communities [19].

To investigate research collaboration between research groups we analyze the structure of a group collaboration network obtained from the enriched co-authorship network. The group collaboration network can be constructed by Algorithm 8.1 based on the following principles:

1. The nodes of the group collaboration network represent research groups identified by a community detection algorithm. Additionally, each node is annotated with its size (the number of researchers in the group), as well as with the number of male and female researchers in the group.
2. Two research groups A and B are connected by an undirected weighted link if at least one researcher from A co-authored at least one paper with at least one researcher from B. The weight of the link connecting A and B is equal to the sum of weights of all co-authorship links connecting researchers from A to researchers from B.

Algorithm 8.1: The algorithm for constructing the group collaboration network for a given co-authorship network.

---

input : $G = (V, E)$ – a co-authorship network
          P – the partition of V according to a community detection algorithm
output: $G^* = (V^*, E^*)$ – the collaboration network of research groups in G

$V^* \leftarrow P$
$E^* \leftarrow \emptyset$
foreach $(p_1, p_2) \in P \times P, p_1 \neq p_2$ **do**
    foreach $(x, y) \in p_1 \times p_2$ **do**
        if $\{x, y\} \in E$ **then**
            if $\{p_1, p_2\} \in E^*$ **then**
                | weight($\{p_1, p_2\}$) $\leftarrow$ weight($\{p_1, p_2\}$) + weight($\{x, y\}$)
            else
                $E^* \leftarrow E^* \cup \{\{p_1, p_2\}\}$
                weight($\{p_1, p_2\}$) $\leftarrow$ weight($\{x, y\}$)
            end
        end
    end
end

It can be noticed that the group collaboration network does not contain parallel links and self-loops. Consequently, the degree of a node A in the network is equal to the number of other research groups with which researchers from A have established research collaboration. Thus, the degree of A reflects the capacity of A to collaborate with other research groups. On the other hand, the weighted degree of A indicates the strength of research collaboration of A with other research groups. To identify the most important groups we rely on the weighted betweenness centrality measure, i.e. betweenness centrality computed according to the shortest paths in a weighted network formed by inverting link weights of the group collaboration network (see Chap. 2, Sects. 2.1 and 2.2.3). The shortest paths in the mirroring network can be determined by the Dijkstra algorithm since link weights in the group collaboration network are positive values.

To separate strong from weak links in the group collaboration network we apply the following scheme. Let L denote the set of all links in the network and let $L(w)$ be a subset of L such that the weight of each link in $L(w)$ is higher than or equal to w.

Definition 8.4 (*Strong links*) We say that $L(w)$ is the set of strong links if $L(w)$ is a minimal set for which the following property holds:

$$\sum_{x \in L(w)} \text{weight}(x) > \sum_{y \in L \setminus L(w)} \text{weight}(y) \tag{8.2}$$

In other words, the set of strong links is a minimal subset of L such that the total weight of strong links is higher than the total weight of weak links.

To investigate differences between researchers from different research groups with respect to their productivity and collaborative behavior we construct and analyze so-called *group superiority graphs*.

Definition 8.5 (*Group Superiority Graph*) Let M be an arbitrary researcher evaluation metric. The group superiority graph corresponding to M is a directed graph without isolated nodes. A link $A \to B$ in the group superiority graph indicates that a research group A is strongly dominant over a research group B with respect to M, i.e. researchers from A strongly tend to have higher values of M compared to researchers from B.

To operationalize the notion of a strong group domination we create link $A \to B$ in a group superiority graph (see Algorithm 8.2) if the following two conditions are satisfied:

- The null hypothesis of the MWU test is rejected for M_A and M_B, where M_A and M_B are the set of values of M for groups A and B, respectively.
- The probability of superiority of A over B is higher than 0.75.

Links in group superiority graphs are transitive: $A \to B$ and $B \to C$ imply $A \to C$. As already mentioned, group superiority graphs do not contain isolated nodes. Therefore, if a research group is neither superior nor inferior to all other research groups with respect to a researcher evaluation metric M then it will not be present in the group superiority graph corresponding to M. An empty group superiority graph corresponding to M implies that strong differences among research groups regarding research performance aspect quantified by M are completely absent. Additionally, we distinguish between two types of nodes in group superiority graphs: superior and inferior groups.

Algorithm 8.2: The algorithm for constructing group superiority graphs

---

input : $G = (V, E)$ – an enriched co-authorship network
P – the partition of V into research groups
M – a researcher evaluation metric
output: $G^* = (V^*, E^*)$ – the group superiority graph of G corresponding to M

$V^* \leftarrow \emptyset$
$E^* \leftarrow \emptyset$
foreach $(A, B) \in P \times P$, $A \neq B$ **do**
   // M_A – the set of M values for group A
   $M_A \leftarrow \{M(a) : a \in A\}$
   // M_B – the set of M values for group B
   $M_B \leftarrow \{M(b) : b \in B\}$

   $p \leftarrow$ apply the MWU test to M_A and M_B

   if $p < 0.05$ **then**
      // compute probabilities of superiority
      $n \leftarrow |M_A| \cdot |M_B|$
      $PS_A \leftarrow \big|\{(m_a, m_b) \in M_A \times M_B : m_a > m_b\}\big| / n$
      $PS_B \leftarrow \big|\{(m_a, m_b) \in M_A \times M_B : m_b > m_a\}\big| / n$

      // create a link in the group superiority graph
      if $PS_A \geq 0.75 \vee PS_B \geq 0.75$ **then**
         $V^* \leftarrow V^* \cup \{A, B\}$
         if $PS_A \geq 0.75$ **then**
            $E^* \leftarrow E^* \cup \{(A, B)\}$
         else
            $E^* \leftarrow E^* \cup \{(B, A)\}$
         end
      end
   end
end

Definition 8.6 (*Superior group*) A group g is a superior node in a group superiority graph S if the in-degree of g in S is equal to zero.

Definition 8.7 (*Inferior group*) A group g is an inferior node in a group superiority graph S if the out-degree of g in S is equal to zero.

To investigate the evolution of an enriched co-authorship network we construct time-ordered snapshots of the network $S = \langle N_y = (V_y, E_y) \rangle$, where N_y denotes the snapshot of the network in a year y. The set of nodes V_y contains all researchers that published at least one article before or during y. Similarly, a link between researchers A and B is present in N_y, $\{A, B\} \in E_y$, if A and B co-authored at least one paper before or during y. Therefore, we have that the evolutionary sequence S satisfies the following property

$$y < z \;\Rightarrow\; V_y \subseteq V_z \wedge E_y \subseteq E_z \tag{8.3}$$

The evolution of some quantifiable property P of the network structure is investigated by analyzing the numerical sequence $P(S) = \langle P(N_y) \rangle$.

One of the most important issues regarding the evolution of a complex network is related to how new nodes integrate into the network. This question can be answered by determining characteristics of old nodes to which new nodes connect when joining the network. Let $N_a = (V_a, E_a)$ and $N_b = (V_b, E_b)$ denote two successive evolutionary snapshots of an enriched co-authorship network such that N_a precedes N_b. We distinguish between two types of nodes in N_b:

1. new nodes (denoted by V_n) – nodes present in N_b but not in N_a ($V_n = V_b \setminus V_a$), and
2. old nodes – nodes present in both N_a and N_b (the set of old nodes is equal to V_a since $V_a \subseteq V_b$).

Additionally, we distinguish between two types of old nodes: *preferential* and *non-preferential* nodes.

Definition 8.8 (*Preferential nodes*) An old node X is a preferential node if there is at least one new node Y such that X and Y are connected in N_b. Otherwise, X is a non-preferential node.

We propose a new algorithm for discovering attachment preferences in enriched co-authorship networks (see Algorithm 8.3). The algorithm is based on the MWU test and frequent itemsets mining using the Apriori algorithm [1]. Let V_p and V_{np} denote the set of preferential and non-preferential nodes in N_a, respectively. Let M be an arbitrary researcher evaluation metric contained in metric vectors annotated to nodes of the network, and let M_p and M_{np} denote the set of values of M for V_p and V_{np}, respectively. In order to determine attachment preferences of new nodes in one evolutionary transition we analyze differences between M_p and M_{np} using the MWU test and probabilities of superiority. For example, we can infer that new researchers tended to establish research collaboration with highly productive researchers in the transition from N_a to N_b if M_p is stochastically superior than M_{np} where M is a metric quantifying research productivity.

For each transition in the evolutionary sequence S, two sets of researcher evaluation metrics are determined:

1. M^p that contains those metrics for which preferential nodes tend to have significantly higher values according to the MWU test, and
2. M^{np} that contains those metrics for which non-preferential nodes tend to have significantly higher values.

Attachment preferences characteristic for one evolutionary transition can be described as an itemset I in which each item is of the form (M, C). M is a researcher evaluation metric and C indicates the category of old nodes (preferential or non-preferential nodes) exhibiting significantly higher values of M. Therefore, we have

$$I = \{(m, \text{preferential nodes}) : m \in M^p\} \cup \{(m, \text{non-preferential nodes}) : m \in M^{np}\}$$

Algorithm 8.3: The algorithm for discovering attachment preferences.

---

input : $S = \langle N_y = (V_y, E_y) \rangle$ – the array of evolutionary snapshots of an enriched
        co-authorship network
        m – minimal support
output: F – frequent itemsets of researcher evaluation metrics

$i \leftarrow 1$
$D \leftarrow \emptyset$ /* an empty itemset */
while $i < \text{length}(S)$ **do**
   $(V_a, E_a) \leftarrow S_i$ /* i-th evolutionary snapshot */
   $(V_b, E_b) \leftarrow S_{i+1}$ /* (i+1)-th evolutionary snapshot */
   $V_n \leftarrow V_b \setminus V_a$ /* the set of new nodes */
   $V_p \leftarrow \{x \in V_a : (\exists y \in V_n) \{x, y\} \in E_b\}$ /* the set of preferential nodes */
   $V_{np} \leftarrow V_a \setminus V_p$ /* the set of non-preferential nodes */
   $I \leftarrow \emptyset$ /* an empty itemset */
   foreach metric M attached to nodes **do**
      $M_p \leftarrow \{M(x) : x \in V_p\}$
      $M_{np} \leftarrow \{M(x) : x \in V_{np}\}$
      $p \leftarrow$ apply the MWU test to M_p and M_{np}
      if $p < 0.05$ **then**
         // compute probabilities of superiority
         $n \leftarrow |M_p| \cdot |M_{np}|$
         $PS_p \leftarrow \big| \{(x, y) \in M_p \times M_{np} : x > y\} \big| / n$
         $PS_{np} \leftarrow \big| \{(x, y) \in M_p \times M_{np} : y > x\} \big| / n$
         if $PS_p > PS_{np}$ **then**
            | $I \leftarrow I \cup \{(M, \text{preferential nodes})\}$
         else
            | $I \leftarrow I \cup \{(M, \text{non-preferential nodes})\}$
         end
      end
   end
   $D \leftarrow D \cup I$
   $i \leftarrow i + 1$
end
$F \leftarrow \text{Apriori}(D, m)$

To discover attachment preferences of the whole evolutionary sequence we do the
following:

- form a set D by aggregating itemsets of all evolutionary transitions ($D = \langle I_{k \rightarrow k+1} \rangle$
  where k goes over all evolutionary snapshots of the enriched co-authorship net-
  works except the last one), and
- apply the Apriori algorithm to detect frequent subsets of items in D.

The algorithm for discovering attachment preferences relies on the Apriori algorithm. Thus, it has an additional parameter, so-called minimal support, required by
Apriori. The support of a subset of items X is the fraction of itemsets in D containing X. X is considered frequent if its support is higher than or equal to the minimal
support.

8.2 Case Study

In this chapter, following the methodology described in the previous section, we investigate the structure and evolution of an enriched co-authorship network reflecting intra-institutional research collaboration at the Faculty of Sciences in Novi Sad, Serbia (FS-UNS). FS-UNS is an educational and scientific institution founded in 1969. It is one of the oldest faculties at the University of Novi Sad (UNS), Serbia. The faculty consists of five research departments:

1. Department of Biology and Ecology (DBE),
2. Department of Physics (DP),
3. Department of Geography, Tourism and Hotel Management (DG),
4. Department of Chemistry, Biochemistry and Environmental Protection (DC), and
5. Department of Mathematics and Informatics (DMI).

The enriched co-authorship network of researchers employed at FS-UNS was constructed from the author and publication records exported from the institutional research information system called CRIS-UNS [9]. CRIS-UNS was developed following the recommendations of the non-profit organization euroCRIS.[1] All researchers employed at FS-UNS are institutionally obligated to have CRIS-UNS profiles and periodically update their bibliographic references. CRIS-UNS is an author-article-centered bibliography database which means that researchers registered in the system have unique CRIS-UNS identifiers appearing in publication records instead of personal names. When recording a new publication, a FS-UNS researcher has to select the rest of the co-authors among the researchers registered in the system. He/she is also able to create a new CRIS-UNS profile for an external researcher (researcher not affiliated with FS-UNS) in the case that the external co-author is not already registered in the system. Then, the recorded publication is automatically associated to all other co-authors.

Each author record in the CRIS-UNS system contains the following information: an unique author identifier, author name, the date of birth, the institution to which the author is affiliated, the organizational unit within the institution in which he/she works, academic rank and gender. Each publication present in the system is described by a record which consists of an unique publication identifier, the complete list of author identifiers, the year of publication, publication title, publication type (journal, conference, monograph, and so on), detailed information about the publication venue and the quantitative evaluation of the publication by the rule book prescribed by the Serbian Ministry of Education, Science and Technological Development. The name disambiguation problem is not present when forming a co-authorship network from CRIS-UNS data since authors are uniquely identified in CRIS-UNS publication records.

The FS-UNS co-authorship network was constructed from 14986 publications authored by 423 FS-UNS researchers and their 5690 external co-authors. The

---

[1]http://www.eurocris.org/.

Table 8.1 The number of researchers, the percentage of male and female researchers and the average age of researchers per departments in the FS-UNS co-authorship network

Department	Researchers	Male [%]	Female [%]	Avg. age
DMI	87	49.43	50.57	45.3
DG	66	57.58	42.42	42.9
DBE	118	25.42	74.58	41.2
DP	57	56.14	43.86	46.5
DC	95	24.21	75.79	42.7

network captures all researchers employed at FS-UNS in 2014 and their collaborations established until that year. The distribution of nodes per FS-UNS research departments is given in Table 8.1.

To extract the FS-UNS co-authorship network from the CRIS-UNS database, a CRIS-UNS module that exports publication and author metadata to two XML files was developed. The first XML file contains metadata about all FS-UNS researchers and their external collaborators. The second XML file contains metadata about publications authored by FS-UNS researchers. We developed a software tool that extracts the FS-UNS enriched co-authorship network from those two XML files. This tool works in five phases:

1. The set of nodes is formed in the first phase by iterating through XML elements in the XML file containing metadata about researchers. For each XML element, a node representing the corresponding researcher is created in the extracted network.
2. The set of links is formed in the second phase by iterating through XML elements in the XML file containing metadata about publications. For a publication p authored by k researchers $k(k-1)/2$ co-authorship links are created connecting each pair of co-authors. A decentralized inverted index mapping authors to their publications is also formed in this phase. The inverted index is decentralized in the sense that the list of all publications of a researchers is directly attached to the corresponding node in the co-authorship network.
3. In the third phase, the weights of co-authorship links are determined according to the Newman weighting scheme [15]. The decentralized inverted index is used to compute the set of joint publications for two researchers connected in the network.
4. In the fourth phase, the number of external co-authors (researchers not affiliated with FS-UNS) and the strength of research collaboration with them are computed for each local node (i.e. nodes representing FS-UNS researchers). Then, nodes representing external researchers are removed from the network.
5. The rest of researcher evaluation metric are computed in the last phase. The network is enriched with three types of researcher evaluation metrics: productivity metrics, collaboration metrics and centrality metrics reflecting institutional importance and social capital of FS-UNS researchers. Table 8.2 shows the complete list of researcher evaluation metrics attached to nodes in the enriched FS-UNS

Table 8.2 Researcher evaluation metrics attached to nodes in the FS-UNS enriched co-authorship network

Metric	Abbreviation	Category
Productivity, normal count	PRON	Productivity
Productivity, fractional count	PROF	Productivity
Productivity, straight count	PROS	Productivity
Serbian Research Competency Index	SRCI	Productivity
The total number of co-authors	COLL	Collaboration
The number of FS-UNS co-authors	LCOLL	Collaboration
The number of external co-authors	ECOLL	Collaboration
The strength of research collaboration with all co-authors	WCOLL	Collaboration
The strength of research collaboration with FS-UNS co-authors	WLCOLL	Collaboration
The strength of research collaboration with external co-authors	WECOLL	Collaboration
Clustering coefficient	CC	Collaboration
The degree of intra-group collaboration	IntraDEG	Collaboration
The degree of inter-group collaboration	InterDEG	Collaboration
The strength of intra-group collaboration	WIntraDEG	Collaboration
The strength of inter-group collaboration	WInterDEG	Collaboration
Betweenness centrality	BET	Importance
Weighted betweenness centrality	WBET	Importance
Closeness centrality	CLO	Importance
Weighted closeness centrality	WCLO	Importance
Eigenvector centrality	EVC	Importance

co-authorship network. Productivity metrics are computed using the decentralized inverted index. Metrics quantifying local collaboration (LCOLL, WLCOLL and CC) and institutional importance (BET, WBET, CLO, WCLO and EVC) are computed from the network itself. The total number of co-authors (COLL) is the sum of the number of external co-authors (the ECOLL metric computed in the fourth phase) and the number of FS-UNS co-authors (LCOLL which is the degree of a node in the FS-UNS co-authorship network; we recall that nodes representing external researchers are removed from the network in the previous phase). Similarly, the total strength of research collaboration of a FS-UNS researcher (WCOLL) is the sum of the strength of research collaboration with FS-UNS co-authors (WLCOLL) and the strength of research collaboration with external co-authors (the WECOLL metric computed in the fourth phase). Finally, metrics quantifying intra-group and inter-group research collaboration (IntraDEG, InterDEG, WIntraDEG and WInterDEG) are computed after identifying the best partition of the network into communities.

Each node in the FS-UNS co-authorship network is enriched with three widely used productivity metrics (normal, straight and fractional counting of publications [11]) and one Serbian-specific research productivity metric known as the *Serbian research competency index* (SRCI). The main idea of the SRCI metric is that the impact factor of a publication venue determines the quality and importance of a scientific publication. The SRCI metric is computed according to the rule book prescribed by the Serbian Ministry of Education, Science and Technological Development. The rule book defines several categories of publication venues and each category corresponds to a certain number of points. The most important categories are:

1. M21a (10 points) – papers published in international journals indexed in Journal Citation Reports (JCR) that are among the top 10% ranked journals in a scientific discipline according to the two-year impact factor.
2. M21 (8 points) – papers published in JCR journals that are between the top 10% and 30% ranked journals.
3. M22 (5 points) – papers published in JCR journals that are between the top 30% and 60% ranked journals.
4. M23 (3 points) – papers published in international journals with impact factor that are not among the top 60% ranked journals.
5. M33 (1 point) – papers published in proceedings of international conferences.

The SRCI of a Serbian researcher is defined as the sum of points of publications he/she authored. CRIS-UNS categorizes publications according to the rules defined in the rule book. The number of points a publication receives is then stored in the CRIS-UNS record of the publication. This means that CRIS-UNS publication records enable the computation of the SRCI metric for individual researchers.

8.3 Network Analysis: Results and Discussion

The FS-UNS co-authorship network contains 2856 links reflecting intra-institutional research collaboration among 423 FS-UNS researchers. In this Section we present the results of the analysis of the FS-UNS network following the methodology described in Sect. 8.1.

8.3.1 Network Structure

The connected components in the network were identified using the BFS (breadth-first search) algorithm. The network consists of 15 connected components where 14 of them are isolated nodes. The isolated nodes in the network correspond to FS-UNS researchers whose production consists entirely of solo-authored publications. All non-isolated nodes are contained in one giant connected component that encompasses a strong majority of FS-UNS employees (96.69% of them). The existence of a giant connected component and an extremely small number of isolated nodes indicates a high degree of institutional cohesion and implies that the scientific output of FS-UNS researchers is not a product of many institutionally separated research groups that do not collaborate among themselves.

The average distance between two randomly selected nodes contained in the giant connected component is equal to 3.32. Consequently, we can say that the FS-UNS network is very compact due to short distances between nodes. The average clustering coefficient of nodes in the giant component is equal to 0.566. On the other hand, the clustering coefficient of a random graph with $N = 409$ nodes and $L = 2856$ links is equal to $2L/N(N - 1) = 0.03$. Therefore, we can conclude that the FS-UNS co-authorship network is a small-world network in the Watts-Strogatz sense [24].

The Newman assortativity coefficient of the network is equal to 0.24 implying that highly intra-collaborative researchers moderately tend to be directly connected among themselves [13]. The values of the P-assortativity coefficients for various productivity metrics (PRON, PROS, PROF and SRCI) are close to zero (see Table 8.3) implying that highly productive FS-UNS researchers tend to have the same number of highly productive and lowly productive institutional collaborators.

Table 8.3 The values of the P-assortativity coefficients for the enriched FS-UNS co-authorship network

P (productivity metric)	PRON	PROS	PROF	SRCI
P-assortativity index	0.0969	0.0351	0.0362	0.0985

Table 8.4 Descriptive statistics of the distributions of researcher evaluation metrics

Metric	Mean	Standard deviation	Skewness	Kurtosis	Max
PRON	73.72	96.76	3.23	16.52	871
PROF	21.21	29.05	3.37	17.58	266.3
PROS	22.23	30.21	2.89	11.32	242
SRCI	116.58	143.88	2.73	10.93	1124.7
COLL	48.66	49.28	1.97	4.94	311
LCOLL	13.97	10.8	1.12	0.94	56
ECOLL	34.69	41.21	2.2	5.98	260
WCOLL	70.15	93.44	3.17	15.51	816
WLCOLL	37.04	46.12	3.04	16.37	442.1
WECOLL	33.11	59.09	5.57	52.24	735.08
BET	474.18	924.68	3.92	22.71	9064.98
WBET	331.98	783.46	4.09	21.63	7044.41

We computed the distributions of researcher evaluation metrics for nodes in the giant connected component of the FS-UNS co-authorship network. The summary of basic descriptive statistics of the distributions is given in Table 8.4. It can be observed that all distributions, except the distribution of LCOLL, are highly skewed to the right and have heavy-tails (skewness > 1 and kurtosis > 3). This means that there are deep inequalities in productivity, non-local collaborative behavior and institutional importance of FS-UNS researchers. For example, the strength of inter-institutional research collaboration (WECOLL) of the best ranked FS-UNS researcher is 22 times higher than the WECOLL of an average FS-UNS researcher.

To analyze relationships between the productivity of FS-UNS researchers and their collaborative behavior, we computed the Spearman correlation between research productivity metrics and metrics of research collaboration/centrality in the co-authorship network. The results are shown in Table 8.5. We can observe that there are moderate to strong positive Spearman correlations between productivity and collaboration/centrality metrics. Therefore, we can say that the most productive FS-UNS researchers tend to be the most collaborative FS-UNS researchers occupying central positions in the network. It can be noticed that the WECOLL metric exhibits the strongest correlation with the productivity metrics implying that a strong inter-institutional research collaboration has a profound impact on the productivity of FS-UNS researchers. We can also observe that the betweenness centrality metric exhibits a stronger correlation with the productivity metrics than the degree centrality metric. Consequently, we can say that a significant brokerage role within the institution has a bigger impact on research productivity than the number of local collaborators.

To investigate whether the network possesses the scale-free property we applied the power-law test proposed by Clauset et al. [6] to the distribution of LCOLL (the

Table 8.5 The values of the Spearman correlation coefficient between metrics of research productivity and metrics of research collaboration/centrality in the FS-UNS co-authorship network

	LCOLL	WLCOLL	ECOLL	WECOLL	BET	WBET
PRON	0.66	0.92	0.87	0.93	0.69	0.68
PROF	0.53	0.83	0.77	0.89	0.66	0.7
PROS	0.51	0.79	0.74	0.85	0.6	0.66
SRCI	0.59	0.86	0.81	0.88	0.68	0.67

Table 8.6 The results of the power-law test. x_m – the lower bound of the power-law region, α – the power-law exponent, p – statistical significance

Metric	x_m	α	p	Power-law plausible
LCOLL	26	5.39	0.0532	No
WLCOLL	93.87	3.79	0.2675	Yes

degree of a node) and WLCOLL (the strength of a node) metrics. The results of the test are summarized in Table 8.6. We can see that the degree distribution of the network cannot be accurately modeled by a power-law ($p < 0.1$), which implies that the unweighted projection of the network does not have a scale-free topology. On the other hand, the tail of the node strength distribution can be modeled by a power-law ($x_m > 1$, $p > 0.1$).

Using the log likelihood ratio test, we compared the power-law fit to the best fits of the exponential probability mass function (PMF) and log-normal PMF in the obtained power-law scaling region of the node strength distribution. Let $R(d_1/d_2)$ denote the log likelihood ratio for two theoretical distributions d_1 and d_2 ("pw" – power-law, "ln" – log-normal, "exp" – exponential). A positive and statistically significant value of R ($R(d_1/d_2) > 0$, $p < 0.1$) indicates that d_1 is preferred over d_2, while a negative and statistically significant value of R indicates exactly the opposite. If the obtained value of R is not statistically significant ($p \geq 0.1$) then both distributions are equally plausible models for the empirically observed distribution. In our case, we obtained the following log likelihood ratio values:

- $R(\text{pw/ln}) = -0.18$ ($p = 0.86$), and
- $R(\text{pw/exp}) = 0.75$ ($p = 0.45$).

Since both p values are higher than 0.1 we can say that log-normal and exponential distributions are also plausible models for the tail of the node strength distribution.

Using the same statistical procedure we also checked which theoretical distribution provides the best fit considering the whole range of node strength values ($x_m = 1$). The obtained results are summarized in Table 8.7. We can see that the best fits of exponential and log-normal distributions are preferred over the best power-law fit. Figure 8.1 shows the complementary cumulative distribution of node strength with fitted theoretical models.

Table 8.7 The results of the power-law test for the distribution of node strength (WLCOLL) through the whole range of node strength values ($x_m = 1$)

d_1	d_2	$R(d_1/d_2)$	p	Preferred model
Power-law	Log-normal	-10.44	$<10^{-4}$	Log-normal
Power-law	Exponential	-2.43	0.0148	Exponential
Log-normal	Exponential	0.135	0.89	Equally plausible

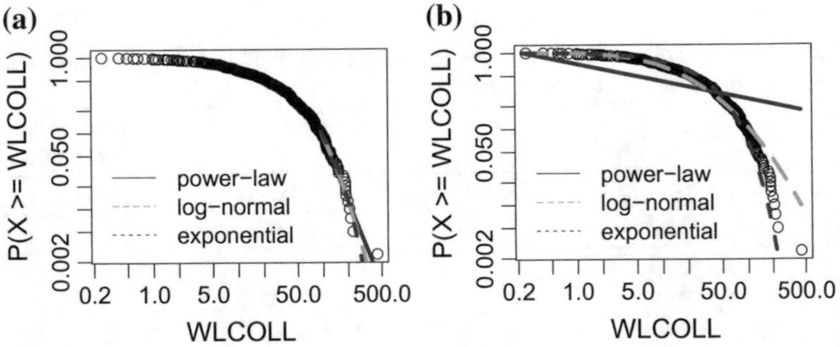

Fig. 8.1 The complementary cumulative distribution of node strength with **a** the best fits of the theoretical distributions in the tail and **b** the best fits of the theoretical distributions through the whole range of values

To summarize, we can conclude that the FS-UNS co-authorship network can not be considered as a scale-free network for two reasons:

1. A power-law is not a plausible model for the degree distribution of the network.
2. The tail of the node strength distribution can be modeled by a power-law, but log-normal and exponential distributions are equally plausible statistical models (see Fig. 8.1a). On the other hand, both alternative models provide significantly better fits through the whole range of node strength values (see Fig. 8.1b).

We applied the k-core decomposition algorithm and determined connected components contained in k-cores in order to check whether the network has a nested core-periphery structure. The network has 19 nested cores all of them being connected subgraphs of the network. This implies that the network has a clear core-periphery structure. The fraction of nodes per k-core and the density of k-cores are shown in Fig. 8.2. We can see that the fraction of nodes in k-cores decays linearly with k, while the density of k-cores grows exponentially with k (a straight line on the semi-log plot), which implies that the network has a strong and balanced core-periphery structure with respect to the number of nodes in each of k-core layers.

The 12-core of the network, which encompasses approximately 32% of FS-UNS researchers, is the minimal core for which the sum of shell index values of nodes in the core is higher than the sum of shell index values of nodes belonging to lower cores. Therefore, FS-UNS researchers whose shell index is equal to or higher than 12 constitute the core of the FS-UNS co-authorship network. We employed the

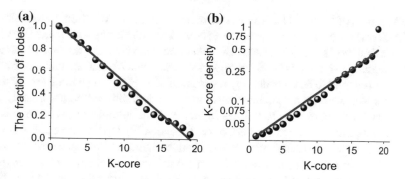

Fig. 8.2 K-core decomposition of the network: **a** the fraction of nodes in each k-core and **b** the density of k-cores

Table 8.8 The results of the comparison between FS-UNS researchers from the core (C) and researchers located at the periphery (P) of the FS-UNS co-authorship network. Avg(C) and Avg(P) denote the average value of the corresponding metric for C and P, respectively. U – the Mann-Whitney U statistic, p – the p-value of U, PS$_1$ – the probability of superiority of C over P, PS$_2$ – the probability of superiority of P over C. The column NHA indicates whether the null hypothesis of the MWU test (no statistically significant differences between groups) is accepted or not

Metric	Avg(C)	Avg(P)	U	p	NHA	PS$_1$	PS$_2$
PRON	124.2576	49.6354	7954	$<10^{-4}$	No	0.78	0.22
PROF	31.7894	16.1697	10003	$<10^{-4}$	No	0.73	0.27
PROS	32.4924	17.3430	10647	$<10^{-4}$	No	0.70	0.28
SRCI	172.8083	89.7819	9684	$<0^{-4}$	No	0.74	0.26
COLL	88.2273	29.8051	4653	$<10^{-4}$	No	0.87	0.13
LCOLL	26.4697	8.0072	784.5	$<10^{-4}$	No	0.98	0.02
ECOLL	61.7576	21.7978	7191	$<10^{-4}$	No	0.80	0.19
WCOLL	120.2045	46.2960	7723	$<10^{-4}$	No	0.79	0.21
WLCOLL	66.9061	22.8109	6641	$<10^{-4}$	No	0.82	0.18
WECOLL	53.2984	23.4851	10089.5	$<10^{-4}$	No	0.72	0.28
BET	813.9461	312.2748	8040	$<10^{-4}$	No	0.78	0.22
CLO	0.3457	0.2897	4622	$<10^{-4}$	No	0.87	0.13
EVC	0.0046	0.0014	849	$<10^{-4}$	No	0.98	0.02
WBET	563.0291	221.8833	12644.5	$<10^{-4}$	No	0.54	0.23
WCLO	0.6649	0.5056	7099	$<10^{-4}$	No	0.81	0.19
CC	0.4659	0.5829	13654	$<10^{-4}$	No	0.37	0.63

MWU test to investigate differences between FS-UNS researchers belonging to the core of the network and researchers at the periphery of the network (i.e. FS-UNS researchers that are not in the 12-core of the network). The results of the MWU test are summarized in Table 8.8.

The results of the MWU test show that there are statistically significant and drastic differences ($PS_1 \gg PS_2$ for all metrics, except for the clustering coefficient for which we have that $PS_2 \gg PS_1$) between researchers from the core and researchers located at the periphery of the FS-UNS co-authorship network. We can see that researchers from the core tend to be considerably more productive by all considered productivity measures. An average researcher from the core has published approximately 124 papers which is more than two times higher than the number of papers published by an average researcher at the periphery. Secondly, FS-UNS researchers from the core tend to have drastically more external collaborators: an average researcher from the core of the network has three times more external collaborators than an average researcher located at the periphery. Both local and external co-authorship links of researchers from the core tend to be significantly stronger than co-authorship links of researchers at the periphery. Additionally, researchers located in the core of the network exhibit a significantly higher institutional importance estimated by both weighted and unweighted centrality measures. The density of ego-networks of researchers at the periphery is significantly higher than the same quantity for researchers from the core. This result implies that researchers from the core have a more significant brokerage role in their ego-networks than peripheral researchers. To summarize, we can conclude that the FS-UNS co-authorship network exhibits a strong core-periphery structure in which researchers from the core strongly tend to be superior to peripheral researchers with respect to all considered indicators of research performance.

8.3.2   Identification of Research Groups

Six different community detection techniques were applied in order to identify the best partition of the FS-UNS co-authorship network into cohesive research groups. The results of the community detection are summarized in Table 8.9. It can be noticed that four algorithms identified community partitions with a high value of the weighted modularity score ($Q > 0.8$) implying that the network has a strong community structure. The community partition obtained by the Louvain algorithm has the highest value of modularity and the lowest ratio between the total weight of inter-community and intra-community links. Consequently, we can conclude that the best partition of the network into cohesive research groups is identified by the Louvain algorithm. Figure 8.3 shows the visualization of the network according to the obtained community partition.

According to the data presented in Table 8.9, we can also order the algorithms considering their performance on the FS-UNS co-authorship network:

$$LV \succ GMO \succ WT \succ IM \succ SOM \succ EB \qquad (8.4)$$

where $A \succ B$ means that algorithm A performs better than B. It is interesting to notice that agglomerative community detection methods (LV, GMO, WT and IM)

Table 8.9 The results of community detection for the FS-UNS co-authorship network. Q – modularity, NC – the number of clusters, w^{intra} – the total weight of intra-community links, w^{inter} – the total weight of inter-community links, r – the ratio of w^{inter} and w^{intra}

Algorithm	Reference	Q	NC	w^{intra}	w^{inter}	r
GMO	[5]	0.8371	18	6919.45	655.66	0.0947
IM	[21]	0.8141	41	6618.53	956.58	0.1445
LV	[3]	0.8466	17	6920.37	654.74	0.0946
WT	[18]	0.8207	37	6873.07	702.04	0.1021
EB	[8]	0.5486	13	5248.49	2326.63	0.4433
SOM	[16]	0.6022	27	6466.84	1108.28	0.1714

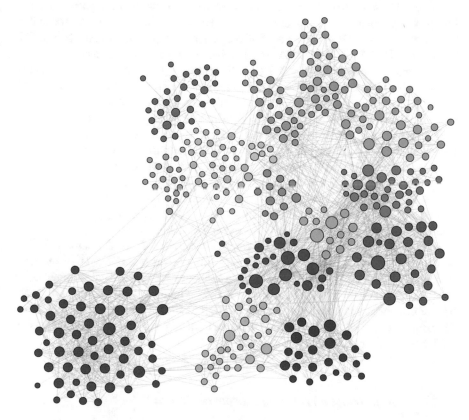

Fig. 8.3 The visualization of the FS-UNS co-authorship network after community detection by the Louvain algorithm. Nodes in the same color belong to the same community. The size of a node is proportional to its degree centrality

perform significantly better than divisive community detection methods (SOM and EB).

The Louvain algorithm identified 17 cohesive research groups of FS-UNS researchers. The basic characteristics of identified research groups are given in Table 8.10. It can be seen that for each group the following property holds: the total strength of research collaboration within a group is significantly higher than the total strength of collaboration with researchers belonging to other groups $(w^{intra} \gg w^{inter})$. Additionally, 12 out of 17 groups are Raddichi strong clusters. This means that each researcher in those 12 groups has established a stronger research collaboration with members of his/her group than with researchers belonging to other research groups.

Table 8.10 also shows the distribution of researchers within a research group per FS-UNS departments. It can be observed that 10 groups are "pure" in the sense that their members come from the same FS-UNS department (groups denoted by C1, C3, C5, C6, C7, C10, C12, C15, C16 and C17). Other 7 groups contain researchers that institutionally belong to different FS-UNS departments, but researchers from exactly one of the FS-UNS departments are in the great majority in each of those 7 groups. The distribution of researchers per FS-UNS departments within a group can be also seen as an indicator of group interdisciplinarity since FS-UNS departments correspond to different general scientific disciplines. We can see that the group C13 possesses a high degree of interdisciplinarity. Namely, C13 encompasses 10 researchers from the Department of Physics, 13 researchers from the Department of Chemistry and 1 researcher from the Department of Mathematics and Informatics. The rest of research groups are homogeneous in the sense that researchers from exactly one FS-UNS department either entirely constitute or strongly dominate in a group. Therefore, such groups exhibit a weak degree of research interdisciplinarity.

Except for the Department of Geography, researchers from the same FS-UNS department tend to be significantly present in more than one of FS-UNS research groups. A large majority of FS-UNS geographers (90% of the total number, 95% of those being non-isolated nodes in the FS-UNS co-authorship network) are members of C15 which is the largest research group identified by the Louvain method. Therefore, we can conclude that the Department of Geography is the most cohesive FS-UNS department.

8.3.3 Collaborations Among Research Groups

We constructed the collaboration network of FS-UNS research groups from the community partition identified by the Louvain algorithm. The collaboration network of FS-UNS research groups contains 79 links (collaborations between different FS-UNS research groups). The average node degree is equal to 9.29 which means that an average FS-UNS research group has established research collaboration with approximately 9 other groups. The visualization of the network is shown in Fig. 8.4. It can

Table 8.10 The list of research groups identified by the Louvain method. DBE, DP, DC, DMI and DG indicate the number of researchers from the corresponding FS-UNS department within a group. DD (dominant department) denotes the FS-UNS department whose members are dominant within a group. w^{intra} and w^{inter} denote the total weight of intra-community and inter-community links for a group, respectively. RS denotes whether a group is a Radicchi strong cluster

ID	Size	DBE	DP	DC	DMI	DG	DD	w^{intra}	w^{inter}	RS
C1	6	0	0	6	0	0	DC	179.33	23.58	No
C2	35	0	1	1	32	1	DMI	670.30	48.75	No
C3	2	2	0	0	0	0	DBE	82.57	2.70	yes
C4	25	23	0	1	0	1	DBE	927.73	260.12	No
C5	24	24	0	0	0	0	DBE	1230.69	117.35	No
C6	32	32	0	0	0	0	DBE	797.62	136.49	No
C7	13	0	0	13	0	0	DC	712.20	131.51	Yes
C8	30	0	1	0	29	0	DMI	843.72	16.83	Yes
C9	26	1	0	25	0	0	DC	1848.67	163.32	Yes
C10	32	32	0	0	0	0	DBE	927.99	73.69	Yes
C11	27	0	21	5	0	1	DP	761.24	68.34	Yes
C12	19	0	0	0	19	0	DMI	299.17	10.23	Yes
C13	24	0	10	13	1	0	DC	537.04	82.45	Yes
C14	35	2	2	31	0	0	DC	948.64	107.32	Yes
C15	59	0	0	0	0	59	DG	1642.18	30.58	Yes
C16	9	0	9	0	0	0	DP	321.13	11.88	Yes
C17	11	0	11	0	0	0	DP	1110.53	24.35	Yes

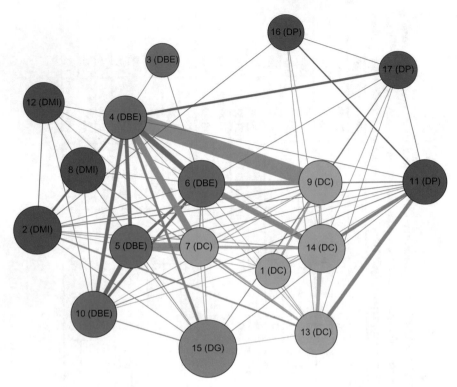

Fig. 8.4 The collaboration network of FS-UNS research groups. The size of a node is proportional to the size of the corresponding research group, while the width of a link is proportional to its weight

be observed that the network does not contain isolated nodes (i.e. FS-UNS groups that have not established research collaboration with any other group) and that it has exactly one connected component. This confirms our previous observation that scientific output of the institution is not a product of many separated research groups that do not collaborate among themselves.

The average link weight in the collaboration network of FS-UNS research groups is equal to 8.28. The complementary cumulative distribution of link weights is shown in Fig. 8.5. We can see that the strength of research collaboration between groups is highly unbalanced: approximately 70% of links in the network have weight less than or equal to 1, but there are also links whose weight is drastically higher than the average link weight (the weight of more than 10% of links is higher than 20). The skewness of the distribution is equal to 3.821, while the coefficient of variation is 1.696 indicating that the distribution is heavy-tailed (the skewness and coefficient of variation of an exponential distribution are equal to 2 and 1, respectively). The strongest research collaboration is between groups C4 (group whose members are dominantly from the Department of Biology and Ecology) and C9 (group whose members are dominantly from the Department of Physics). The weight of the link

Fig. 8.5 The complementary cumulative distribution of link weights in the collaboration network of FS-UNS research groups

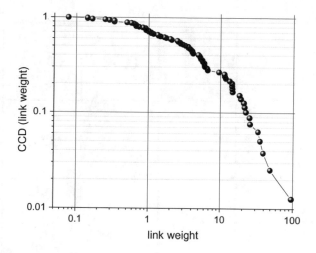

connecting those two groups is equal to 95.81 which is more than 10 times higher than the average link weight.

Exactly three research groups have the highest degree in the collaboration network of FS-UNS research groups: one group associated with the Department of Physics (C11) and two groups associated with the Department of Biology and Ecology (C4 and C6). Those three groups have established research collaboration with 13 other groups (approximately 75% of the total number of FS-UNS research groups). The highest weighted degree has C4 which means that it has established the strongest research collaboration with other groups. It is important to emphasize that C11 and C6, although having the largest degree in the network, do not have the second largest weighted degree – group C9 (one of groups associated with the Department of Physics) has a smaller degree compared to C11 and C6, but a larger weighted degree. In other words, C9 has established stronger intra-group collaborations than C11 and C6 in spite of having a smaller number of partners. To identify the most important research groups within FS-UNS, we computed weighted betweenness centrality of nodes in the collaboration network of FS-UNS research groups. The most central group in the network is C4 (as already mentioned, this group also has the highest degree and the highest weighted degree in the network). Exactly 9 research groups (C2, C4, C5, C6, C7, C9, C11, C13 and C14) have non-zero weighted betweenness centrality, while the weighted betweenness centrality of the rest of groups is equal to zero. Therefore, we can conclude that nearly half of FS-UNS research groups are located at the periphery of the group collaboration network. We computed Spearman's correlations between group size and node centrality metrics. Figure 8.6 shows the obtained correlation matrix. We can observe the following:

- The size of a research group significantly correlates only with its node degree. This means that larger FS-UNS research groups tend to collaborate with a larger number of other research groups, which is quite expected. However, the strength

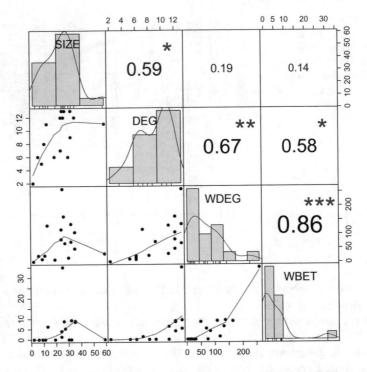

Fig. 8.6 The Spearman correlation matrix of the size, degree (DEG), weighted degree (WDEG) and weighted betweenness centrality (WBET) of nodes in the collaboration network of FS-UNS research groups

of inter-group research collaboration and the importance of a group within the network are independent of group size.

- There are strong positive correlations between weighted node degree and weighted betweenness centrality – groups that stronger collaborate with other research groups tend to be more important in the group collaboration network (Fig. 8.4).

We determined the set of strong links in the collaboration network of FS-UNS research groups according to Definition 8.4. The network contains 9 strong links which is 11.39% of the total number of group collaboration links. Figure 8.7 shows a reduced collaboration network of FS-UNS research groups that contains only strong links. Isolated nodes in the reduced network correspond to research groups that have established weak research collaborations with other research groups. We can see that 8 FS-UNS research groups are isolated nodes in the reduced network:

- Research groups associated with the Department of Mathematics and Informatics and the Department of Geography are isolated nodes suggesting that those two FS-UNS departments do not stimulate inter-group and inter-disciplinary research collaboration.

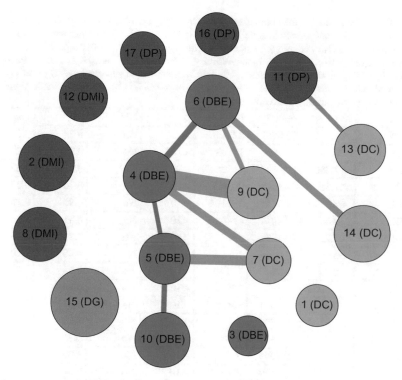

Fig. 8.7 The reduced collaboration network of FS-UNS research groups that contains only strong links

- Two research groups associated with the Department of Physics are also isolated nodes in the network. However, the largest research group from this department (C11) has established a strong inter-disciplinary research collaboration with C13 (the members of C13 are dominantly from the Department of Chemistry).

The vast majority of research groups associated with the Department of Biology and Ecology and the Department of Chemistry have established strong research collaborations among themselves. Only the smallest research groups from those two FS-UNS departments are isolated nodes in the reduced network. Therefore, we can conclude that those two departments tend to stimulate both inter-group and inter-disciplinary research collaboration. The large number of isolated nodes in the reduced network and the large fraction of weak links in the collaboration network of FS-UNS research groups imply that both inter-group and inter-disciplinary research collaboration at FS-UNS could be considerably improved, especially in the case of the Department of Geography and the Department of Mathematics and Informatics.

We can distinguish between two groups of researchers with respect to inter-group research collaboration:

Table 8.11 The results of the comparison between researchers involved in inter-group research collaborations (G_1) and researchers collaborating only with colleagues from their own research groups (G_2). Avg(G_1) and Avg(G_2) denote the average value of the corresponding metric for G_1 and G_2, respectively. U – the Mann-Whitney U statistic, p – the p-value of U, PS_1 – the probability of superiority of G_1 over G_2, PS_2 – the probability of superiority of G_2 over G_1. The column NHA indicates whether the null hypothesis of the MWU test (no statistically significant differences between groups) is accepted or not

Metric	Avg(G_1)	Avg(G_2)	U	p	NHA	PS_1	PS_2
PRON	98.6367	26.8662	8221.5	$<10^{-4}$	No	0.78	0.21
PROF	26.5373	11.1954	11043	$<10^{-4}$	No	0.71	0.29
PROS	27.8801	11.6127	11325	$<10^{-4}$	No	0.69	0.28
SRCI	151.3667	51.1648	9037.5	$<10^{-4}$	No	0.76	0.24
COLL	65.1873	17.5845	5301	$<10^{-4}$	No	0.86	0.14
LCOLL	17.4569	7.4014	6744.5	$<10^{-4}$	No	0.81	0.16
ECOLL	47.7303	10.1831	5932	$<10^{-4}$	No	0.84	0.15
WCOLL	95.1948	23.0563	7651.5	$<10^{-4}$	No	0.80	0.20
WLCOLL	48.7996	14.9347	8520.5	$<10^{-4}$	No	0.78	0.22
WECOLL	46.3951	8.1216	8065	$<10^{-4}$	No	0.79	0.21
IntraDEG	10.3408	7.4014	12475.5	$<10^{-4}$	No	0.64	0.30
InterDEG	7.1161	0.0000	0	$<10^{-4}$	No	1.00	0.00
WIntraDEG	43.8952	14.9347	9557	$<10^{-4}$	No	0.75	0.25
WInterDEG	4.9045	0.0000	0	$<10^{-4}$	No	1.00	0.00
BET	687.4335	73.2130	5683	$<10^{-4}$	No	0.84	0.14
CLO	0.3291	0.2676	3142	$<10^{-4}$	No	0.92	0.08
EVC	0.0031	0.0013	6730	$<10^{-4}$	No	0.82	0.18
WBET	457.4900	95.9977	12196.5	$<10^{-4}$	No	0.51	0.16
WCLO	0.6060	0.4651	8647.5	$<10^{-4}$	No	0.77	0.23
CC	0.4866	0.6553	11824	$<10^{-4}$	No	0.30	0.68

- G_1 – researchers collaborating with researchers from other research groups (researchers whose inter-cluster degree is non-zero), and
- G_2 – researchers collaborating only with colleagues from their own research groups (i.e. researchers whose inter-cluster degree is equal to zero)

We employed the Mann-Whitney U test to compare those two groups of researchers with respect to their productivity, collaboration and institutional social capital. The results of the statistical testing are presented in Table 8.11.

The null hypothesis of the MWU test (no statistically significant differences between two groups of researchers) is rejected for all researcher evaluation metrics. It can be seen that for each researcher evaluation metric, except for the clustering coefficient, we have that $PS_1 \gg PS_2$, where PS_1 is the probability of superiority of G_1 over G_2, while PS_2 is the opposite probability of superiority. Therefore, it can be concluded that researchers involved in inter-group research collaboration tend to be drastically more productive (by all considered productivity metrics), drastically more collaborative considering all forms of research collaboration (intra- and inter-institutional, intra- and inter-group) and drastically more institutionally important (by all considered node centrality metrics) compared to researchers who solely collaborate only with colleagues from their own research groups. For example, the average value of the Serbian research competency index for researchers from G_1 is nearly three times higher than the average value of the same metric for researchers from G_2. Moreover, the probability of superiority of G_1 over G_2 with respect to this metric is 2.71 times higher than the opposite probability of superiority. On the other hand, ego networks of researchers from G_2 tend to be significantly more dense than ego networks of researchers from G_1. Therefore, we can conclude that researchers from G_1 tend to have a more significant brokerage role within their ego networks than researchers from G_2, which additionally signifies their institutional importance.

8.3.4 Comparison of Research Groups

In order to determine whether there are strong differences in productivity and collaboration of researchers from different FS-UNS research groups we formed and analyzed the structure of group superiority graphs corresponding to researcher productivity (PRON, PROF, PROS and SRCI) and collaboration metrics (COLL and WCOLL). Three FS-UNS groups that have less than 10 members (C1, C3 and C16) are excluded from the analysis since their extremely small size cannot guarantee statistically accurate results. The basic characteristics of obtained group superiority graphs are presented in Table 8.12. For each considered researcher evaluation metric, the table shows the number of nodes and links in the corresponding group superiority graph, the number of superior research groups (nodes having zero in-degree), the number of inferior research groups (nodes having zero out-degree) and whether the group superiority graph has a bipartite structure. It should be noticed that all non-empty group superiority graphs have a bipartite (two-layered) structure which means that they do not contain nodes having both non-zero in-degree and out-degree (such nodes correspond to research groups that are superior to some groups, but inferior to some other groups). For example, the group superiority graph corresponding to the PRON research productivity metric is shown in Fig. 8.8. We can see that there are three FS-UNS research groups associated with departments of chemistry and physics that have significantly higher productivity (measured by PRON) compared to four FS-UNS research groups associated with departments of mathematics and biology.

All group superiority graphs corresponding to research collaboration metrics are non-empty. Consequently, we can conclude that there are significant differences between researchers from different FS-UNS research groups with respect to their collaborative behavior. We can also observe that group superiority graphs corresponding to productivity metrics are smaller than group superiority graphs corresponding to collaboration metrics. Therefore, we can say that differences in research collaboration between FS-UNS researchers from different groups are more drastic

Table 8.12 The basic characteristics of group superiority graphs

Metric	Nodes	Links	Superior groups	Inferior groups	Bipartite structure
PRON	7	7	3	4	Yes
PROF	0	0	/	/	/
PROS	0	0	/	/	/
SRCI	11	10	2	9	Yes
COLL	13	24	9	4	Yes
WCOLL	10	12	4	6	Yes

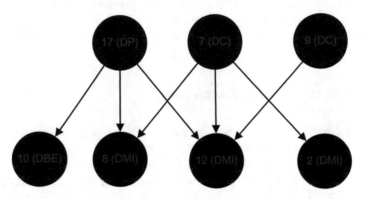

Fig. 8.8   The group superiority graph corresponding to the PRON research productivity metric

than differences in research productivity. Thirdly, it can be seen that group superiority graphs corresponding to PROF and PROS are empty, while group superiority graphs corresponding to PRON and SRCI are non-empty. This means that there are strong differences in productivity of different FS-UNS research groups when research productivity is measured by PRON and SRCI, but strong differences are absent when research productivity is measured by PROF and PROS. The principal difference between PROF and PROS on the one side and PRON and SRCI on the other side is that the first two researcher productivity metrics take into account the number of authors of a publication. Therefore, the absence of strong differences in research productivity measured by PROF and PROS indicates that SRCI and PRON are biased measures of research productivity. The links contained in group superiority graphs corresponding to PRON and SRCI are shown in Table 8.13. It can be seen that two research groups from the Department of Chemistry (C9 and C7) and one research group from the Department of Physics (C17) are superior nodes in both group superiority graphs. Researchers from mentioned groups tend to have significantly higher PRON/SRCI than researchers from groups associated mainly with the Department of Mathematics and Informatics. Consequently, we can conclude that PRON and SRCI are biased towards researchers from physics and chemistry research groups.

Table 8.13 The links in group superiority graphs corresponding to the PRON and SRCI research productivity metrics

Metric	Superior group	Superior to
PRON	C9 (DC)	C12 (DMI)
	C17 (DP)	C12 (DMI), C8 (DMI), C10 (DBE)
	C7 (DC)	C12 (DMI), C8 (DMI), C2 (DMI)
SRCI	C7 (DC)	C8 (DMI)
	C17 (DP)	C4 (DBE), C8 (DMI), C6 (DBE), C5 (DBE), C11 (DP),
		C10 (DBE), C15 (DG), C2 (DMI), C12 (DMI)

Table 8.14 The links in the group superiority graphs corresponding to the COLL and WCOLL research collaboration metrics

Metric	Superior group	Superior to
COLL	C17 (DP)	C2 (DMI), C12 (DMI), C8 (DMI), C10 (DBE)
	C11 (DP)	C2 (DMI), C12 (DMI), C8 (DMI)
	C7 (DC)	C2 (DMI), C12 (DMI), C8 (DMI)
	C9 (DC)	C2 (DMI), C12 (DMI), C8 (DMI)
	C14 (DC)	C2 (DMI), C12 (DMI), C8 (DMI)
	C15 (DG)	C2 (DMI), C12 (DMI), C8 (DMI)
	C5 (DBE)	C2 (DMI), C12 (DMI), C8 (DMI)
	C6 (DBE)	C12 (DMI)
	C13 (DC)	C12 (DMI)
WCOLL	C17 (DP)	C6 (DBE), C15 (DG), C8 (DMI), C2 (DMI),
		C12 (DMI), C10 (DBE)
	C7 (DC)	C8 (DMI), C2 (DMI), C12 (DMI)
	C9 (DC)	C8 (DMI), C12 (DMI)
	C5 (DBE)	C12 (DMI)

The links present in group superiority graphs corresponding to collaboration metrics (COLL and WCOLL) are listed in Table 8.14. It can be observed that researchers belonging to all 3 research groups from the Department of Mathematics and Informatics tend to have a significantly lower number of collaborators than researchers belonging to seven other FS-UNS research groups. We can also see that 4 groups that are superior nodes in the WCOLL group superiority graph are also superior nodes in the COLL superiority graph, but conversely is not true. Researchers from five FS-UNS research groups (C6, C11, C13, C14 and C15) tend to have significantly more collaborators than researchers from mathematical research groups, but those collaborations do not tend to be stronger than research collaborations established by FS-UNS mathematicians.

The group C17 is the most superior research group regarding both the number of collaborators and the strength of research collaborations. It is interesting to notice that

C17 is superior regarding the strength of research collaboration to a larger number of groups than with respect to the number of collaborators. Namely, there are no strong differences in the number of collaborators between researchers from C17, C6 and C15, but collaborations established by researchers from C17 tend to be significantly stronger than collaborations established by researchers from C6 and C15. Finally, it can be seen that the COLL and WCOLL group superiority graphs do not contain links connecting research groups belonging to the same FS-UNS department. This means that there are no strong differences in collaborative behavior of FS-UNS researchers from different groups working in the same scientific field.

8.3.5 Gender Analysis of Research Groups

The number of male and female researchers in each FS-UNS research group is shown in Fig. 8.9. The figure also shows the fraction of male researchers within each group. It can be observed that a majority of research groups contain more female than male researchers. Only in 5 research groups (C8, C12, C15, C16 and C17) a majority of researchers are of the male gender. Large gender gaps regarding the number of male and female researchers are present in the following three research groups:

1. C3 – this is the smallest FS-UNS research group and it is associated with the Department of Biology and Ecology. The group contains only two female researchers whose research collaboration is especially strong.

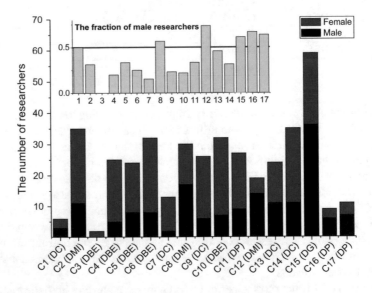

Fig. 8.9 The total number researchers and the number of male/female researchers per FS-UNS research groups. The inset shows the fraction of male researchers in each group

Table 8.15 FS-UNS research groups exhibiting significant gender differences

Group	Metric	Avg(M)	Avg(F)	U	p	PS_M	PS_F
C8 (DMI)	COLL	22.7	18.31	59.5	0.033	0.72	0.26
	CC	0.5	0.76	52.5	0.013	0.21	0.73
C15 (DG)	InterDEG	2.08	0.48	245.5	0.005	0.54	0.13
	WInterDEG	0.75	0.15	248	0.005	0.56	0.15
	BET	550.13	187.74	270	0.026	0.68	0.37
	CLO	0.32	0.29	215	0.002	0.74	0.26

2. C7 – this is the fourth smallest FS-UNS research group and it is associated with the Department of Physics. The group encompasses 11 female and only 2 male researchers.
3. C4 – this is the ninth largest FS-UNS research group and it is associated with the Department of Biology and Ecology. The group contains 20 female and 5 male researchers.

Using the MWU test we compared the productivity, collaboration and social importance of male and female researchers within FS-UNS research groups that have more than 10 male and 10 female members. Those groups are C15, C2, C14 (the three largest FS-UNS research groups associated with the Department of Geography, the Department of Mathematics and Informatics and the Department of Chemistry, respectively), C8 (group associated with the Department of Mathematics and Informatics) and C13 (group associated with the Department of Chemistry). The results of statistical testing are summarized in Table 8.15. The table shows only those groups and researcher evaluation metrics for which the null hypothesis of the MWU test was rejected. We can observe that there are no statistically significant gender differences in C2, C14 and C13 regarding all examined researcher evaluation metrics. Secondly, in all five groups male researchers do not tend be more or less productive compared to female researchers.

Significant gender differences were detected in two research groups: C8 and C15. Male researchers from C8 tend to have significantly higher number of collaborators than female researchers from the same group. On the other hand, ego networks of female researchers tend to be more cohesive. Regarding gender differences in C15, male researchers from this group tend to occupy more central positions in the FS-UNS co-authorship networks compared to female researchers. Additionally, male researchers from C15 tend to have a significantly higher number of collaborators from other FS-UNS groups and their intra-group research collaboration tend to be more stronger than intra-group research collaboration of female researchers. Therefore, it can be concluded that male researchers from the largest FS-UNS research group tend to be more institutionally important than female researchers due to their collaborations with FS-UNS researchers from other research groups.

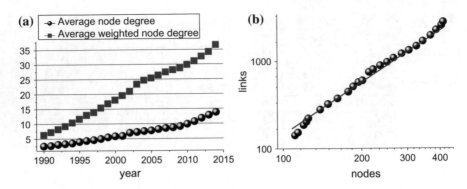

Fig. 8.10 The evolution of intra-institutional research collaboration at FS-UNS: **a** the evolution of the average (weighted) node degree in the FS-UNS co-authorship network, and **b** the plot of the number of links versus the number of nodes in the evolutionary snapshots of the network (log-log scales)

8.3.6 Network Evolution

We examined the evolution of the FS-UNS co-authorship network in the period from 1990 to 2014 (inclusive) at the yearly level. The network evolved from 111 nodes and 114 links in 1990 to 423 nodes and 2856 links in 2014. Figure 8.10a shows the evolution of the average node degree in the examined time period. It can be seen that the average node degree and the average weighted node degree exhibit a growing trend. Therefore, we can conclude that intra-institutional research collaboration at FS-UNS is *densifying* in time since the number and strength of intra-institutional collaboration links grow at a faster rate than the number of active FS-UNS researchers. Figure 8.10b shows the log-log plot of the number of links versus the number of nodes in the network. Each point in the plot represents one evolutionary snapshot of the FS-UNS co-authorship network such that the x coordinate is the number of nodes, while the y coordinate is the number of links. We can see that the evolution of the FS-UNS co-authorship network, similarly to many other co-authorship networks whose evolution is studied in the literature, obeys a densification power-law [10] which manifests as the straight line on the log-log plot shown in Fig. 8.10b. The observed increase of intra-institutional research collaboration at FS-UNS can be explained by two factors:

1. The fraction of single-authored papers authored by FS-UNS researchers steadily decreased from 13% in 1990 to only 3% in 2014.
2. In the same period, the average number of FS-UNS authors per publication increased from 1.61 in 1990 to 3.03 in 2014.

The evolution of the number of isolated nodes and the size of the largest connected component are shown in Fig. 8.11. We can see that the fraction of isolated nodes continuously decreases in time from 18.9% in 1990 to only 1% in 2014. Secondly, it can be noticed that the giant connected component of the network emerged very

Fig. 8.11 The evolution of connected components in the FS-UNS co-authorship network

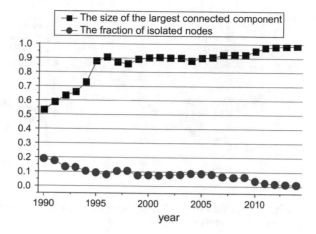

Fig. 8.12 The average value of the metrics of research productivity and inter-institutional research collaboration over time

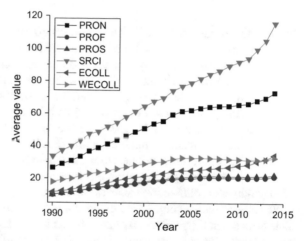

early (in 1995). Since 1995, the largest connected component in each evolutionary snapshot encompasses more than 85% of active FS-UNS researchers. Additionally, from 2011 all non-isolated FS-UNS researchers are located in the giant connected component of the network. Consequently, it can be concluded that the degree of FS-UNS institutional cohesion has steadily increased over time.

The evolutionary dynamics of four research productivity indicators and two metrics of inter-institutional research collaboration are shown in Fig. 8.12. It can be observed that two research productivity indicators, PRON and SRCI, exhibit a clear growing trend. On the other hand, PROF and PROS have significantly different evolutionary behavior than PRON and SRCI: the average productivity of FS-UNS researchers estimated by PROF/PROS increased in the period from 1990 to 2005, but in 2005 the growth stopped and stabilized. Taking into account that PRON and SRCI are biased indicators of research productivity towards researchers from FS-UNS physics and chemistry research groups (see Sect. 8.3.4), we can conclude that

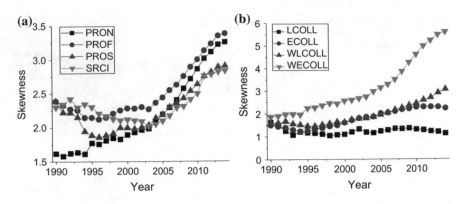

Fig. 8.13 The evolution of the skewness of the distributions of **a** research productivity metrics and **b** research collaboration metrics

the average productivity of FS-UNS researchers did not significantly changed in the last years of the examined time period. We can also notice two different evolutionary trends with respect to inter-institutional research collaboration. The average number of external collaborators (ECOLL) shows a continuously growing trend during the whole examined period. However, the average strength of inter-institutional collaboration (WECOLL), after a period of initial growth, stabilized in 2004 and remained constant afterwards. Therefore, we can conclude that in the last years FS-UNS researchers acquired more external collaborators than in previous years, but the intensity of inter-institutional research collaboration has not increased.

Figure 8.13 shows the evolution of the skewness of the distributions of research productivity and collaboration metrics. We can notice that the skewness of all research productivity metrics (Fig. 8.13a) exhibits a continuous growth starting from 2003. This means that the productivity gap between highly and lowly productive FS-UNS researchers increased very rapidly in the last 10 years of the examined period. A similar evolutionary behavior can be observed for the metrics quantifying the strength of intra-institutional and inter-institutional research collaboration. On the other hand, the skewness of the degree distribution (LCOLL) is evolutionary stable, it oscillates with small changes around a constant value throughout the whole examined period. Secondly, we can see that the degree distribution of the network exhibits a small skewness in all evolutionary snapshots, smaller than the skewness of an exponential distribution, which implies that there are no large gaps in intra-institutional collaborative behavior of FS-UNS researchers. This result additionally signifies that the FS-UNS co-authorship network does not belong to the class of scale-free networks (the skewness of the degree distribution of a scale-free network increases in time).

The FS-UNS co-authorship network contained 114 links in 1990. The total number of links in the network in 2014 is equal to 2856. Therefore, 2714 new collaboration links were established in the period from 1990 and 2014. A majority of new links (73.2%) are links between old nodes, 25.5% of new links are links connecting new to old nodes, while 1.3% of new links are links between new FS-UNS researchers.

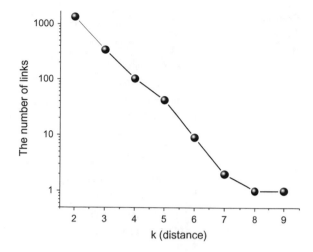

Fig. 8.14 The number of links created to nodes k hops away

The most interesting question regarding new links connecting old nodes is related to the distance between old nodes before they become connected. Firstly, it should be noticed that two old nodes may belong to different connected components of the network. In our case, 7.7% of links between old nodes are links that connected researchers belonging to different connected components of the network. A large fraction of such links was established in the year when the giant connected component emerged (1995). The distribution of locality of old links (the number of links created to nodes k hops away) is shown in Fig. 8.14. We can see that a vast majority of new links between old nodes are short-range links: 73% of the links connecting old nodes are links between FS-UNS researchers that had at least one common collaborator before they established research collaboration. Additionally, we can observe that the probability that two FS-UNS researchers establish collaboration decreases exponentially with their distance in the FS-UNS co-authorship network.

We applied the algorithm for mining attachment preferences (Algorithm 8.3) to the evolutionary sequence of the FS-UNS co-authorship network in order to determine how new FS-UNS researchers tend to integrate into the FS-UNS co-authorship network. The algorithm identified 7 frequent 1-itemsets, 8 frequent 2-itemsets and 2 frequent 3-itemsets for the minimal support equal to 0.5. The frequent itemsets are shown in Table 8.16. An item of the form (A, pref) indicates that preferential nodes (the nodes to which new nodes attach when integrating into the network, see Definition 8.8) tend to have significantly higher values of a metric A than non-preferential nodes, while an item of the form (A, non-pref) implies exactly the opposite. We can observe that the algorithm has not identified any item of the form (A, non-pref) which means that new FS-UNS researchers do not tend to integrate into the co-authorship network by establishing research collaboration with lowly productive and lowly collaborative FS-UNS researchers.

From the results shown in Table 8.16, it can be inferred that new FS-UNS researchers, when integrating into the FS-UNS co-authorship network, tend to con-

Table 8.16 The results of the algorithm for mining attachment preferences

Itemset size	Itemset	Support
1	{(PRON, pref)}	0.625
	{(SRCI, pref)}	0.667
	{(COLL, pref)}	0.667
	{(LCOLL, pref)}	0.583
	{(WCOLL, pref)}	0.708
	{(IntraDEG, pref)}	0.583
	{(EVC, pref)}	0.625
2	{(PRON, pref), (SRCI, pref)}	0.625
	{(PRON, pref), (WCOLL, pref)}	0.625
	{(SRCI, pref), (WCOLL, pref)}	0.625
	{(COLL, pref), (LCOLL, pref)}	0.583
	{(COLL, pref), (WCOLL, pref)}	0.625
	{(LCOLL, pref), (WCOLL, pref)}	0.583
	{(WCOLL, pref), (IntraDEG, pref)}	0.583
	{(WCOLL, pref), (EVC, pref)}	0.583
3	{(PRON, pref), (SRCI, pref), (WCOLL, pref)}	0.625
	{(COLL, pref), (LCOLL, pref), (WCOLL, pref)}	0.583

nect to FS-UNS researchers that are highly productive (measured by the PRON and SRCI metrics), highly collaborative (the COLL and WCOLL metrics) and that have a significantly higher institutional importance estimated by the eigenvector centrality metric. Preferential FS-UNS researchers also tend to have a significantly higher number of both total FS-UNS and intra-group collaborators compared to non-preferential researchers. Finally, the obtained frequent 2-itemsets and 3-itemsets indicate that new FS-UNS researchers tend to attach to highly productive FS-UNS researchers that have established a strong research collaboration with institutional colleagues.

8.4   Conclusions

In this chapter we proposed the methodology to analyze the structure and evolution of enriched co-authorship networks. The main idea of the methodology is to use community detection techniques and graph clustering evaluation metrics to determine research groups within an analyzed enriched co-authorship network, and then to employ non-parametric statistical tests to compare identified groups with respect to researcher evaluation metrics attached to nodes. Non-parametric statistical procedures are also utilized to perform gender-based analysis of research groups and investigate differences between (1) researchers from the core and researchers located at the periphery of the network and (2) researchers involved in inter-group research collaboration and researchers whose collaboration is bounded to their own research groups. Regarding the evolution of enriched co-authorship networks, we proposed the algorithm for mining attachment preferences that takes into account researcher evaluation metrics attached to nodes.

We applied the proposed methodology on the enriched co-authorship network reflecting intra-institutional research collaboration at FS-UNS (the Faculty of Sciences, University of Novi Sad, Serbia). The network was extracted from the database of the FS-UNS institutional research information system. We exploited all the benefits the database provides: (1) researchers are uniquely identified in the database and, consequently, there are no name disambiguation problems when extracting the co-authorship network, and (2) the database contains the categorization of research publications according to the rule book prescribed by the Serbian Ministry of Education, Science and Technological Development enabling us to enrich the nodes in the co-authorship network with the national research competency index used by the government to evaluate researchers.

The results of the analysis showed that the enriched FS-UNS co-authorship network has a giant connected component encompassing all non-isolated nodes. Secondly, the number of isolated nodes is extremely small. Therefore, it can be concluded that researchers employed at FS-UNS form a highly cohesive research community. The giant connected component exhibits the small-world property in the Watts-Strogatz sense, but it is not scale-free. It also has a strong core-periphery structure in which researchers from the core tend to be superior to peripheral researchers regarding productivity and inter-institutional collaboration.

The application of six different community detection algorithms revealed that the giant connected component of the FS-UNS co-authorship network also exhibits a strong community structure. The best partition of the network into cohesive research groups was identified using the Louvain algorithm. This algorithm identified 17 highly cohesive research groups at FS-UNS and 13 of them are Radicchi strong clusters. The analysis of the network reflecting inter-group research collaboration at FS-UNS showed that the strength of research collaboration between FS-UNS research groups is highly unbalanced indicating that inter-group and interdisciplinary research collaboration at FS-UNS could be drastically improved. Secondly, we observed that the strength of inter-group research collaboration and the importance of a research group within the institution are independent of group size. Finally, the analysis of inter-group research collaboration at FS-UNS revealed that FS-UNS researchers involved in inter-group research collaboration tend to be significantly more productive, collaborative (both locally and externally) and institutionally important compared to FS-UNS researchers collaborating only with colleagues from their own research groups.

Using the proposed methodology we also made an in-depth comparison of FS-UNS research groups. The obtained results indicate statistically significant differences in research productivity when it is estimated by metrics which do not take into account the number of authors of a publication (the normal counting scheme and Serbian research competency index). However, statistically significant differences in productivity are absent when research productivity is measured by the fractional and straight counting schemes, suggesting that the normal counting scheme and Serbian research research competency index are biased indicators of research productivity. By analyzing the group superiority graphs of those two productivity metrics, we found that they tend to favor researchers from chemistry and physics research groups. Significant differences between researchers from different FS-UNS research groups are also present regarding their collaborative behavior. However, strong differences in collaborative behavior of researchers belonging to different research groups working in the same scientific field are absent. The gender-based analysis of research groups revealed that significant gender differences in collaborative behavior and institutional importance are present in two FS-UNS research groups.

The proposed methodology can also be used to investigate the evolution of enriched co-authorship networks. We examined the evolution of the FS-UNS co-authorship network in a period of 25 years (from 1990 to 2014). The obtained results indicate that the network is densifying in time, i.e. the number of intra-institutional co-authorship links grows at a faster rate than the number of FS-UNS researchers. On the other side, the average productivity of FS-UNS researchers does not show an increasing trend in the last 10 years of the examined period. However, in the same period the gap between highly productive and lowly productive researchers increased very rapidly. Our analysis also showed that the probability that two FS-UNS researchers establish research collaboration decreases exponentially with their distance in the FS-UNS co-authorship network. Finally, our algorithm for discovering attachment preferences revealed that new FS-UNS researchers when integrating into the network tend to establish research collaboration with FS-UNS researchers that have both a high productivity and a high strength of intra-institutional research collaboration. All previously mentioned findings indicate that our methodology enables an in-depth analysis of research collaboration, as well as mutual relationships between research collaboration and other indicators of research performance.

References

1. Agrawal, R., Srikant, R.: Fast algorithms for mining association rules in large databases. In: Proceedings of the 20th International Conference on Very Large Data Bases, VLDB '94, pp. 487–499. Morgan Kaufmann Publishers Inc., San Francisco, CA, USA (1994)
2. Barabasi, A.L., Albert, R.: Emergence of scaling in random networks. Science **286**(5439), 509–512 (1999). https://doi.org/10.1126/science.286.5439.509
3. Blondel, V.D., Guillaume, J.L., Lambiotte, R., Lefebvre, E.: Fast unfolding of communities in large networks. J. Stat. Mech. Theory Exp. **2008**(10), P10,008 (2008)
4. Boccaletti, S., Latora, V., Moreno, Y., Chavez, M., Hwang, D.: Complex networks: structure and dynamics. Phys. Rep. **424**(45), 175–308 (2006). https://doi.org/10.1016/j.physrep.2005.10.009
5. Clauset, A., Newman, M.E.J., Moore, C.: Finding community structure in very large networks. Phys. Rev. E **70**, 066,111 (2004). https://doi.org/10.1103/PhysRevE.70.066111
6. Clauset, A., Shalizi, C., Newman, M.: Power-law distributions in empirical data. SIAM Rev. **51**(4), 661–703 (2009). https://doi.org/10.1137/070710111

7. Erceg-Hurn, D.M., Mirosevich, V.M.: Modern robust statistical methods: an easy way to maximize the accuracy and power of your research. Am. Psychol. **63**(7), 591–601 (2008). https://doi.org/10.1037/0003-066X.63.7.591

8. Girvan, M., Newman, M.E.J.: Community structure in social and biological networks. Proc. Natl. Acad. Sci. **99**(12), 7821–7826 (2002). https://doi.org/10.1073/pnas.122653799

9. Ivanović, D., Milosavljević, G., Milosavljević, B., Surla, D.: A CERIF-compatible research management system based on the MARC 21 format. Program: Electron. Libr. Inf. Syst. **44**(3), 229–251 (2010)

10. Leskovec, J., Kleinberg, J., Faloutsos, C.: Graph evolution: densification and shrinking diameters. ACM Trans. Knowl. Discov. Data **1**(1) (2007). https://doi.org/10.1145/1217299.1217301

11. Lindsey, D.: Production and citation measures in the sociology of science: the problem of multiple authorship. Soc. Stud. Sci. **10**(2), 145–162 (1980)

12. Mann, H.B., Whitney, D.R.: On a test of whether one of two random variables is stochastically larger than the other. Ann. Math. Stat. **18**(1), 50–60 (1947). https://doi.org/10.2307/2236101

13. Newman, M.E.J.: Assortative mixing in networks. Phys. Rev. Lett. **89**, 208,701 (2002). https://doi.org/10.1103/PhysRevLett.89.208701

14. Newman, M.E.J.: Analysis of weighted networks. Phys. Rev. E **70**, 056,131 (2004). https://doi.org/10.1103/PhysRevE.70.056131

15. Newman, M.E.J.: Who is the best connected scientist? A study of scientific coauthorship networks. In: Ben-Naim, E., Frauenfelder, H., Toroczkai, Z. (eds.) Complex Networks. Lecture Notes in Physics, pp. 337–370. Springer, Berlin (2004). https://doi.org/10.1007/978-3-540-44485-5_16

16. Newman, M.E.J.: Finding community structure in networks using the eigenvectors of matrices. Phys. Rev. E **74**, 036,104 (2006). https://doi.org/10.1103/PhysRevE.74.036104

17. Perc, M.: Growth and structure of Slovenia's scientific collaboration network. J. Informetr. **4**(4), 475–482 (2010). https://doi.org/10.1016/j.joi.2010.04.003

18. Pons, P., Latapy, M.: Computing communities in large networks using random walks. J. Graph Algorithms Appl. **10**(2), 191–218 (2006)

19. Radicchi, F., Castellano, C., Cecconi, F., Loreto, V., Parisi, D.: Defining and identifying communities in networks. Proc. Natl. Acad. Sci. **101**(9), 2658–2663 (2004). https://doi.org/10.1073/pnas.0400054101

20. Redner, S.: Citation statistics from 110 years of physical review. Phys. Today **58**(6), 49–54 (2005). https://doi.org/10.1063/1.1996475

21. Rosvall, M., Bergstrom, C.T.: Maps of information flow reveal community structure in complex networks **105**(4), 1118–1123 (2007). https://doi.org/10.1073/pnas.0706851105

22. Savić, M., Ivanović, M., Dimić Surla, B.: Analysis of intra-institutional research collaboration: a case of a Serbian faculty of sciences. Scientometrics **110**(1), 195–216 (2017). https://doi.org/10.1007/s11192-016-2167-z

23. Seidman, S.B.: Network structure and minimum degree. Soc. Netw. **5**(3), 269–287 (1983). https://doi.org/10.1016/0378-8733(83)90028-X

24. Watts, D.J., Strogatz, S.H.: Collective dynamics of small-world networks. Nature **393**, 440–442 (1998). https://doi.org/10.1038/30918